UNITED STATES DEPARTMENT OF THE INTERIOR, Douglas McKay, *Secretary*

BUREAU OF RECLAMATION, W. A. Dexheimer, *Commissioner*

DAMS AND

CONTROL

WORKS

THIRD EDITION : 1954

REPRINTED 1993

UNITED STATES GOVERNMENT PRINTING OFFICE, WASHINGTON : 1954

FOREWORD

This third edition of Dams and Control Works presents an outline and summarization of the Bureau of Reclamation's experience in the design and construction of dams and control works. Articles on individual structures, written by members of the Bureau's engineering staff, were selected to exhibit the wide range of sizes, types, and designs of the dams, spillways, and outlet works that the Bureau has used under varied topographic, foundation, and climatic conditions, materials availability, and water need and use.

In contrast with the earlier editions, the presentation has been simplified to widen the range of interest and at the same time to increase the ease with which major design features may be studied. Detailed design drawings and site layouts have been replaced by highly simplified, functional representations. A special article on the over-all phases of dam design covers the general considerations and problems facing the dam designer. Another describes the types and usage of gates and valves; and a brief résumé describes the scope of Reclamation engineering.

CONTENTS

❯❯

RECLAMATION ENGINEERING

By Everett H. Larson and David L. Goodman

By the end of 1952—slightly more than one-half century after passage of the Reclamation Act on June 17, 1902—Federal reclamation had made water available to more than 6,500,000 acres of irrigable lands in the 17 Western States. It had completed 195 dams, nearly 19,000 miles of canals and laterals, 47 power plants with a combined capacity of more than 4,000,000 kilowatts, 354 pumping plants, 200,000 irrigation structures, and 9,400 miles of transmission lines. In addition, numerous associated works had been constructed, including railroads and highways, bridges, and towns complete with water, sewerage, and communication systems.

The dams and the multitude of associated hydraulic structures constructed during Reclamation's first half-century have transformed large areas of wilderness into lands of productivity and wealth. The accumulated value of agricultural production from these lands within this 50-year period totaled $7,984,000,000, nearly four times the cost of project construction. In 1951 alone, farmers on irrigated farms served by the Bureau harvested crops valued at nearly $822,000,000. In that year also, Bureau hydroelectric power plants produced about 23 billion kilowatt-hours of energy for farms, cities, and industries. Of most importance to the Nation, Federal reclamation has resulted in the establishment of thousands of homes and farms and has afforded new opportunities for industrial and agricultural enterprise.

SCOPE OF RECLAMATION ENGINEERING

Reclaiming arid western lands and controlling the rivers for agricultural and power purposes has impelled continuous progress in the science of irrigation engineering. The first irrigation projects were simple, consisting usually of a diversion dam, headworks, canals, and suitable turnouts. These structures involved no special problems other than those peculiar to each site. Similarly, the first storage projects were simple and were built following design and construction procedures generally known to the engineering profession.

But maximum utilization of the West's limited water supplies required the development of projects of far greater scope of operation. To obtain maximum storage capacities, dams of increasing heights were required, and their construction introduced new problems to the engineer. The function of power was added early in Reclamation experience. The many storage dams built for irrigation provided excellent opportunity for generating power for construction purposes and for pumping water to higher lands, and in 1906 Congress authorized sale of surplus power generated at Reclamation projects.

In 1928, by passage of the Boulder Canyon Project Act, Congress inaugurated a new era in the conservation and utilization of western water. The Boulder Canyon project on the Colorado River, with Hoover Dam as its principal structure, embraced for the first time the multiple-purpose concept, which has since been the keystone of Bureau of Reclamation activities. This and other major undertakings such as the Central Valley project in California, the Columbia Basin project in Washington, the Colorado-Big Thompson project in Colorado, and the Missouri River Basin project in the Great Plains area are multiple purpose in scope, and they function for maximum utilization of water and land resources in large areas or entire river basins. These projects not only provide for the conservation, control, and utilization of water for irrigation and power production, but may also provide for flood control, silt control, navigation improvement, river regulation, municipal water supply, fish and wildlife preservation, recreation, salinity repulsion, and other functions.

The many functions involved in the Bureau's multiple-purpose projects are usually interrelated and require careful consideration through all phases of planning, design, construction, and operation. They have vastly increased the scope, variety, and complexity of

1

Reclamation engineering. At the inception of a project, and with congressional approval, Reclamation engineers investigate the economic, agricultural, and engineering feasibility of land and water resources. Lands are surveyed, mapped, and classified according to their agricultural potential. Rivers and streams are measured and their hydrologic characteristics—both quantitative and qualitative—are determined. Snow surveys and forecasts of river run-off are made for use in irrigation, storage, power generation, and flood-control designs. The proposed development is studied for conformance with water laws, local and state riparian rights, interstate compacts, and international treaties. Power generation sites and electrical transmission systems are studied and selected, and the economic feasibility of power markets is determined.

In the planning stage, engineers and geologists study the geologic formations and special foundation conditions at construction sites, study seepage losses in potential reservoir basins, and explore ground-water conditions. Geologists and materials specialists assemble data on availability of construction materials for concrete and earth structures. Drainage of water-logged lands and salt-water reclamation methods are also considered in project feasibility studies.

From these investigations and project plans, engineers develop preliminary designs and estimates for consideration by the Congress. After Congress has appropriated funds for construction, detail drawings and specifications are prepared for the great variety of structures. Designs of concrete and earth storage dams and their outlet works, and diversion dams and their headworks, are planned and executed. Power and pumping plants and related hydroelectric features such as transmission lines, switchyards, and substations are integrated into the project scheme as required. Canals and their many auxiliary works are designed, laid out, and built. Other features include tunnels, pipelines, water-supply and sewage-disposal systems, houses and administrative buildings for project personnel, and telephone systems for project use.

When required, desilting works are designed and built to prevent sediment deposits from entering canal headworks. To protect migratory fish in rivers on which dams are built, fish screens and fish ladders are designed and constructed. When it is determined that a contemplated Reclamation reservoir will flood an existing highway or railroad, the highway or railroad is relocated and constructed around the site. Where there is a possibility of flash floods damaging Bureau works, protective dikes and levees are included in the project designs.

Reclamation projects are dynamic enterprises which require capable administration. The irrigation works on today's Reclamation developments are turned over to water users' organizations for operation and maintenance as soon after completion as feasible. Experienced Bureau engineers manage, operate and maintain the projects before the transfer is made and assist in the operation and maintenance of the project after the transfer. They continually survey individual works, and recommend rehabilitation and betterment as required. Operation and maintenance engineers also make studies and survey economic needs for existing projects, and assist in soil and moisture conservation programs.

Assisting the planning and designing engineers as well as the operating personnel, research engineers develop improved methods and procedures, test new construction materials, and find more efficient ways of using existing materials. Structural and hydraulic model studies are made to verify the results obtained by mathematical methods. Using special laboratory techniques, research engineers also solve many problems not amenable to mathematical solution.

This, then, is the scope of Reclamation engineering—planning the projects, translating these plans into designs and specifications, building the structures, and operating and maintaining the completed systems.

THE ENGINEERING CENTER

The hub of the Bureau of Reclamation's technical activities is the Reclamation Engineering Center—the design offices and laboratories where engineers, scientists, and technicians concertedly attack the wide diversity of problems encountered in reclaiming the arid and semi-arid regions of the West. Here the technical skills and diversified talents of the civil, electrical, and mechanical engineer are teamed with those of the physicist, the chemist, and geologist. The hydraulic, concrete, soils, chemical, and other laboratories located at the Center are among the most complete in the world and are invaluable aids in the solving of water resources problems.

The Engineering Center may be considered as an engineering clinic where the manifold complexities of Reclamation engineering may each be examined by specialists of high qualification. It is a repository of experience, which forms the basis for sound engineering judgment and a guide to progressive improvement in engineering practice. The Engineering Center is located at the Federal Center near Denver, Colo.

Through technical assistance and training programs for foreign engineers at the Engineering Center and by the Bureau's engineers traveling abroad to render first-hand assistance, Reclamation's technology has helped to promote human and economic welfare of scores of countries. These undertakings are financed by non-Reclamation funds.

DESIGN ACTIVITIES

Although they are unified developments, Bureau of Reclamation projects comprehend a great diversity of structures such as dams, power plants, canals, and many other associated features. These structures must be planned and built economically and for competency, serviceability, and safety under widely varying climatic, topographic, and foundation conditions. Structures in great number have been designed and built to operate under extremes of climate varying from the intense winter cold of Alaska to the subtropic summer heat of southern Arizona. Reclamation works have been designed for service in rugged mountainous regions, for the uplands of the Pacific Northwest, and for the vast Great Plains area.

From the earliest days of Reclamation organization, its engineers have given intensive study to designing for minimum cost. Many examples of design and construction economy will be found in the works of even the earliest projects. Following World War II, Reclamation programs were greatly expanded to meet the ever-increasing agricultural and industrial needs of the West, and these expanding programs together with steadily mounting price and cost levels increased the pressure for construction economy.

The increased scope and pace of the expanding programs engendered new and complex engineering problems in addition to rising costs. Through solution of these problems and those of past decades, Reclamation engineers established an impressive record of contributions to engineering skill and ingenuity of design. These contributions for three major types of works are briefly discussed in the following paragraphs.

Dams and Control Works

Reclamation engineering is perhaps best known for its contributions to design of dams. From the crucible of varied experience Reclamation has produced a succession of dams of unprecedented heights and proportions. At five different times in its history, Reclamation was called upon to design and build a dam higher than any then in existence in the world.

As concrete dam heights and sizes increased, existing methods of design and construction became inadequate and new methods were devised through progressive design and laboratory research. To remove the tremendous heat generated by the hardening of concrete in the interiors of large dams, Bureau engineers evolved the technique of cooling the concrete by means of circulating water through embedded pipe systems. To facilitate the construction of such dams, the block method of construction was adopted as standard.

Trial-load methods were devised for analyzing stresses in concrete arch and gravity dams, which permitted great savings through selection of more economical dam sections. In a continuing effort to improve economy of design while retaining structural adequacy, dependability, and safety, engineers of the Bureau established new criteria for determining temperature effects, earthquake loads, uplift forces and ice pressures, foundation deformations, and physical properties of construction materials for dams.

To adapt rolled-earth construction to higher and higher dams, the Bureau has analyzed characteristics of stability and pore pressure of earth embankments, developed solutions for the stresses in earth dams, and evolved criteria for moisture control, material selection, and embankment placement.

For further information on these advances in design, the reader is referred to the articles on dams—such as those on Grand Coulee, Hoover, and Hungry Horse Dams—and the special article on design of dams, in this volume.

The design of hydraulic equipment to control the release of water from dams has paralleled progress in dam design. Butterfly valves were developed for emergency closure of penstocks and outlet pipes. Slide gates were adapted to higher pressures by the addition of ring followers and wheels to withstand the high thrust and to reduce friction. Gates of this type were designed for 120-inch conduits and for pressures greater than 260 pounds per square inch.

Higher pressures also required improved designs of valves which were represented first by a needle-type valve, followed by the interior differential needle valve, the tube valve, and finally the present hollow-jet and jet-flow valves. Sizes have been built up to 108 inches in diameter and for pressures as great as 200 pounds per square inch.

Reclamation's innovations in the techniques of energy dissipation of falling water have similarly influenced successive improvements in design of water control structures. Spillways of large capacities ranging up to 1,000,000 cubic feet per second, and penstocks, intakes, draft tubes, forebays, and tailwater channels in great number and variety bear evidence of these design improvements.

Further information on reservoir outlets and spillways may be obtained in the two special articles at the end of this volume.

Power Plants and Systems

The Bureau's large multiple-purpose projects include power plants of unprecedented size, and the successful design of these plants with their associated control, protective, and transmission systems has added greatly to the knowledge of hydroelectric power development. An

Design and construction.

outstanding example is the Grand Coulee power plant, a vital link in the comprehensive program of power development in the Columbia River Basin, which began initial operation in April 1942. This unique plant, housing 18 huge generators rated at 108,000 kilowatts each, produces more than 2,000,000 kilowatts of power. A single generator, weighing about 1,000 tons, is 34 feet in diameter and equal in height to a 2½-story building. Each of the two powerhouses is about 730 feet long, 100 feet wide, and 150 feet high. Design of the Grand Coulee power plant presented problems of the first magnitude, from determining the best size and number of generating units to designing facilities to handle the unprecedented equipment loads during installation and repair. Foundation and structure loads were beyond the scope of previous experience in powerplant design.

Other major hydroelectric power installations have added to Reclamation's engineering experience. In time of initial installation, Grand Coulee was preceded by Hoover power plant, of 1,332,000 kilowatts ultimate generating capacity, on the lower Colorado River. It was followed, in order, by Shasta power plant, of 375,-000 kilowatts capacity, in the Central Valley, Calif.; Davis power plant, of 225,000 kilowatts capacity, also on the lower Colorado River; and the 285,000-kilowatt Hungry Horse power plant in northern Montana.

Protection of power centers has received continued attention by Reclamation engineers. The Bureau, with the assistance of the electrical manufacturing industry, pioneered in the development of the large circuit breakers guarding against lightning and other interruptive forces. As the result of thorough design studies and field tests, the circuit breakers stand sentinel over the Bureau's large power-generating facilities to assure delivery of vital power.

Electrical transmission lines extend great distances over prairie lands, foothills, and mountainous regions to bring power to load centers and to other points of use. The lines—with voltages as high as 287,500—and their associated structures have been designed for maximum efficiency of operation under all types of weather and service conditions. Many Bureau transmission lines have been integrated into extensive transmission systems linking major sources of power. An a. c. network analyzer permits Bureau engineers to study in miniature the planning, design, and operating problems of these complex power systems, with great savings over mathematical methods.

Pumping Plants

The Bureau's hydroelectric power plants often provide economical sources of energy for pumping irrigation water to otherwise inaccessible areas. In certain localities there has been an increasing need to bring

water to these higher lands, and pumping plants of capacities ranging from 2 to 16,000 cubic feet per second (ultimate) have been designed, constructed, and placed in operation. For the larger installations, investigations embracing many phases of design and analytical and laboratory research were required to assure reliable pump performance and uninterrupted service.

The Grand Coulee pumping plant, on the Columbia Basin project, is the largest in the world and will consist ultimately of 12 units, each capable of lifting water 285 feet at a capacity of 1,350 cubic feet per second. The Tracy pumping plant, a major feature of the Central Valley project, can lift water at the rate of 4,600 cubic feet per second against a pumping lift of 197 feet. The Granby pumping plant, on the Colorado-Big Thompson project, can pump water a height of 186 feet at the rate of 600 cubic feet per second.

The unusual size of these three plants and their special conditions of operation intensified the usual problems of design and introduced new ones, and the successful solutions of these problems represented contributions to engineering skill and knowledge. At Grand Coulee pumping plant, for example, special methods were devised to eliminate or control vibration in the penstocks induced by pressure oscillations in the large centrifugal pumps. While Granby pumping plant is not nearly so large as the other two, a fluctuation of 94 feet in the inlet head, the necessarily long inlet pipe, and the low barometric pressure at the plant's 8,180-foot elevation in the Rocky Mountains prompted the use of subterranean construction. The center of the pump is about 107 feet below the level of the surrounding ground, and the building frame had to be designed to resist high hydrostatic pressure as well as earth pressures.

These three large pumping installations are parts of systems which take surplus waters to arid lands distant from natural stream flows. As such, they represent the beginning of a new era in American agriculture. About 2¾ million acres of irrigable lands will receive new or supplemental irrigation water when the pumping plants are in full operation.

An interesting feature on the Colorado-Big Thompson project, the Flatiron pumping-power plant is believed to be the first such installation in the United States. This plant is so designed that during periods of low power demand pumping units can be used to pump water from the Flatiron Reservoir into Carter Lake Reservoir, and during periods when extra power is needed the same units can be operated in reverse as hydroelectric generators, using water released from Carter Lake Reservoir.

Canals and Associated Facilities

Thousands of miles of canals and laterals and their associated structures convey water to irrigated farm

Power.

lands on Bureau projects throughout the West. Building of these vital arteries to bring and distribute water from stream bed or reservoir to arid and semi-arid lands is a principal responsibility of Bureau engineers. The ever-increasing demand for maximum use of the country's limited water supplies has been a compelling force to improve water conveyance and distribution works and has had a pervasive influence upon the design of modern-day canals and their far-flung distributaries.

Outstanding on Bureau projects are the All-American, Friant-Kern, Delta-Mendota, East Low and West Canals. These are virtually rivers, carrying large volumes of irrigation water to hundreds of thousands of acres of fertile lands.

The All-American Canal, a principal feature of the All-American Canal system in southern California, is 80 miles long. This huge waterway traverses the rich Imperial Valley and carries water at a maximum capacity (before lateral diversions) of 15,000 cubic feet per second. The Coachella Branch of the canal is 145 miles long, its capacity is 2,500 cubic feet per second. Friant-Kern Canal on the Central Valley project in California is 153 miles long, one of the longest man-made waterways in the world. The canal carries water at a maximum capacity of 5,000 cubic feet per second, serving farm lands in the San Joaquin Valley. Delta-Mendota Canal, another major canal of the Central Valley project, is 117 miles long and carries water at a maximum capacity of 4,600 cubic feet per second to extensively irrigated farm developments on the project. The 130-mile-long East Low Canal, carrying 4,500 cubic feet per second, and the 88-mile-long West Canal, 5,100 cubic feet per second, are major irrigation features on the great Columbia Basin project.

Design advances contributed greatly to the completion of these canals and their related structures. Among the more important of these advances are developments in soil mechanics technology, in the control of temperature and settlement cracking in concrete canal structures, in the sealing of joints to prevent leakage, in the control of turbulence of water, and in the theory of hydraulic forces involved in flowing water; and numerous improvements in the application of established hydraulics principles. Laboratory investigations gave vital information on the materials and forces involved.

Intensive studies by Reclamation engineers have brought out much information on the reduction of the cost of suitable canal linings. Under the Lower Cost Canal Lining Program, initiated in 1946, unreinforced concrete, pneumatically placed mortar, buried asphalt membrane, compacted earth, soil-cement, and other types of lining have been introduced and successfully applied as economical substitutes for reinforced concrete linings in Bureau canals. These efforts not only have made possible new economies in canal construction, but they have also assured the saving of thousands of acre-feet of precious water annually.

In addition to the development of major canal structures, irrigation systems on Reclamation projects have required the design of a great number and variety of structures to carry the water properly, control it, and protect the canal systems from failures. Desilting works, chutes, drops, culverts under railroads and highways, flumes, inverted siphons, tunnels, closed conduits, turnouts, weir structures, drainage inlets, and wasteways illustrate the diversity of associated canal features.

Reclamation engineers also design extensive underground concrete-pipe distribution systems for irrigation, and they have introduced many innovations in the design of pipe lines for the systems. Steel-reinforced and steel-lined concrete pipes to withstand high pressures have also been built from Bureau designs.

CONSTRUCTION ACTIVITIES

Administration of the Bureau of Reclamation's construction programs—which in the years following World War II required the translation of an annual investment of about one-quarter billion dollars into useful works—is a continuing responsibility of Reclamation engineers. To discharge this responsibility, the Chief Engineer of the Bureau and the Bureau's construction engineers—both at the Engineering Center and at the individual projects—must see that construction requirements are met, that adequate standards are observed, and that design conceptions achieve reality in the form of completed works.

After a short period of construction by contract, construction in the early years was largely carried out by Government forces because of the contractors' limited experience in reclamation undertakings. Since 1925, however, nearly all Bureau structures have been built by contractors' forces. The progressive expansion of Reclamation programs since that time has been paralleled by the growth of the skills of the contracting industry. Much of the success of Reclamation project development can be attributed to the capabilities and resourcefulness of modern contracting organizations.

Assembled at the Engineering Center is a group of engineers who contribute in large part to the expeditious and economical functioning of Reclamation construction programs. These engineers are specialists in estimating costs of construction work, in contract administration, in construction administration, and in the coordination of activities in the field and design offices.

In recent years contractual relationships between the Bureau and contractors have been strengthened by periodic conferences on mutual problems. This endeavor and the concerted efforts of construction engineers have been reflected in reduction of cost of con-

Irrigation canals.

struction on Reclamation projects and increased rate of performance.

RESEARCH ACTIVITIES

Research plays a dominant role in the program of Reclamation engineering. The designing and building of safe and efficient dams, power plants, canals, and other structures necessitates continued investigation of engineering applications and diligent inquiry into new materials and methods. An effective stimulus of research is the stringent requirement for building Reclamation works economically and within the water users' ability to repay the costs of construction, as established by Reclamation law. Research has not only given impetus to development of new engineering applications, introduction of new materials, and amplification and strengthening of existing procedures, but it has also produced major economies in design and construction.

The Bureau's engineering laboratories at the Reclamation Engineering Center have made major contributions in virtually every phase of water engineering advancement. In the fields of hydraulic research, structural testing, and investigations of construction materials, laboratory effort has improved design and construction methods.

Hydraulic research embraces a widely diversified scope of activities. Many specialized problems not amenable to analytical solution may be solved by hydraulic laboratory techniques. Hydraulic structures such as dams, spillways, canal chutes and drops, pipelines, outlet works, stilling basins, desilting works, and hydraulic machinery and appurtenances are studied in model form; and such model tests have led to improvements in hydraulic operating characteristics, safety in operations, and economy of design. The solution of specific problems has often contributed to the solution of future problems.

The erosion effects of flowing water increase as the head, and hence the velocity, increases. Erosion therefore becomes of serious concern in spillways, outlet valves, turbine runners, and other water passages associated with modern high-head dams. Hydraulic research techniques have identified the causes of erosion and have led to corrective design measures for reducing costly repairs to structures and machinery. Discharge capacities of spillways and other hydraulic structures have been increased through laboratory studies. Economies have been effected through reduction in the size of water passages when laboratory studies showed that such modifications were practical. Design dimensions of many spillway chutes have thus been decreased, with resultant improvement in operation of these features.

The hydraulic laboratory has applied electronics to the study of certain hydraulic phenomena. Such application has been valuable in the study of flow through a complex network of fresh water channels into tidal estuaries, in studies of pressure surges in pipe lines, and in investigations of sedimentation in Bureau reservoirs.

A primary requisite to sound design and construction is the structural research conducted for the accurate measurement of stress, strain, deflection, and other physical effects developed in structures. Tests on models of dams, bridges, power and pumping plants, gates, and other structures or structural parts are invaluable adjuncts to safe and economical designs. Structural research is also applied profitably to the study of foundations, characteristics of materials, particularly stress-strain relationships; and the effects of dynamic shock or impact on structural members. Structural studies employ a variety of special equipment, including a 5,000,000-pound capacity universal testing machine, a triaxial testing machine, and special optical, mechanical, and electrical measuring devices in the laboratory.

Significant contributions of the Bureau's materials research activities are the development of tests to determine the suitability of natural construction materials, the intensive inquiry into the properties of cements, the development of procedures for evaluating concrete mixes, and the introduction of new construction materials. Especially noteworthy advances have been made in concrete technology, in which a continuous effort is exerted to decrease the cost of concrete mixes while retaining or improving strength, durability, workability, and other desirable characteristics. Research by others—furthered by the Bureau—disclosed that entrainment of small amounts of air in concrete considerably increases the durability, as measured by resistance to freezing and thawing. Air entrainment also increases the workability of concrete mixes and allows substantial reductions in cement and water content, with resultant savings in construction costs. Research on pozzolans disclosed that the introduction of fly ash or other suitable pozzolanic materials makes possible the use of leaner concrete mixes, thereby reducing the heat of hydration of the cement and minimizing the requirements for artificial cooling. Pozzolans also reduce destructive chemical reactions between the alkalies in cements and certain aggregates and improve workability of concrete mixes. While pozzolans are not used primarily to reduce costs, savings often accompany their use.

From research and experience have come basic criteria for control of mixing, transporting, placing, and curing of concrete. Progress has also been made in identifying and preventing the destructive reaction between concrete aggregates and the alkalis of certain cements, and in the studies of physical and chemical properties of concrete.

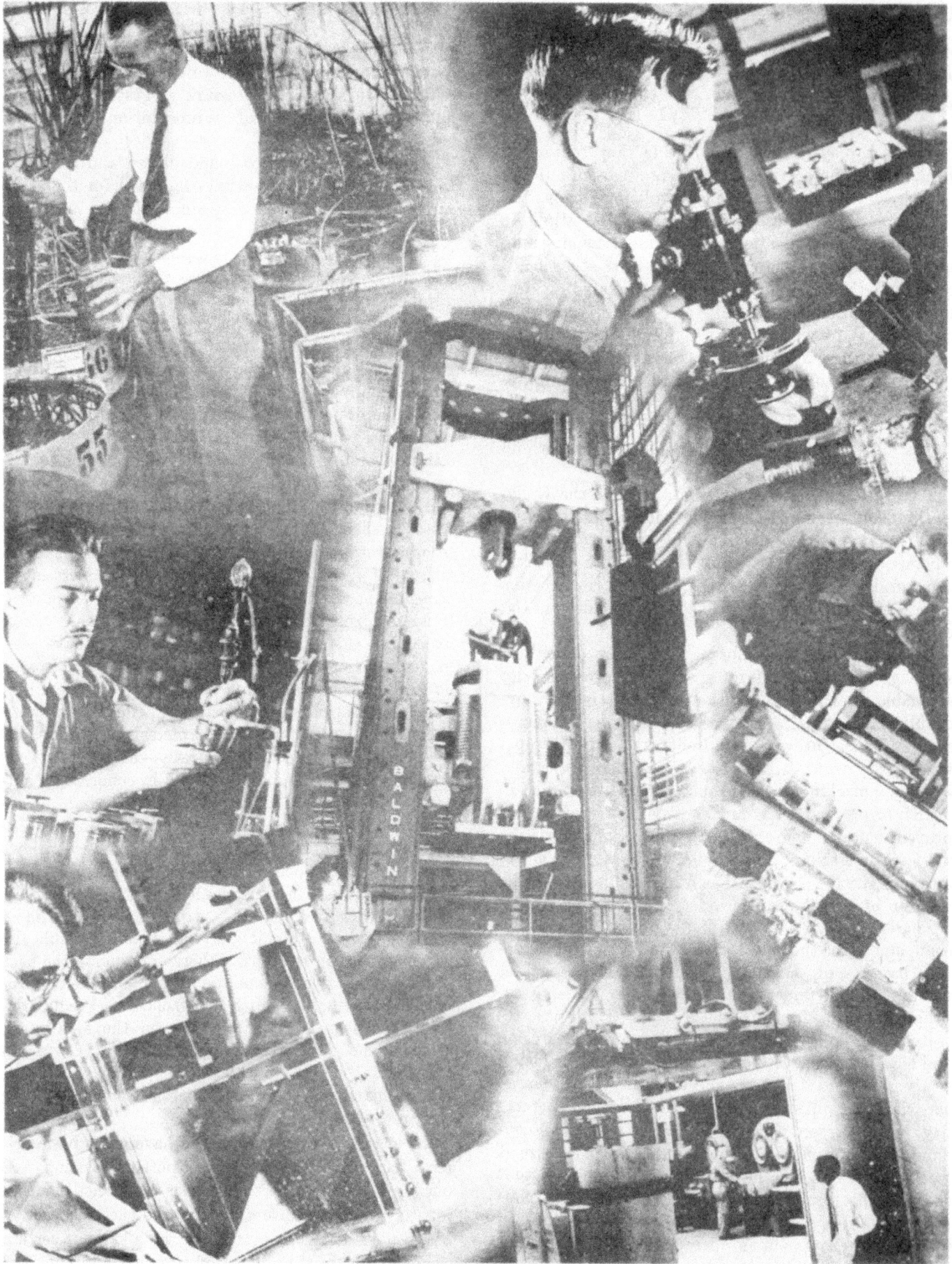

Research.

To permit most effective use of available soils as economical construction materials and for foundation purposes, laboratory research engineers have undertaken basic studies of soils. The investigation and construction control techniques thus developed now aid Reclamation engineers to build many large earth dams safely and economically on foundations formerly considered inadequate for large structures. The introduction of heavy, compacted-earth linings for canals—one of the most promising types of canal linings developed in recent years—received impetus through laboratory investigations. Foundation and canal subgrade soils are now analyzed to determine load bearing capacities and to insure that expansive soils or soils subject to high settlement are properly considered in design or construction practices. In cooperation with other agencies, Bureau engineers have standardized the classification of soils. Unique tests for determining physical, petrographic, and chemical properties of soils have been established in the engineering laboratories.

The Bureau's search for better protective coatings has resulted in coatings of increased durability, lower cost, and easier application. For example, the coal-tar enamel now specified will give protection to water piping under a wider range of conditions than was previously possible, and a service life of at least 40 years can now be expected for this material. The vinyl-resin paints developed in recent years for use on the numerous flow-control gates in canals, dams, and other structures will last at least three times as long as the coatings formerly used. Another phase of the Bureau's research on corrosion prevention is "cathodic protection" of metal irrigation structures. In this electrochemical process the structure is protected from excessive corrosion by use of an "expendable anode" which takes the destructive corrosion itself and frees the structure—as the cathode—from corrosion damage.

Through its weed control research program, in cooperation with the Bureau of Plant Industry, Department of Agriculture, the Bureau has made much progress in the development of effective and economical weed killers. As low-cost substitutes for more expensive mechanical removal, aromatic solvents such as coal-tar naphtha with an emulsifying agent have been developed to control aquatic weeds. Recent laboratory research has utilized radio-isotope techniques to study plant life and the effect of weed killers on aquatic and land-type weeds.

A variety of field programs are carried out jointly by project offices and the engineering laboratories to improve structural performance and to further economies in design, construction, and operation. The Lower Cost Canal Lining Program, previously mentioned, is one of these cooperative endeavors. Considerable success has been realized from the application of laboratory research in asphaltic materials to the construction of low cost canal linings. A joint water-measurement standardization program has added to the knowledge of performance of measuring devices and has led to improved operational methods in irrigation practices. In other directions, cooperative research effort has been important in the study of electrical methods for inhibiting corrosion of submerged metalwork, in evaluating protective coatings, in gathering data on the effect of ice pressure on structures, in the measurement of seepage from canals, in the study of ground water, and in the operation, maintenance, and rehabilitation of existing structures.

*   *   *   *   *

The gradual evolution of Reclamation from the first simple diversion dam and canal to the modern multiple-purpose projects has brought ever more challenging problems to the Reclamation engineer. The West is about at the halfway point in developing irrigation resources available under present economic standards. Only about 9,000,000 of the estimated 50,000,000 kilowatts of hydroelectric power have yet been harnessed. The more difficult water diversion, storage, and conveyance works and power developments are yet to be undertaken. Full utilization of Western water resources will require the exchange of water through multiple-basin exportation systems. Also, full development of many irrigable areas will require an economical means of demineralizing saline or brackish waters to make them suitable for irrigation.

Concepts in planning will continue to be broadened, and techniques improved for designing and constructing the complex developments required by the ever-increasing demands for irrigation water and electrical power.

Aerial view looking upstream of river showing Hungry Horse Dam, power plant, and reservoir.

HUNGRY HORSE DAM

HUNGRY HORSE PROJECT, MONTANA

By Charles L. Townsend

Hungry Horse project in Montana is a combination power and flood control project that was contemplated as a result of a survey of the area by the United States Geological Survey in 1934–35. World War II and the "Battle of Flathead Lake" finally induced congressional action authorizing the Hungry Horse Dam and power plant and designating the Bureau of Reclamation as the agency to prepare plans and specifications and to supervise construction. The renowned "battle" occurred in June 1943 when wartime need for power caused serious consideration of a plan to raise the level of Flathead Lake, about 7 miles downstream from Kalispell, Mont., and thereby increase the capacity of existing power plants through added units and more firm power. Inasmuch as this would have inundated a large part of the Flathead Valley and flooded the streets of Kalispell, residents of the area protested so vigorously that the Hungry Horse project was approved instead. The name of the project, incidentally, comes from Hungry Horse Creek, a small tributary entering the river about 2 miles upstream from the dam. This creek obtained the name when several half-starved horses were found in the snow-filled valley during the early days of the region.

Hungry Horse Dam is located on the south fork of the Flathead River about 9 miles southeast of Columbia Falls, Mont. The area is mountainous and heavily timbered and ranges in elevation from 3565 at the top of the dam to about elevation 9200 at the headwaters of the south fork. The basin above Hungry Horse Dam is 80 to 85 miles long and drains an area of approximately 1,640 square miles. Annual precipitation in the area ranges from 22 inches per year at Columbia Falls to about 35 inches in the mountains.

Hungry Horse Dam and its power plant constitute a key unit in a comprehensive, long-range program to insure full control and utilization of the Columbia River for agricultural and industrial development of the Pacific Northwest. Extensive studies were made in cooperation with the Bonneville Power Administration to determine the economic height of the dam, and the cost of the project was balanced against the benefits to be derived from power generated at the power plant. As of the date of preparation of this article, the dam and power plant are essentially completed. The plan for operation contemplates full capacity power generation during the winter months when power is needed most and water releases can be used to the best advantage downstream. In spring and summer months, when ample power is available at the downstream power plants, the project will store water and generate a minimum amount of power. This plan lends itself ideally to flood control of the Columbia River. If snow surveys and winter climatic data indicate an abnormal run-off, the reservoir can be lowered early so that it can retain the high water during the flood season.

Hungry Horse Dam is also expected to play an important part in the future Kalispell project which will irrigate 85,000 acres of rich valley soil. Several plans for this project are under consideration, the foremost of which calls for pumping water from the middle fork of the Flathead River with low cost power from Hungry Horse power plant.

RESERVOIR

The gross storage capacity of Hungry Horse Reservoir is about 3,500,000 acre-feet and 2,980,000 acre-feet of this is active storage to be used for power generation, flood control, and future irrigation needs. When full, the reservoir extends upstream about 35 miles and covers about 24,000 acres.

Contracts for clearing the reservoir area were awarded in several schedules totalling over $7,000,000. An estimated 70,000,000 board feet of merchantable

More than 750,000 acre-feet of water was in storage in Hungry Horse Reservoir in December 1952. The dam was finished, except for minor clean-up tasks, and was functioning as an integral part of the system of dams and reservoirs controlling the Columbia River and its tributaries.

timber was involved in the reservoir clearing operations and, in order to salvage as much of this as possible, the Government awarded separate logging contracts to precede the clearing operations by about 4 months. Payment for the removal of the trees was made on a stumpage basis, the loggers purchasing the timber as it was removed. The money received for this lumber was then applied against the cost of the reservoir clearing.

To facilitate the rapid clearing of the dense forest area, the clearing contractors developed a method which involved some rather unusual equipment. The method depended primarily on a 2-inch steel cable which was dragged by two heavy tractors to snag and uproot brush and small trees on the steep sides of the reservoir. A 4½-ton hollow steel ball, 8 feet in diameter, whose axle was attached by swivels to the middle of the cable, weighed down the cable and supported it at the most effective height. The cable was equipped with swivels every 50 feet to prevent twisting and varied in length from 400 to 800 feet. The width of the strip cleared was varied by the tractor operators depending on the size and number of trees between the tractors. About

4,000 acres were cleared in this manner and, in some instances, as many as 200 acres were cleared in 4 hours.

THE DAM

Ranking with the highest and largest dams in the United States, Hungry Horse Dam is a variable-thickness, concrete arch dam approximately 2,100 feet long at the crest with a maximum height of 564 feet. In height, Hungry Horse Dam is exceeded only by Hoover Dam (726 feet) and Shasta Dam (602 feet). Containing approximately 2,900,000 cubic yards of concrete, it ranks fourth in the Nation in volume, exceeded only by Grand Coulee, Shasta, and Hoover Dams. Four 162-inch-diameter penstocks and three 96-inch-diameter outlet pipes extend through the dam. The penstock entrances on the upstream face of the dam are protected by individual trashracks and can be closed by hydraulic gates. The outlet pipe entrances on the upstream face of the dam are protected by two trashracks and can be closed by a bulkhead type gate lowered from the top of the dam. The main features of the dam and their relative locations are shown in the drawing.

Supported by giant steel balls, 8 feet in diameter, cables were dragged through the forest to clear the reservoir area of trees and brush. Merchantable timber was salvaged, the remainder was burned.

Division of the mass concrete into blocks by vertical contraction joints was necessary because of the size of the dam. Transverse (radial) joints were continuous from upstream to downstream face and spaced at 80-foot intervals on the upstream face of the dam. As the base thickness of the dam is over 320 feet, one longitudinal joint was placed in the longer blocks to prevent development of continuous, wide cracks across the blocks. The longitudinal joint was offset in adjacent blocks and so placed that the longest single block would be 186 feet. The vertical contraction joints thus divide the dam into blocks, the largest of which is 80 by 186 feet and contains about 3,700 cubic yards of concrete in each 5-foot lift.

To assure monolithic action of the structure, the contraction joints were filled with cement grout under pressure through a system of piping after the concrete was cooled to the desired minimum temperature of 38° F. and joints were opened to their maximum width. To obtain the desired temperature and joint openings during the construction period, it was necesary to cool the concrete by means of an embedded pipe system through which river water was circulated. This artificial cooling system was then used, at very slight increase in cost,

to control cracking by lowering the temperature rise of the concrete during the early age. This, in turn, permitted the wider spacing of contraction joints than would otherwise have been permissible. The cooling system consisted of 1-inch diameter thin-wall steel tubing spaced 3 to 5½ feet horizontally and placed on the foundation rock and on the top of each 5-foot lift of concrete. No special measures for cooling the aggregate, sand, cement, or water were required. Cooling the concrete and grouting the contraction joints followed progressively as the level of construction rose, with the contraction joints being grouted as soon as possible after initial cooling was completed.

One of the most extensive laboratory investigations undertaken by the Bureau laboratories for a single project covered the studies of concrete to be placed in Hungry Horse Dam. Over 1,000 concrete cylinders and bars were tested and analyzed during the course of the work to determine the effects of various concrete mixes on compressive strengths, permeability, and durability. One result of this comprehensive program was the requirement that the contractor use fly ash, or any pozzolanic material that would produce equivalent results, in the concrete. Fly ash is a finely divided dust precipitated from the stacks of steam power plants which burn pulverized coal and is well known as an active pozzolan. (Pozzolans are materials which, though not cementitious in themselves, have the ability to combine with lime in the presence of water to form compounds which possess cementing properties.) Fly ash proved to be the least costly and most readily available of several acceptable pozzolanic materials that were tried. Fly ash made possible the use of leaner mixes, with a consequent reduction in heat of hydration. Other advantages included increased workability and durability, reduced segregation of the fresh concrete, and incidental monetary savings.

SPILLWAY

The outstanding hydraulic appurtenance of Hungry Horse Dam is the shaft-and-tunnel or "morning-glory" type of spillway. Overflow water from the reservoir plunges over the crest of the spillway near the right end of the dam and drops about 480 feet, discharging through the spillway outlet structure 550 feet downstream from the powerhouse. The entire tunnel is lined with concrete and is capable of passing a flow of 50,000 cubic feet per second. Excavation of the 1,100-foot-long tunnel required removal of about 30,000 cubic yards of rock. The major portion of the spillway tunnel has a circular section which varies from 24½ feet in diameter to about 35 feet in diameter. Then, in order to make use of a portion of the tunnel used for diversion of the river during construction of the dam, the circular

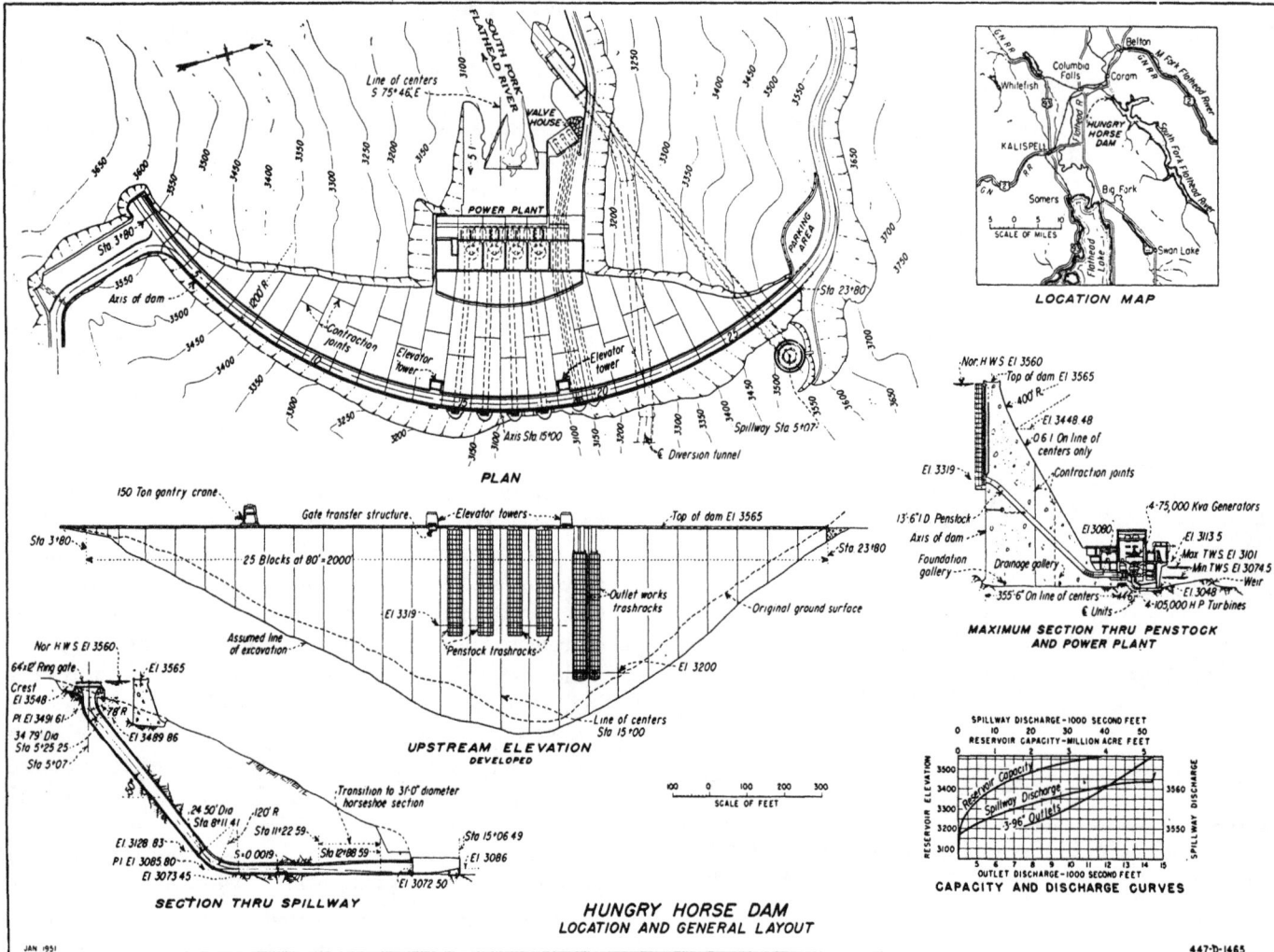

HUNGRY HORSE DAM
LOCATION AND GENERAL LAYOUT

section connects to the horseshoe-shaped section of the diversion tunnel.

The circular spillway crest is controlled by a 64-foot diameter ring gate which can be raised or lowered as desired through a distance of 12 feet. The ring gate is a closed, watertight, annular ring which is raised by buoyant force. The shape of the top of the gate and of the interior of the concrete crest structure was determined by hydraulic model studies.

The spillway outlet structure maintains the 36-foot width of the spillway tunnel. The downstream end of the spillway floor is raised 13½ feet by a long-radius curve and forms a vertical deflector which raises the high-velocity jet above the river tailwater and allows the jet to drop into the river a safe distance from the structure.

OUTLET WORKS

Three 96-inch diameter outlet pipes were provided to regulate more closely the flow of water to meet downstream requirements than would be possible through the

spillway and turbine discharges. In addition, they provide a means for rapidly lowering the reservoir in the face of an impending flood. The outlet pipes extend radially through the dam to a valve house located on the right bank of the river channel about 200 feet downstream from the power plant. Discharge through the outlet pipes is controlled by hollow-jet valves in the valve house and ring-follower gates located in the foundation of the power plant. A bulkhead-type gate is also provided for emergency or maintenance work and can be lowered over the upstream end of the outlet pipes by a 150-ton gantry crane located on the crest of the dam. The maximum capacity of the three pipes is 14,400 cubic feet per second with the reservoir surface at elevation 3560.

POWER PLANT

The powerhouse is located at the toe of the dam in the original channel and accommodates four main generating units. Each unit consists of a 75,000-kv.-a. generator driven by a 105,000-horsepower turbine. Four

Trashracks for the power penstocks were under construction on the upstream face of the dam. Trash bars were already in place on the lower portions of the outlet works trashracks, submerged in the rising reservoir.

162-inch diameter penstocks, each about 450 feet long and embedded in the dam except for a short exposed section are provided to pass the water from the reservoir to the turbines. To aid in the erection and maintenance of the generators and turbines, two 250-ton, overhead traveling cranes are provided.

GEOLOGY AND FOUNDATION TREATMENT

The foundation rock is a blue-gray limestone which contains appreciable quantities of clay, silica, and magnesia. The rock is somewhat superior to pure limestone for the foundation of an engineering structure since the impurities have increased its hardness and greatly lessened its solubility. The bedding planes of the rock are regular and range in thickness from a few inches to several feet. The planes all dip in one direction so that the stripped foundation followed rather closely the planes on the left abutment, while in the bottom and on the right side the foundation excavation cut across the planes.

Several vertical faults were uncovered during the excavation of the left abutment and river channel. These zones were effectively sealed by cut-off shafts at the upstream and downstream toes of the dam. The shafts were washed with air and water using high-velocity jets, backfilled with concrete, and sealed by high-pressure cement grouting.

The foundation was grouted to prevent seepage of water through the seams and cracks of the foundation rock. Low-pressure shallow grouting over most of the foundation was carried out prior to the placement of any concrete in the dam. Final high-pressure grouting

of the foundation to depths of from 50 to 200 feet awaited placement of sufficient concrete to prevent any displacement of the foundation rock.

RIVER DIVERSION

River diversion required a 36-foot diameter horseshoeshaped tunnel about 1,050 feet long through the right abutment to carry the flow of the river during construction of the dam. This tunnel was constructed by separate initial contract. The actual diversion and care of the river during construction of the dam was the responsibility of the general contractor, who constructed cofferdams upstream and downstream from the dam site and diverted the water through the diversion tunnel. Under the contractor's original plan, flood flows in excess of 12,500 cubic feet per second would flood the construction area and cause some delay in the construction program. During 1949, a speedup in the construction program was adopted, the diversion plan was revised, and the upstream cofferdam was raised to permit flood flows up to 40,000 cubic feet per second to pass through the diversion tunnel. By the end of the 1950 concreting season, the dam was high enough

Left abutment stripping was completed and excavation nearly finished in the river bed. The stockaded opening in the foreground is the entrance to the spillway tunnel, high on the right abutment.

In preparation for the next lift of concrete, the top of each block was cleaned, necessary form work and reinforcement for galleries placed, and the cooling coils laid and connected to the cooling headers.

so that there was no further danger of overtopping by flood flows. At the rate of concrete placement in 1951, the dam was completed to about elevation 3400 and was able to store any water coming down the river in excess of the capacity of the river outlets. The diversion tunnel was closed by placing the concrete tunnel plug during the 1951–52 winter season, after which storage began in the reservoir.

CONSTRUCTION

Construction on the Hungry Horse project commenced on September 26, 1946, when construction was initiated on the upper access road to the dam site. The first construction activity at the dam site itself was work on the diversion tunnel during the 1947–48 winter season. The first work on the dam and power plant occurred in June 1948 when the foundation excavation was started. The first concrete was placed in September 1949 and 50,000 cubic yards of concrete were placed before cold weather stopped placement in November. Concreting was resumed in March 1950 and 950,000 cubic yards were placed that year. At the end of 1951, 2,401,000 cubic yards were in place, and the last concrete in the dam was placed October 4, 1952, one year ahead of schedule. The first generator in the power plant went into service in October 1952 and the second during December.

Concrete was placed from a cableway system patterned after that used at Shasta Dam. The overhead system consisted of three cableways radiating from one fixed head tower on the left abutment to three movable

The concrete mixing plant perched on the right abutment, downstream from the excavated area. Round-the-clock activity raised the dam in the valley in a race with winter. At the lower left is the outlet of the diversion tunnel, later to become part of the spillway outlet.

tail towers operating in a single runway on the right abutment and a separate, fourth cableway which served the powerhouse area. The system of movable tail towers and fixed head towers thus formed a blanket coverage of the entire area from the spillway upstream to the power plant downstream. The main head tower, anchored to the south or left wall of the canyon, was 200 feet high and the tail towers were 52 feet high. The single cableway towers were 85 feet high and 102 feet high for the head and tail towers respectively. Each of the steel cables of the main cableway were 3 inches in diameter and about 2,400 feet long. Concrete placement was made from 8-cubic-yard capacity buckets on the main cableways and a 6-cubic-yard capacity bucket on the single cableway.

The concrete mixing plant was located on the right abutment. The plant equipment consisted of five 4-cubic-yard concrete mixers with a peak capacity of about 400 cubic yards of concrete per hour, a batching plant, an aggregate rescreening plant, and cement and pozzolan storage silos. Cement and pozzolan were hauled in special 150-barrel capacity trucks and trailers from the railhead at Coram, 6 miles away.

Sand and gravel were hauled to the gravel plant in 22-cubic-yard capacity bottom-dump trailers from the aggregate pit located about 5 miles downstream from the dam. The gravel plant was located on the flat immediately downstream from the dam and occupied an area about 200 feet wide by 2,000 feet long. The plant had a capacity of 700 tons of graded sand and gravel per hour and was equipped with crushers and a rod mill so that any deficiencies in smaller size aggregate could be overcome.

Climatic conditions were such that concrete placement was stopped during the winter months. From about the middle of November to the middle of March each year, all construction activities were reduced to a minimum with only drilling, repair, and maintenance, and other incidental work carried on. During this time the contractor established a rule of no work when the temperature at 8 a. m. was below 0° F. As an example of the effect of this rule, during January 1949 the job was completely shut down about half of the time.

The total cost of the project will be over $100,000,000. Of this total, the largest single contract was for $43,431,000 for the construction and completion of Hungry Horse Dam, power plant, and the roads and parking areas in the immediate vicinity. Furnishing and installing the power facilities is estimated at about $9,100,000. Materials furnished by the Government will amount to about $25,700,000. The remainder of the project costs include relocation of forest service roads which are to be inundated by the reservoir, temporary construction and facilities, and administration and design.

Concrete aggregate in various sizes was stock-piled along the river bank, to be withdrawn as needed and transported to the mixing plant by belt conveyor.

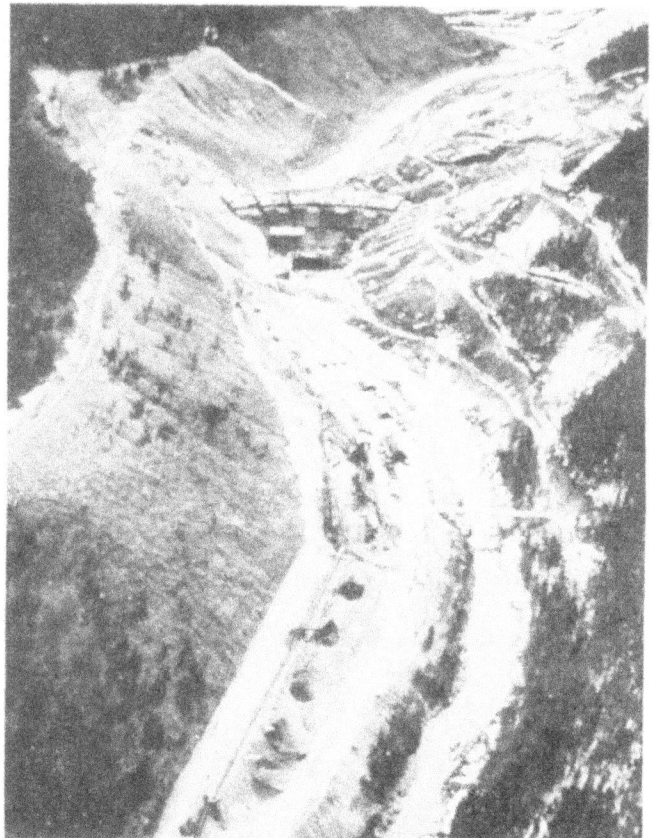

Although the severe winters of northwestern Montana compelled almost complete cessation of concrete placement, construction went forward at Hungry Horse with the erection of structural steel for the powerhouse.

GRAND COULEE DAM

COLUMBIA BASIN PROJECT, WASHINGTON

By Fred A. Houck

⌄⌄

Pioneers homesteaded much of the Columbia Basin area of eastern Washington in the latter part of the last century. A succession of several dry years drove most of them from the region and the hardy few who remained existed largely through their own efforts to irrigate with water from deep artesian wells. In 1904, the United States Reclamation Service first investigated the possibility of irrigating an area along the Columbia River in eastern Washington. In 1918, local interests suggested a plan to build a dam on the Columbia River high enough to divert water into the famed Grand Coulee—the river bed of the Columbia during the ice ages—which would carry the water by gravity for more than 50 miles to the northern end of the irrigable area at Soap Lake. The Washington State legislature subsequently created the Columbia Basin Survey Commission to study these possibilities. In a report published in 1920, the Commission discussed a dam and pumping plant on the Columbia at the head of the Grand Coulee, but concluded that the dam would be impracticable—perhaps impossible. The report favored an all-gravity plan which would take water from the Pend Oreille River in Northern Idaho.

In 1932 a comprehensive report prepared by the United States Corps of Engineers concluded that the development of the Columbia River for water power, irrigation, and flood control would best be served by construction of the Grand Coulee Dam and several other downstream dams. Subsequently, the Bureau of Reclamation reviewed all the Columbia River investigations and issued a report recommending development of the Columbia Basin project substantially as it exists today.

In 1933, the President approved the Bureau's review report and allocated funds for construction of the Grand Coulee Dam. The first construction contract was awarded on July 16, 1934.

LOCATION

In its generally southwesterly course from Canada to the Pacific Ocean, the Columbia River in central Washington forms a half loop known as the Big Bend, between the mouths of the Spokane and Snake Rivers. During one of the ice ages, a glacier extended across the present river channel, diverting the flow across the open eastern side of the Big Bend loop. By the time the ice age had passed, this diverted water had eroded what is now known as the Grand Coulee, so that when the river returned to its present channel the Coulee was left as a potential high-elevation irrigation storage reservoir—waiting only to be sealed at each end and filled with water. The Grand Coulee Dam and power plant are located in the present Columbia River Channel adjacent to the upper end of the Grand Coulee, about 90 miles northwest of Spokane, Wash.

FUNCTIONS OF GRAND COULEE DAM

Grand Coulee Dam is the key to the development of power on the Columbia River—the greatest potential source of energy among the rivers of America. Present plans contemplate, ultimately, 10 dams on the Columbia between the Canadian border and the mouth of the river to utilize 92 percent of the fall of the river in the United States. Grand Coulee Dam accounts for 27 percent of the total drop, providing the energy for the Grand Coulee power plant which, with its 18 generating units, is rated as the largest single power producer in the world today. From Lake Roosevelt—the reservoir backed up by Grand Coulee Dam—the Grand Coulee pumping plant, containing the largest pumping units ever built, lifts the water required for irrigation 280 feet into the storage reservoir formed by sealing off both ends of the dry gorge of the Grand Coulee.

Aerial view shows the pumping plant at the dam in foreground, the silver-painted discharge pipes which carry the water uphill from the world's largest pumps inside the plant, and the feeder canal on the hill, where the water discharges from the pipes.

This perspective of the irrigated area of the Columbia Basin project illustrates the key position of Grand Coulee Dam in the development of the project. Here a million acres of farmland will be brought under cultivation and supplied with water delivered from the elaborate system of canals depicted.

In addition to its irrigation and power functions, Grand Coulee Dam is a potent factor in controlling the floods on the Columbia River.

RESERVOIR (LAKE ROOSEVELT)

Extending from Grand Coulee Dam upstream 151 miles to the Canadian Border, Lake Roosevelt has a shore line of 600 miles, a surface area of 82,000 acres, a total storage capacity of 9,517,000 acre-feet and an active or usable storage capacity of 5,220,000 acre-feet. The drainage area contributing to the reservoir is 74,100 square miles, which in the period prior to existence of the dam produced a maximum annual discharge of 103,000,000 acre-feet and a maximum momentary discharge of 725,000 cubic feet per second. Since completion of the dam, a maximum momentary discharge of 637,800 cubic feet per second has been recorded.

DESIGN

Original designs were prepared for a low dam and a complete power structure with provisions for subsequent enlargement to a high dam. Various types of dams were studied—slab and buttress, massive head buttress, and multiple arch—but it was concluded that only a massive, gravity-type structure could accommodate itself to the diversion requirements during construction and to the ultimate development planned.

This low dam would have been approximately 3,500 feet long, 350 feet high from the lowest foundation, and would have impounded 250,000 acre-feet of water with the water surface 150 feet above the low water level in the river. Because the low dam would have been eventually consolidated within the high dam, the original design omitted regulating gates and bridge from the 1,800-foot spillway section.

Water pours smoothly down the spillway of the completed dam. Franklin D. Roosevelt Lake, behind the dam, extends 151 miles to the Canadian border.

The original design contemplated completion of the west end section of the powerhouse, some distance downstream from the left abutment of the low dam, and installation of three main generating units and two service units capable of producing 520,000 kilowatts. For future generating units, penstocks were to be embedded in the concrete of the low dam and temporarily capped. A permanent concrete cofferdam would have been placed downstream in the correct location to form the toe for the high dam.

In the hectic days of 1933, the construction of the low dam started under the $63,000,000 initial P. W. A. allotment. Meanwhile, politics and policy had injected other factors into the Northwest Federal relief-works program; other important projects were under construction in the region which would temporarily remove the demand for power from Grand Coulee; and severe drought conditions had become widespread, causing large numbers of people to vacate their farms. The low dam under construction could afford but limited relief to agriculturists, owing to the inadequate

amount of storage and the cost of pumping water 500 feet above the reservoir elevation to a comparatively small acreage.

It was recognized by the designers from the outset that certain objectionable features were involved in building the high dam in two stages. One of these was the difficult task of securing a tight joint between the low dam and the additional mass concrete to be placed in enlarging it to the high dam so that accumulation of dangerous hydrostatic pressure in the joint would be prevented. Another was the uneconomical adaptation of the turbines in the power plant from low- to high-head conditions. Finally, extensive slides within the foundation area during the excavation operations for the low dam had established the likelihood of recurring slides and the desirability of placing the entire concrete base for the high dam to avoid duplication of the excavation work.

The disadvantages of completing the low dam in accordance with the original plans were so cumulative that an order for changes was issued to the construction

Hemispherical bulkheads closed the upstream ends of the power penstocks for the right powerhouse until turbines and generators were installed.

THE DAM

Grand Coulee Dam is a straight-gravity type, massive concrete structure consisting of a river overflow spillway section, 1,650 feet long, and adjoining abutment sections which give the dam a total crest length of 4,173 feet. It is founded on excellent granite from the lowest point of which the dam rises to a height of 550 feet. The dam has a mass of 10,585,000 cubic yards of concrete. More detailed information of technical interest may be found on the accompanying drawings.

In the central portion of the length of the dam the horizontal loads are carried by gravity action to the foundation. In the vicinity of the abutments, part of the horizontal load is distributed to the foundation and the remainder is carried horizontally to the abutments by a twisting action. To accommodate this twisting action in the locality of the abrupt change from the horizontal planes of the rock floor to the sloping planes of the rock abutments, transverse vertical slots, three on the east side and two on the west side, each 6 feet wide and extending the full height of the dam, were left at intervals in the vicinity of each abutment where the effect of twist was most pronounced. Heavily reinforced concrete bulkheads were cantilevered longitudinally across the ends of the openings with their exterior faces forming the upstream and downstream faces of the dam. Flexible joints were provided to permit independent deflection between the two sections thus connected. The slots were filled with sand prior to grouting all contraction joints. Later, when the structure had deflected as a result of the filling of the reservoir, the sand was removed by jetting and the slots filled with concrete.

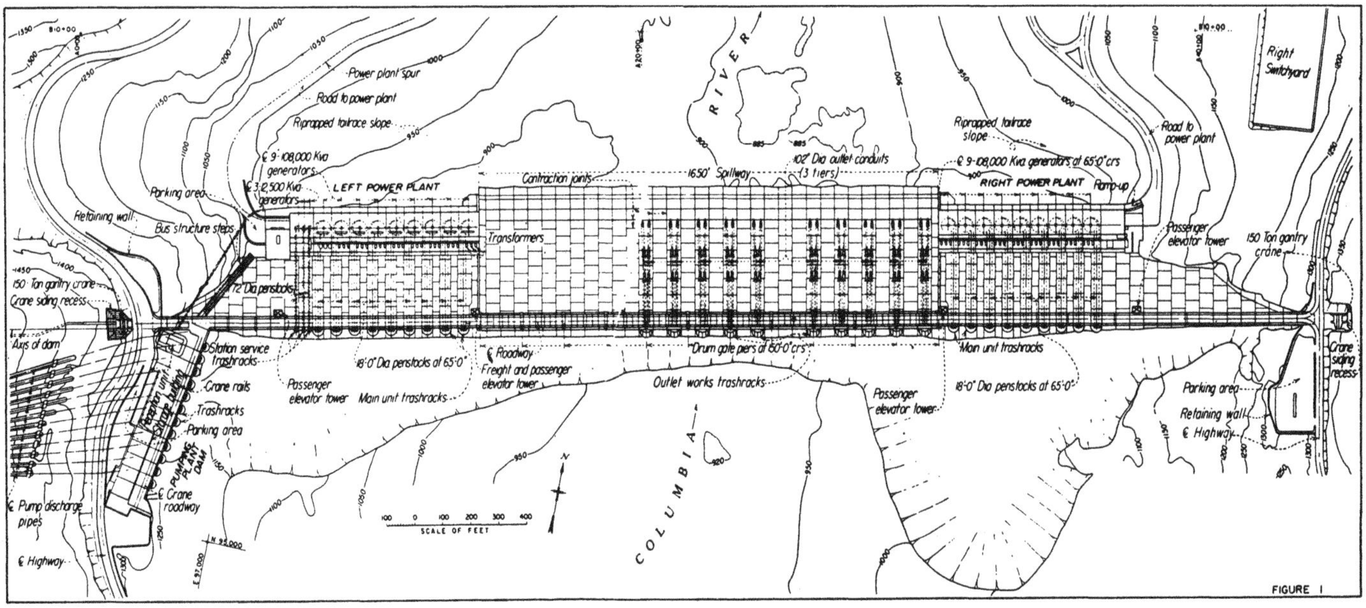

contractor on June 5, 1935. The order provided for prosecuting construction in accordance with designs which eliminated the original concept of an initial low dam and partial power plant. Under the changed initial construction contract, all the river bed and channel excavation for the high dam was performed and the lower portion of the high dam was constructed. Under a subsequent contract, the dam was completed to full height and the powerhouse constructed.

Grand Coulee power plant as constructed consists of two powerhouses, each housing nine generators with rated capacities of 108,000 kilowatts. The total installed nameplate capacity is therefore 1,974,000 kilowatts, but the plant has operated for long periods at a capacity of about 2,200,000 kilowatts.

FIGURE 1

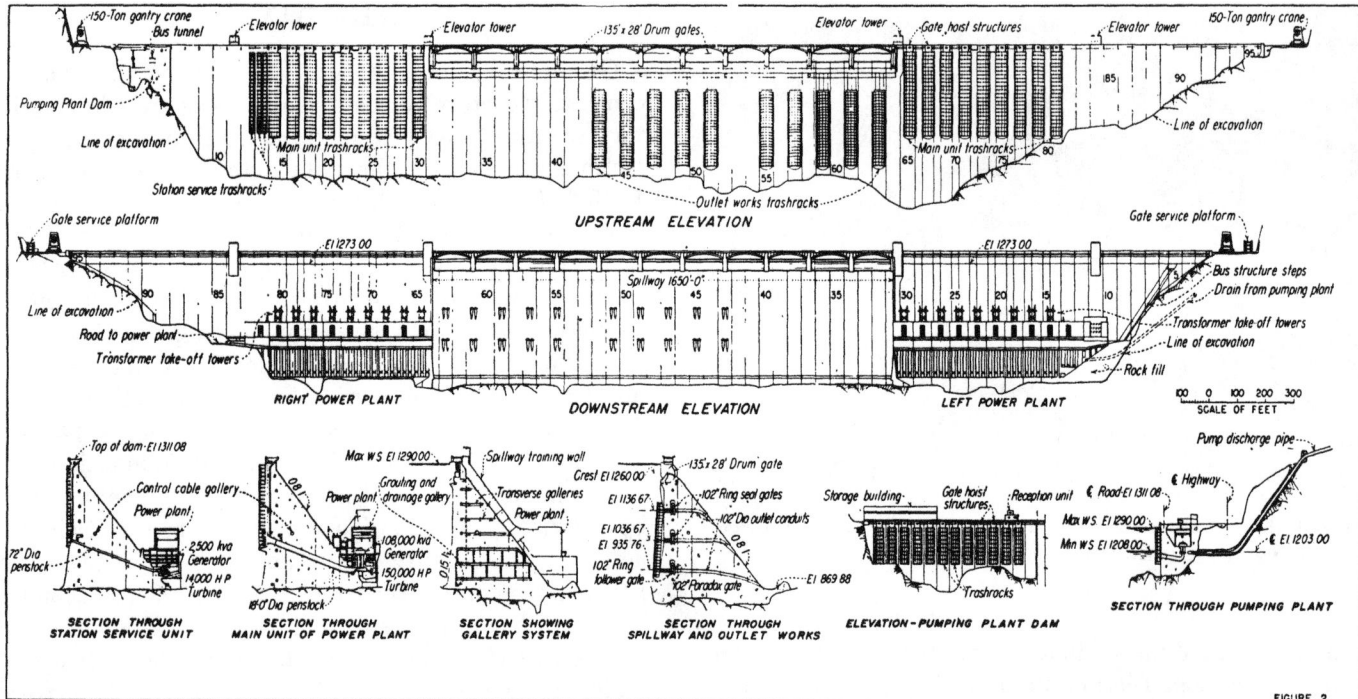

FIGURE 2

SPILLWAY

The spillway is designed to discharge 1,000,000 second-feet with an effective head on the crest of 30.5 feet when the drum gates are down. The spillway consists of eleven 135-foot openings, separated by ten 15-foot-thick concrete piers. A structural steel drum gate, 28 feet high and 135 feet long, is mounted on the crest of each opening to regulate the reservoir level between spillway crest elevation 1260 and the raised position of the gates at elevation 1288.

Gate movement is effected hydraulically by filling the recessed crest chamber with water to raise the gate, or by emptying the chamber to lower the gate, both operations being performed automatically by a control mechanism which raises or lowers the gates through a predetermined range to correspond with an increase or decrease in the reservoir water surface elevation. Complete lowering of the gates from the extreme raised position to obtain the maximum discharge is manually controlled. The gate operating chambers and controls are located in the piers near the upstream face, the chambers being interconnected by a longitudinal operating gallery.

Tests on 14 hydraulic models of various types, representing designs from the embryonic to the final stage, were made in evolving the adopted design of the spillway. The studies and tests were responsible for eliminating the apron type of energy dissipator in favor of the roller-bucket type having a continuous lip rather than dentated sills. An apron would have involved tremendous amounts of additional concrete and rock excavation, and the studies showed that it would seriously interfere with the tailrace discharges from the power plants.

RIVER OUTLETS

Thirty pairs of outlet conduits, each 102 inches in diameter, arranged in groups of 10 pairs in 3 tiers which differ in elevations by about 100 feet, are located in the spillway section to regulate reservoir outflow during periods of normal operation. This arrangement results in a more uniform distribution of the discharging jets over the downstream area, and also permits the use of a single trashrack structure to serve three pairs of outlets and a single gate operating chamber to serve a pair of outlets. The arrangement and control is better visualized by referring to the plan, elevations, and sections.

The upstream section of each 102-inch diameter outlet conduit is lined with steel for a distance of 51 feet from the upstream face. The outlet control gates are included within this length. Beyond this point the conduits are formed in the mass concrete, which is heavily reinforced around the openings. The conduits emerge from the downstream face of the dam with the direction of discharge such that the water flows on the spillway surface rather than springing clear.

The discharge from each conduit is controlled by a motor-operated service gate and a hydraulically-operated emergency gate, installed in tandem near the up-

The dam was nearly completed and the great drum gates for the spillway were being erected. The piers between the gates are surfaced with electrically heated plates to prevent ice formation.

stream end of the conduit. The gates are designed to operate under a maximum head of 250 feet, providing for a total maximum discharge of 275,000 cubic feet per second with all gates open and the reservoir water surface at elevation 1184. With the reservoir surface above this elevation, the gates of the lower tier of conduits are closed and the discharge correspondingly reduced. The intermediate tier of gates operate under a maximum head of 250 feet. The upper tier of gates operate under a head of 153 feet with maximum water surface at elevation 1290.

CONCRETE COOLING

The problem of dissipating the heat generated by the setting of the cement in the mass concrete of Grand Coulee Dam was solved by means of an embedded artificial cooling system similar to the one used for Hoover Dam. The cooling plant consisted of a system of 1-inch metal tubing embedded in the concrete as construction progressed. Circulation of unrefrigerated river water through the system proved satisfactory in reducing the temperature of the concrete to 45° F.—the mean annual air and water temperature to which the dam is exposed.

The pipes were laid on the top of each 5-foot lift of concrete and spaced 5½ feet apart. The embedded pipes approximate a combined length of 2,000 miles.

Cooling of the concrete progressed upward. When the concrete served by a circuit was cooled sufficiently, the header system was dismantled and reinstalled at a higher elevation. This helped limit the total requirement to approximately 500 tons of header piping, valves, and fittings, for the entire cooling program.

GROUTING AND DRAINAGE

Prevention of seepage under the dam was undertaken by an extensive program which had as its objective the filling of all seams or crevices in the foundation area with a grout mixture of cement and water. The mixture varied according to the receptivity of the rock, and pressures used to force the grout into holes drilled in the rock varied from 100 to 500 pounds per square inch. Great care had to be exercised throughout the program not to displace the foundation rock or concrete that had been placed.

Prior to placing any concrete, the upstream portion of the foundation to an average depth of 30 feet along a strip 100 feet wide, parallel with the axis of the dam, was grouted at low pressures to consolidate the surface rock into a blanket preparatory to subsequent high-pressure grouting. Other areas in the immediate vicinity which indicated the presence of seams or fractures were likewise grouted at low pressures.

After the dam foundation concrete was in place and before the upstream toe of the dam was submerged, inclined holes at 20-foot spacings were drilled approxi-

Outlet pipes pass through the dam in the spillway section. Flow through these outlets is controlled by high-pressure gates which were installed in tandem, providing a service and an emergency gate for each outlet.

mately 50 feet into the foundation rock through pipes embedded in the concrete fillet at that point. These holes were grouted at moderate pressures.

The final foundation grouting was performed from a longitudinal grouting and drainage gallery located near bedrock, 17.5 feet downstream from the axis. From this gallery a curtain of holes averaging 10 feet apart and 150 feet deep was drilled through the intervening concrete foundation and into the bedrock. The grouting of these holes was done at high pressures.

After completion of the final curtain grouting, a line of drain holes at 10-foot centers was drilled from the same gallery approximately 50 feet into the foundation to minimize the possibility of uplift pressures against the base of the dam resulting from water which might accumulate despite the careful grouting. Drainage is effected by pumping from the foundation gallery sumps into galleries at higher elevations, whence the water is wasted by gravity.

The dam was constructed in unit blocks 50 feet square to allow for controlled contraction of the concrete as a result of cooling. Copper grout stops were installed across the joints between adjoining blocks in tiers to confine the grout to the joints, and additional metal stops were installed across the transverse joints at the upstream and downstream faces to prevent escape of the grout or the entrance of water. After the concrete had been cooled, the joints were grouted at moderate pressures from the galleries in the dam.

Five-inch diameter porous concrete drain tile was installed vertically at 10-foot centers, 13 feet in from the upstream face of the dam. These drains run into the longitudinal galleries within the dam and serve as interceptors for water that may seep into the concrete.

PRELIMINARY CONSTRUCTION

Somewhat in the background but of vast importance in building the dam were several "smaller" tasks, each one, however, a good-sized job in itself. Before the dam construction could effectively get under way, access, power, and housing and working facilities had to be provided. The Columbia River was bridged, first with a temporary pile trestle bridge and then with a permanent bridge. An access railroad, 30½ miles long, was built from the nearest existing railroad through the Grand Coulee to deliver the vast quantities of materials that went into the construction of the dam. Two 30-mile-long, hard-surfaced highways to the site were built and power was brought in on a high-tension transmission line of equal length. Two modern and complete towns were built at the dam site—Mason City, the contractor's town, on the east side of the river and Coulee Dam, the Government town, on the west side.

DIVERSION OF RIVER

The general plan was to unwater the two shore ends of the dam site by cofferdams, excavate to bedrock, and place concrete within the enclosures up to a predetermined height, leaving the river channel for final closure. The scheme was based on a possible flood of 550,000 cubic feet per second with upstream cofferdam at elevation 1000 and the downstream at 990.

The west or left side cofferdam was built first and was completed prior to the flood season of 1935. The cofferdam enclosed a 60-acre area of dam site and was formed of timber cribs faced with sheet-steel interlocking piles and a series of gravel-filled cells of sheet-steel piling along the river. Inside this cofferdam, the two blocks (blocks 39 and 40) closest to the river and the major portion of the dam adjacent to the abutment were constructed to elevation 1000 with the intervening portion of 350 feet left at a lower elevation to become the bypass channel during the next stage of diversion.

Concurrently with construction of the west portion of the dam, the east abutment portion of the dam was constructed behind the protection of a cofferdam somewhat lower than the west cofferdam. The east cofferdam was overtopped once, with delay the only damage.

After the west end of the dam base was completed, the flow of the river in its natural channel was stopped by a downstream cross-river cofferdam extending from the east bank to the west cofferdam. The river channel portion of this cross-river cofferdam consisted of timber cribs, built to fit the contour of the river bottom. The cribs were floated into place, sunk by loading with gravel, and protected with a facing of sheet-steel piles and fill. A similar upstream cross-river cofferdam, extending to the end of the west side concrete structure (blocks 39 and 40), completed the enclosure of a 55-acre area which included the natural river channel. In constructing the cross-river cofferdams, about 9,000,000 board feet of lumber, 950,000 cubic yards of fill, and 2,200 tons of sheet-steel piling were used.

With the west cofferdam breached to allow the river to flow over the low blocks of the west portion of the dam, unwatering within the east cofferdam was started January 3, 1937. The impounded water—approximately 80,000,000 gallons—was pumped out in 6 days. Placing of concrete in the center section of the dam was begun in May of 1937. Removal of the cross-river cofferdams was begun in the fall of the same year and in November the river was flowing through low blocks of the spillway section of the dam. By January 1938, when the first construction contract was completed, the cofferdams had disappeared from the river channel, forebay, and tailraces, and the results of over 4 years of work were almost concealed by the waters of the Columbia.

The river poured through low blocks left for the purpose in the spillway section of the dam. In the canyon wall in the background appear the tunnel openings for the pump outlets.

Diversion continued through low blocks of the dam during completion of the dam under the contract awarded January 28, 1938. Alternate blocks were raised in sequence so that the river passed through the intervening slots. Closure of each slot was effected behind a huge gate 35 feet high and 52 feet wide. The gates, weighing 75 tons each, were handled and floated into position at the upstream face by means of a large barge.

Final closure of that last diversion slot was made September 20, 1939, by which time a sufficient number of outlets had been installed to pass the flow of the river during the remainder of construction.

EXCAVATION

One of the most gigantic tasks in the construction program was the excavation and disposal of approximately 15,000,000 cubic yards of foundation overburden and 2,000,000 cubic yards of rock from the dam site.

The lack of suitable areas near the dam site sufficiently large to accommodate these volumes of waste plus the problem of a 540-foot difference in elevation between the canyon rim and construction site resulted in installation of an elaborate belt conveyor system.

The material was excavated by large electric shovels and loaded on fleets of dump trucks and tractor-drawn trailers, varying in capacity from 8 to 20 cubic yards. The material was dumped by the trucks and trailers onto four feeder conveyors which converged to a hub at the main conveyor. A surge hopper and feeder at the hub protected the main conveyor from overloading. When incoming material began piling up at the surge feeder, all tributary belts were stopped until the overload disappeared.

The length of the main 60-inch-wide conveyor was approximately 1 mile. A stacker at the end of the belt permitted uninterrupted delivery as the end was moved along the face of the dump or extended outward with the advancing fill. The belt was designed to transport 2,500 cubic yards of material per hour and achieved a maximum daily performance of 50,700 cubic yards.

In the geologic history of the dam site, the entire mass of glacial overburden had been disturbed many times to a greater or lesser degree from the original bedding. Consequently, it was not unexpected when several major slides occurred during the progress of the excavation. Early in the job, a slide of approximately 2,000,000 cubic yards of earth moved across the road and railway leading to the dam site operations. Another minor but troublesome slide in the east forebay area was controlled by the contractor who ingeniously froze the creeping mass into an arch dam.

The necessity for rapid filling of the cells for the west cofferdam in advance of the 1935 flood season was responsible for the development of a shuttle conveyor system, another of the unique features for handling materials on this huge job. The fill was transported by belt conveyor to a hopper located midway of the length and about 200 feet shoreward from the row of cells. The hopper straddled a track 1,600 feet long on which an 870-foot-long traveling conveyor could travel 800 feet in either direction with some portion of the belt always under the hopper. Two 200-foot boom arms were mounted at each end of the traveling conveyor and discharged the fill directly into the cells.

AGGREGATES

Aggregates for the concrete were sorted into four coarse sizes of gravel and three sizes of blended sand. The pit was located 1½ miles downstream from the dam, and 900 feet above the river on the east side. Power shovels transferred material from the pit directly to a belt conveyor system, and the belt conveyer delivered it to the washing and screening plant which

Man's ingenuity in overcoming obstacles was never more strikingly illustrated during the progress of this biggest of all construction jobs than in the temporary frozen cofferdam shown here. A mud slide into the construction area was controlled by driving pipes into the ground at close intervals, circulating brine through them, and freezing the ground until the concrete was finished above this point.

had a rated capacity of 1,250 tons of finished aggregate per hour.

Another belt conveyor system transported the finished aggregate from the plant to the live storage piles near the east end of the dam site, whence other conveyors supplied the concrete mixing plants.

CONCRETE PLANT

The mixing plants set up for this job established the then world's record of 20,684 cubic yards of concrete in 24 hours. During the month of October 1939, 536,264 cubic yards of concrete were manufactured and placed in the dam.

The first contractor built a concrete mixing plant on each side of the Columbia River, conveniently located for placing the foundation concrete of the dam. The second contractor moved the two plants and reassembled them into one plant high on the east abutment. Each mixing plant, constructed of structural steel except for reinforced concrete supporting columns and mixing floor, was octagonal in plan, 42 feet in diameter,

With the river prevented from entering the excavation area by the cofferdam paralleling the stream, excavation and rock stripping proceeded apace. An elaborate system of spoil disposal by means of conveyor belts sprawled over the hillside.

Rising concrete blocks encased the legs of the trestles. Hammerhead and Whirley cranes spotted concrete buckets at a number of different points.

and 102 feet high. The top of the plant was devoted to aggregate and cement storage space. From here the materials passed downward to the weighing batchers and thence into a battery of four 4-cubic-yard mixers which discharged the concrete into 4-cubic-yard, bottom-dump buckets. These were hauled in units of four on a flat car by Diesel-electric locomotives to the point of placement via three steel trestles, two with three standard-gage railroad tracks and the third with four. The foundation portions of the trestles are embedded within the dam.

The trestles also carried revolving traveling cranes with a reach of 165 feet. The cranes handled the buckets for placing the concrete in the forms, moved the forms, and placed penstocks, gates, and other heavy items which went into the building of the dam and power plants.

OPERATION AND MAINTENANCE

Two features of construction were deferred to such time as their need should develop and their requirements should be better understood. Actually, the features are items of maintenance of vast practical importance to the operation of this great plant.

For design purposes, the hydraulic models indicated the proper type of energy dissipator for the spillway but could not reveal all future actions of the prototype when in operation. Early in 1943, the first underwater inspections of the spillway bucket revealed that severe erosion in the form of potholes and torn-out chunks had occurred in many places on the curved surface and lip of the bucket and on the lower portion of the sloping face of the spillway. The damage was predominantly the result of abrasive grinding or impact of large volumes of rocks, gravel, sand, and other debris from the streambed and cofferdam deposits carried into the bucket by the vicious river currents during diversion and subsequent power plant operation.

The difficult problem of making repairs to the spillway bucket was solved by a special caisson shaped to fit the lower face of the spillway and the curve of the bucket. The caisson is floated out to position, moored, sunk, and anchored to the downstream face during periods of the year when the discharge over a portion of the spillway can be stopped. The repair of the damaged concrete surfaces by placing new concrete is accomplished within the caisson. When the section covered by the caisson has been repaired, the caisson is refloated and moved to another section of the bucket or anchored in a specially prepared drydock on the right bank of the river.

In designing the dam, it was realized that there was

Concrete placement began while excavation clean-up was still in progress. The construction trestles were extended along the length of the dam as concrete placement proceeded.

The floating caisson can be moved to the foot of the spillway, where it is sunk in place, providing working space to permit repair of the spillway bucket.

no way to determine in advance how much downstream bank protection should be provided for spillway and power plant discharges or for a large flood. The flood of June 1948 was the second largest known flood in the river's history and it did considerable damage to the riprap along the riverbanks below the dam. From the knowledge of experience, the main river channel downstream from the spillway bucket and the right and left powerhouse tailrace areas were dredged and all underwater obstructions in the main river channel downstream from the spillway bucket were removed; the left bank of the river immediately downstream from the left powerhouse for a distance of approximately 2,500 feet was excavated and resloped and covered with a blanket of riprap and armor rock weighing as much as 8 tons; underwater riprap, armor rock, and concrete blocks were placed on the right tailrace slopes; and drainage facilities were constructed in unstable riverbank areas or, where possible, the unstable areas were unloaded by removing overburden material.

On a dam of this size the normal continuing maintenance operations assume the proportions of major construction activities, from which it may be concluded that Grand Coulee Dam will always be a scene of interest.

HOOVER DAM

BOULDER CANYON PROJECT, ARIZONA-CALIFORNIA-NEVADA

By Arthur W. Arlt

Hoover Dam, an undertaking of unprecedented size and scope, is the monument of the first successful control of the mighty Colorado River. Several attempts were made to control the river and put it to beneficial use before the building of Hoover Dam, but all ended sooner or later as costly failures. As early as 1901 an attempt was made to use the Colorado to irrigate the Imperial Valley of California, which lies below sea level. For 4 years the only difficulties encountered were those caused by silting, but in 1905 the Colorado burst its banks, inundating the Imperial Valley and its flourishing communities. For 2 years the flood raged, making a waste of the land and leaving as a permanent memento a huge lake nearly 300 square miles in area, which is today known as the Salton Sea. In 1909 a million-dollar levee was built to avert a similar disaster. Shortly after completion of this levee the river crashed through in a new direction, and was kept in this course for 10 years by another expensive levee system. In the meantime, deposition of silt by the river raised the river bed and increased the cost of maintaining the levees. Besides the constant threat of flood, there were years when the river flow was so reduced that it was not possible to provide sufficient water for survival of crops and livestock. The necessity for the permanent control of the Colorado River became imperative.

INVESTIGATIONS

Investigations on the Colorado River which finally led to the construction of Hoover Dam were started by the Reclamation Service in 1904.

After a prolonged examination of available storage sites in the upper part of the main valley a reconnaissance was conducted on the river below the mouth of the Virgin River. The preliminary studies resulted in the concentration of work on the better dam sites in Boulder and Black Canyons in 1918. From 1918 until contracts for construction were awarded in 1931, investigations, surveys, foundation explorations, geological examinations, and engineering studies on feasibility and cost were carried forward for the Boulder Canyon and alternative sites. The Weymouth report, submitted by the Chief Engineer of the Bureau of Reclamation to the Secretary of the Interior in 1924, contains much of the data derived from these investigations.

Congress, by joint resolution (45 Stat. 1011) of May 1928, directed the Secretary of the Interior to appoint a board of five engineers and geologists to review the plans and estimates for the Boulder Canyon project and report on the safety, the economic and engineering feasibility, and the adequacy of the proposed structure and incidental work. Following a favorable report by this board in November 1928, the Boulder Canyon Project Act was passed by Congress and signed by the President on December 21, 1928.

The purpose of Boulder Canyon project is defined in the act as follows: "Controlling the floods, improving navigation and regulating the flow of the Colorado River, providing for storage and for the delivery of the stored waters thereof for reclamation of public lands and other beneficial uses exclusively within the United States, and for the generation of electrical energy as a means of making the project herein authorized a self-supporting and financially solvent undertaking."

RESERVOIR

The drainage area above the dam amounts to 167,800 square miles and the average annual run-off is 12,-945,000 acre-feet, varying from a maximum of 25,200,000 acre-feet in 1909 to a minimum of 4,186,000 acre-feet in 1934.

Lake Mead, created by Hoover Dam, has a maximum capacity of 31,047,000 acre-feet of water, at which stage the water surface area is 157,700 acres. The lake extends 115 miles up the Colorado River, through Boulder Canyon, Virginia Canyon, Iceberg and Tra-

Hoover Dam today.

Dark waters of Lake Mead contrast sharply with the concrete of the dam and the surrounding terrain.

vertine Canyons, and into the lower end of Grand Canyon. The width varies from several hundred feet in the canyons to a maximum of 8 miles.

A total of 9,500,000 acre-feet of storage capacity is reserved primarily for flood-control purposes, with incidental use for the production of secondary electrical energy. The remaining capacity of 21,642,000 acre-feet is used for regulation of water for irrigation, for firm power production, and as dead storage for creating power head and providing a silt pocket.

DAM SITE

Hoover Dam is located in Black Canyon about 30 miles southeast of Las Vegas, Nev. The most char-

acteristic features of the canyon are reddish-black cliffs and exceedingly precipitous slopes. The relatively open arroyos of the uplands approaching the main canyon become narrow ravines or gorges through which the streams from occasional rains plunge to the river over successive falls. The canyon walls at the dam site rise several hundred feet above the river.

The rocks forming the foundation and abutments for the dam are of volcanic origin. They are of excellent character, hard, strong, and practically impermeable. Such faults as occur in the region are not active or likely to become so. Neither are they associated with any soft, crushed, or broken material such as would cause trouble in connection with the dam.

PRELIMINARY CONSTRUCTION

Preparations for the construction of Hoover Dam involved many diversified operations of considerable magnitude. It was necessary to provide access to the site for transportation of supplies and equipment. Housing and municipal facilities had to be constructed to accommodate a population of 5,000 in the middle of a desert. Power had to be provided for construction and domestic purposes.

The Union Pacific Railroad Company constructed a branch line, 22.7 miles in length, from a point near Las Vegas, Nev., to the site for Boulder City. Under a contract with the Lewis Construction Company of Los Angeles, Calif., the Government extended this line 10½ miles from Boulder City to the rim of Black Canyon, directly above the dam site, at a cost of $653,-551.03. Taking off from the end of the Government line, Six Companies, Incorporated, general contractors for construction of the dam, built 26 miles of railroad, one branch extending to the Arizona gravel deposits just above river elevation approximately 8 miles upstream from the dam site, and another to the Nevada side of the gorge at the dam site.

The State of Nevada constructed a new highway between Las Vegas and Boulder City, a distance of 23 miles. This highway was extended by the Government to the rim of Black Canyon under a contract with the General Construction Company of Seattle, Wash., providing for a gravel-base, oil-surfaced highway, 22 feet wide and 8.3 miles in length, including side roads. From the Government highway, Six Companies, Incorporated, constructed a road which descended to the lower portals of the Nevada diversion tunnels and a steel suspension bridge across the river to provide access to the Arizona tunnels. Several miles of additional roads also were constructed by the contractor to reach various parts of the work.

The contract for furnishing power was awarded to the Southern Sierras Power Company. A 222-mile transmission line, to deliver power from generating stations near San Bernardino, Calif., was constructed to the dam site. The line, together with a substation near the Nevada rim of the canyon, was constructed by the power company at a cost of approximately $1,500,000. It now serves as a permanent installation to transmit power from the dam. Auxiliary power lines were erected from the substation to various parts of the work, and a 7-mile line from the substation to Boulder City was constructed by the Government.

High temperatures at the dam site during the summer months necessitated the selection of a town site where more comfortable living quarters could be provided. Boulder City was located on the summit of the divide about 7 miles from the dam site, at an eleva-tion 2,500 feet above sea level and 1,855 feet above the river.

DIVERSION TUNNELS

To unwater the dam site, four huge tunnels, two on each side of the river, were driven through the canyon walls to carry the flow of the river during construction. These tunnels, circular in cross section, were excavated 56 feet in diameter and lined with 3 feet of concrete to make a finished section 50 feet in diameter. The combined length of the four tunnels was 15,946 feet and required the removal of approximately 1,500,000 cubic yards of rock. The rock proved to be of excellent quality and no support was required for any portion of the length. The general plan of tunnel advancement was to drive a 12- by 12-foot pioneer heading at the top of the ultimate cross section. The enlargement followed the pioneer bore to the full width of the tunnel, but only 41 feet high, leaving 15 feet in the invert to be excavated at a later period.

A truck-mounted drill carriage with platforms at four levels, carrying a total of 30 drills and wide enough to drill one-half of the bench in one operation, was devised to carry on the drilling operations.

Mucking was carried on in the tunnel enlarging operations by 100-ton electric shovels, equipped with 3½-cubic-yard buckets. The muck was loaded on 10-ton trucks and hauled up roads along the canyon walls to the disposal areas in side canyons.

For removal of the invert section the same drilling carriage with the top part removed and with two folding wings built on either side was used, and the whole of the invert section was drilled in one operation.

For the purpose of applying concrete lining, the circular section was divided into three parts: the lower or invert section of 74°, to be placed first; the side-wall portions of 88° each, above and on either side of the invert, to be placed second; and above the side-wall sections the roof or arch section of 110°, to be placed last.

Concrete operations were started at the upstream portals and carried progressively toward the downstream ends of the tunnels. All concrete was mixed at the low-level mixing plant located a short distance upstream from the tunnel inlets on the Nevada side of the river. Truck transportation was used exclusively, consisting of either two 2-cubic-yard form buckets carried on platform trucks, or 4-cubic-yard agitator bodies.

Invert concrete was placed by a 10-ton, electrically operated gantry crane. For the side-wall portion of the concrete lining, a huge structural steel framework or jumbo, weighing about 270 tons for an 80-foot section, was provided to support the wall forms. The jumbo was completely equipped to handle concrete; an electric crane operated on top of the jumbo and delivered concrete through a system of chutes.

The forms and concrete placing equipment for the arch portion included an 80-foot length of form supported by a structural steel framework or jumbo, a separate gun carriage containing two 2-cubic-yard pneumatic concrete guns with the necessary hoisting equipment, and a separate traveler connected to the gun jumbo to support two placement pipes.

Low-pressure grouting was performed in the arch section of the tunnels to fill the voids between the lining and rock. High-pressure grouting, using pressures from 100 to 500 pounds per square inch, was accomplished through rings of eight holes, drilled radially into the rock to a depth of 24 feet, holes in successive rings being staggered. Drainage holes were drilled after grouting was completed, to drain water from back of the lining. A total of 122,000 linear feet of grout holes were drilled in the diversion tunnels and more than 200,000 cubic feet of grout injected.

The river was diverted through the two Arizona tunnels on November 13, 1932, less than 2 years after excavation of the tunnels was commenced in June 1931, and almost a year in advance of the date contemplated at the time the contract was awarded.

COFFERDAMS

With the river diverted through the tunnels during the winter season of low flow, work was concentrated on the construction of the cofferdams to complete the river diversion program before the arrival of the spring floods. The upstream cofferdam, an earth- and rock-fill structure was located 600 feet below the diversion tunnel portals. Approximately 250,000 cubic yards of river silt and loose deposit were removed to secure an adequate foundation of consolidated sand, gravel, and boulders. The structure as completed was 98 feet from base to crest, which was 30 feet higher than the top of the diversion tunnels. The dam was 450 feet in length, 750 feet thick at the base, and contained 516,000 cubic yards of earth and 157,000 cubic yards of rock. The upstream face, of 3 to 1 slope, was protected by 6-inch concrete paving, laid on a 3-foot thickness of rock blanket; and the downstream slope of 4 to 1 was covered by a heavy rock fill.

The downstream cofferdam was a rolled earth-fill structure, 66 feet high, 350 feet long, and 550 feet thick at the base. It contained about 230,000 cubic yards of earth and 63,000 cubic yards of rock. This structure was also founded on consolidated material of sand, gravel, and cobbles, after the loose riverbed material had been removed. The downstream slope was 5 to 1, and the upstream slope 2 to 1. The downstream slope was protected by a thick rock blanket.

A rock barrier, 54 feet high, 375 feet long, 200 feet thick at the base, and containing 98,000 cubic yards of rock, was provided about 365 feet downstream from the lower cofferdam, to protect the earth fill from the backwash of the river during flood discharges. The cofferdams, rock barrier, and the Nevada diversion tunnels were completed in March 1933, in advance of the spring flood flows.

FOUNDATION

With the river out of the canyon, work was concentrated on the excavation of the dam site. Men lowered from the top of the cliffs in rope slings stripped the canyon walls of loose rocks, using bars, jackhammers, and explosives, beginning from the top of the canyon walls and working down along the required excavation line.

Crawler-mounted electric shovels, loading directly into trucks, were used in excavating the foundation for the dam and exposing the bedrock in the area between the limits of the cofferdams. This involved removal of more than 800,000 cubic yards of gravel and 100,000 cubic yards of rock. Excavation of the river bed confirmed the conditions indicated by the exploratory drilling. An inner gorge about 75 to 80 feet deep, and bordered by rock benches on either side of the canyon, was uncovered in the bottom of the main canyon.

FOUNDATION GROUTING

Wherever concrete was to be poured against rock, advance provisions were made for forcing a mixture of cement and water into rock fissures and seams, at intervals and depths consistent with the nature of the rock formation and its location with respect to the dam. The main grout curtain or cut-off under the dam included three systems of holes drilled into the rock foundation. The first, or "B" hole grouting system, was designed to provide a leakproof layer under the upstream part of the dam, below which the deeper high-pressure grouting could be performed. These holes, spaced 20 feet on centers, were drilled 30 to 50 feet in depth and grouted at pressures ranging from 150 to 400 pounds per square inch before any concrete was placed. More than 5,300 linear feet of "B" holes were drilled under the dam, which took a total of about 6,400 sacks of cement.

As a part of the "B" hole grouting, a total of 1,418 feet of holes was drilled along two faults situated just upstream from the dam, and 1,064 sacks of cement were applied under pressures ranging up to 400 pounds per square inch.

The "A" line grouting system, which forms the principal cut-off curtain under the dam, was not started until at least 100 feet of concrete in the dam had been placed over the site of the hole, the concrete had cooled,

and the joints had been grouted. The following maximum pressures governed the introduction of the grout:

	Pounds per square inch
Below elevation 800	1,000
From elevation 800 to 1000	750
Above elevation 1000	500

Drilling was carried to a maximum depth of 150 feet
from the main grout gallery, which closely follows the
rock foundation. Before concrete was placed in the
dam, holes were drilled 5 feet into the rock and pipes
leading to the gallery were embedded in the concrete
on 5-foot centers. A total of 54,400 linear feet of holes
was drilled and 60,024 sacks of cement applied as grout
through these pipes.

The "C" line grouting system, to provide a supplemental cut-off under the heel of the dam, followed the
upstream outline of the dam from the lowest point of
the foundation to elevation 775 on either abutment.
Holes were drilled on a 10-foot spacing to a maximum
depth of 100 feet and were inclined downstream. Grout
was placed under pressures varying from 300 to 750
pounds per square inch. A total of 7,325 linear feet of
holes was drilled and 7,106 sacks of cement applied.

The grout curtain was extended into the abutments

Jumbo used in abutment grouting from the diversion tunnels. Jumbos
and forms of unprecedented size were used in these 50-foot diameter
tunnels for drilling the rock, forming the sidewalls, and grouting the
abutments.

by deep drilling and grouting from the diversion and
penstock tunnels.

THE DAM

Hoover Dam is a concrete arch-gravity structure in
which water load is carried both by the massive weight
of the structure on the foundation and by arch action to
the abutments. A total of 3,250,000 cubic yards of
concrete was used to form the dam, which has a maximum height of 726.4 feet above the lowest point of the
foundation, a crest length of 1,282 feet, and a thickness
ranging from 45 feet at the crest to 660 feet at the base.
An additional 1,150,000 cubic yards of concrete were
required to construct the power plant and appurtenant
works.

As a part of the preliminary design studies for the
dam, numerous detailed stress analyses by the trial-
load method were made for the structure to determine
magnitudes and distribution of stresses due to loads and
conditions imposed by water, temperature, and earthquakes. The action of the dam and resulting stress
conditions under water and temperature loads for the
final design were checked by detailed experimental investigations on two models of the structure. Results
obtained from the model tests furnished very satisfactory confirmations of the trial-load analyses and special
stress studies, and demonstrated the adequacy of the
design.

The large volume of concrete to be placed, and the
unprecedented problems involved, justified a concrete
research program to insure a satisfactory mass concrete
structure. The comprehensive program which was in-
stituted covered studies and tests of ultimate compressive stress, permeability, Poisson's ratio, modulus of
elasticity, sliding friction, bond at horizontal construction joints, variations of strength with age and curing
temperature, proper gradation of fine and coarse aggregates, effects of vibrating fresh concrete, proper mix
proportions, and numerous problems of a thermal nature. In the latter category, the most important and
difficult problems were the generation of heat in a large
mass of concrete due to hydration of cement, and the
volumetric changes occurring in mass concrete due to
temperature changes and other causes. Numerous tests
and studies were made to determine the chemical com-
position of a cement which would have the lowest heat-
generating qualities and greatest durability characteristics.

The dam was divided into vertical columns or blocks
by radial and circumferential contraction joints. The
blocks ranged in size from 25 by 30 feet at the downstream face at the base of the dam to 50 by 60
feet at the upstream face. Vertical joints were interlocked by keys, formed to provide maximum cross-sec-

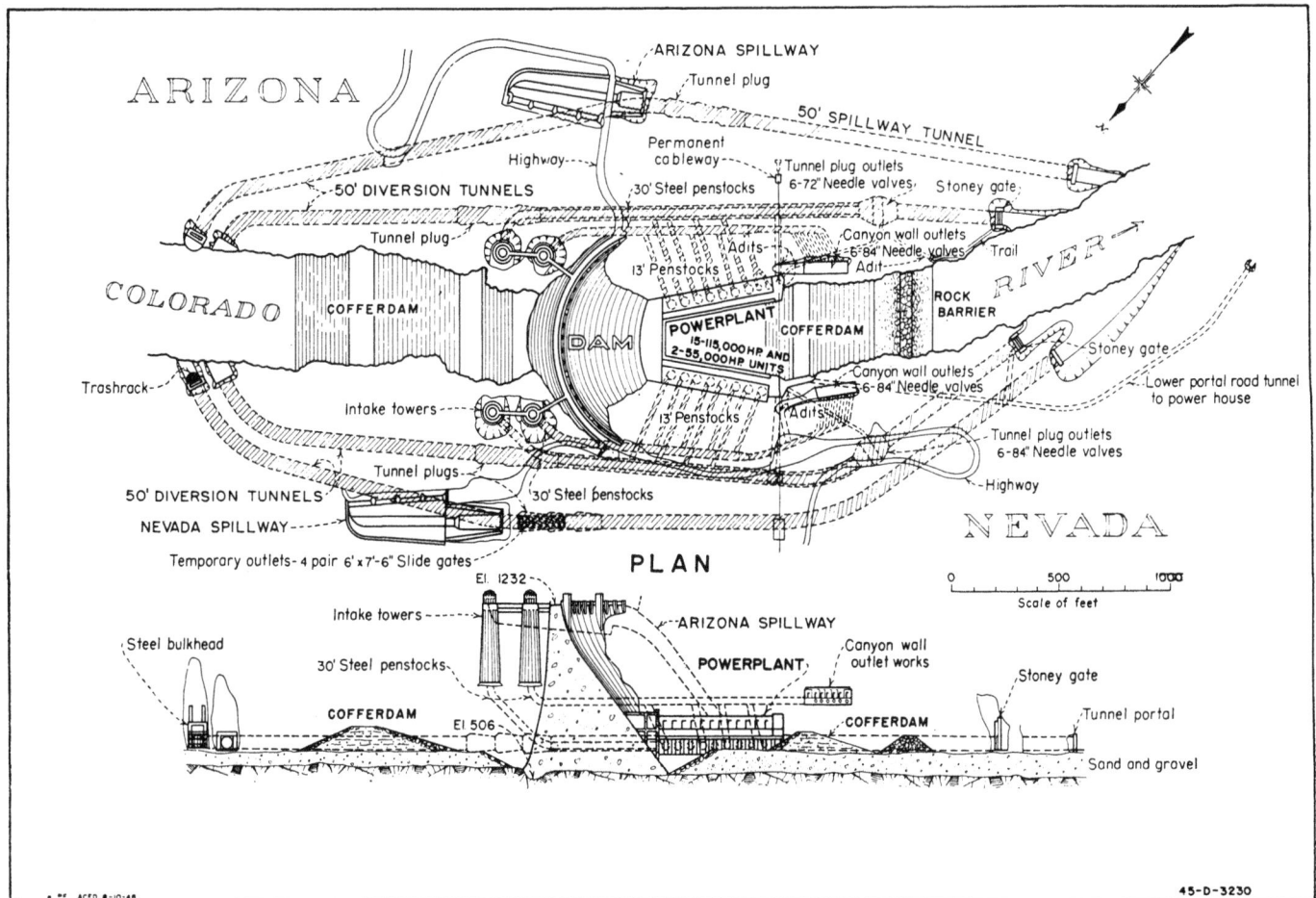

PLAN

45-D-3230

tional area for resistance to shear after the joints were grouted. Concrete was placed in the blocks by 8-cubic-yard bottom discharge buckets operating from an overhead cableway.

Each lift of concrete was finished with horizontal keys, spaced 10 feet apart, and the surface was washed with an air and water jet under high pressure, accompanied by any brushing or chipping required to furnish satisfactory bond with the next lift. Curing was effected on the sides of the blocks by sprinkling from perforated pipes attached to the forms. The top surface of each block was kept wet by hose sprinkling.

To prevent volume change in the large mass of concrete, an artificial cooling system was provided to remove the heat developed during the hardening process and to concentrate the cooling and shrinking to a relatively short period of time. The cooling system consisted of 1-inch outside diameter, 14-gage tubing buried in the concrete. Cooling was carried on in two stages: first, by circulation of air-cooled water through the pipe system; and, second, by circulation of refrigerated water through the same system. An atmospheric type of cooling tower supplied the air-cooled water for the first stage of cooling. The low humidity in the region and

the natural draft in the canyon combined to make this method of cooling very effective. Refrigerated water for the second stage of cooling was provided by an ammonia compression system, similar in most details to the systems used in making ice.

An 8-foot slot in the center of the dam, extending through the dam to the upstream face, was provided for the cooling system header pipes. From 6-inch, cork-insulated header pipes, each loop or coil of 1-inch cooling pipe extended circumferentially to the abutment and then back to the slot. More than 570 miles of tubing were embedded in the dam. The cooling pipes were laid directly on top of each 5-foot lift of concrete after the concrete had hardened, and were anchored by wire loops buried in the concrete while the concrete was still plastic. All sections of the pipe were connected with special couplings which permitted movement.

The temperature of the concrete was obtained by electrical resistance thermometers buried in the concrete or inserted in the ends of cooling pipes at the slot. For cooling tower operations, the average temperature difference for water entering and leaving the concrete was 7.3° F., and for refrigeration plant operations, 11.1° F.

This stiffleg derrick was used to place concrete in some of the upstream blocks of the dam and to transfer concrete buckets from the low level railroad terminus to points where they could be picked up by the overhead cableways.

CONTRACTION JOINT GROUTING

To insure monolithic action of the dam and the desired stress distribution in the structure, the contraction joints in 50-foot lifts were grouted as rapidly as the concrete was cooled and the cooling pipe header slot concreted. Metal grout stops were placed across the vertical radial contraction joints at both the upstream and downstream faces of the dam and across the circumferential joints at each junction with a radial joint. Horizontal grout stops were placed in every vertical joint at 50-foot intervals.

For the injection of grout into the joints, a system of pipes with an outlet for every 30 to 50 square feet of joint area was embedded in the concrete adjacent to the contraction joints. The outlet pipes were connected to headers extending through the grouting galleries in the dam and to both faces of the dam. The joints were filled from the headers by gravity flow, pumping being used only to drive off surplus water and thin grout and to consolidate the film of grout in the joint when the joint was practically filled. Pressures up to 300 pounds per square inch were used in some joints to secure the desired results, but a pressure of 100 pounds per square inch was ordinarily sufficient to drive the water into the concrete on either side of the joint and produce a grout film of satisfactory density.

Cement for grout was screened through a 200-mesh screen to increase the penetration of the grout into fine seams. An average of one bag of cement was required for 90.79 square feet of radial contraction joint area and 129.09 square feet of circumferential contraction joint area, a total of 33,425 bags of cement being used. The approximate average thickness of the radial contraction joint grout films was 0.126 inch and of the circumferential joints 0.087 inch.

SPILLWAYS

Many studies, designs, and model tests were completed before final selection of the side-channel type of spillway with drum-gate crest. Two identical spillways, each with a clear crest length of 400 feet, one on either side of the canyon, discharge into inclined tunnels, 50 feet in diameter and 600 feet in length, which lead into the outer (away from the river) diversion tunnels on each side of the river.

The weir section for each spillway is actually a gravity dam of overflow profile with maximum heights for the Nevada side of 75 feet, and for the Arizona side, 85 feet. At the quarter points on the crest of these overflow sections, piers divide the crest into 100-foot sections for the structural steel drum gates. More than 600,000 cubic yards of rock were removed for the construction of the spillways.

Each spillway channel has a uniform bottom width of 40 feet and the floor slopes on a 12 percent grade from a depth of 75 feet at the upper end to a depth of 128 feet at a point opposite the lower end of the crest. From this point, a 55-foot length of channel ending in a 36-foot-high vertical step-up improves the hydraulic characteristics and reduces the disturbance at the entrance to the transition leading to the inclined tunnel.

The sides and bottom of the spillway channels and transitions were lined with an average thickness of 24 inches of concrete. The lining was secured to the rock by hooked, 1½-inch-square bars grouted into holes drilled into the rock a minimum distance of 5 feet. Concrete from the high-level mixing plant was delivered in 4-cubic-yard agitators on trucks, and either moved directly by cableway to the point of placement

Water discharging into the Arizona spillway looks almost gentle, and the placid lake gives no hint of the great value in water for farms and cities, and the electric power for industry and agriculture locked up within its depths.

Cleanup work was still in progress following completion of Hoover Dam when this picture was taken. The irregular lines snaking across the hillsides are retaining walls intended to prevent rock from falling on structures below.

or transferred to 2-cubic-yard buckets for more convenient handling. More than 127,000 cubic yards of concrete were placed in the spillways.

Each spillway has four structural steel drum gates, each 100 feet long, 16 feet high, and weighing approximately 500,000 pounds. In the lowered position, the top face of the gates provides a curved surface conforming to the profile of the weir crest. Hinged at the top and upstream side of the concrete weirs, the gates float in recesses in the weir section and are designed to allow flow of water over them to a maximum depth of about 8 feet while the gates are in the fully raised position.

Automatic control with optional operation by manual control is provided for raising and lowering the gates. When in raised position, a gate may be held continuously in that position by the pressure of water against its bottom until the water surface of the reservoir rises above a fixed point when, by action of a float, the gate is automatically lowered. As the flood peak decreases, the gate rises automatically. Aside from the automatic control, the gate can be operated by manual control so as to gradually empty the flood control portion of the reservoir without creation of flood conditions downstream.

INTAKE TOWERS

Release of water from the reservoir under normal conditions is made through the power plant and through outlet works. Four towers, two on either side of the canyon upstream from the dam, are the intakes for the water releases. The towers, located symmetrically with respect to the dam, are founded on rock benches or shelves excavated in the canyon walls. Each tower consists of an inner barrel with a nominal inside diameter of 29 feet 8 inches, surrounded by 12 radial buttresses or fins to accommodate the trashrack sections and afford structural support for the barrel. The distance between faces of opposite fins varies from 82 feet at the base of the towers to 63 feet 8⅞ inches at the top of the parapets, 342 feet above the bases of the structures.

Because the great height and relative slenderness of the intake towers made them unusually vulnerable to earthquake shocks, an unusually large quantity of reinforcement steel, amounting to more than 160 pounds per cubic yard of concrete, was incorporated in each structure. (A normal figure is 100 pounds of reinforcing steel per cubic yard of concrete.) Two sets of openings are provided in each barrel, 12 openings at the base of the tower and 12 more 150 feet above the base. Closure of each set of openings is provided for on the inside of the barrel by cylinder gates 32 feet in diameter and 11 feet in height operated by electrically operated screwstem hoists. Normal travel of the gates is approximately 10 feet and the hoisting mechanisms are arranged to open or close either or both gates in 52 minutes. A set of 12 bulkhead gates is provided on the outside of the barrel for each set of gate openings for unwatering the towers when required for inspection and maintenance of the cylinder gates and seats.

The steel trashrack sections for the towers are constructed of vertical trash bars, 5 inches by ¾ inch in section, and horizontal spacer bars 3 inches by ⅝ inch in section. Two plate-girder type bridges, one on each side of the river, connect the towers with the dam.

Concrete for the intake towers was produced at the high-level mixing plant and handled by derricks located between the towers and near the top of each canyon wall. The concrete was conveyed to a hopper erected immediately above the center of the tower and flowed through short chutes to the barrel, fins, and trashrack beams. A total of 93,674 cubic yards of concrete was placed in the towers.

PENSTOCK AND OUTLET TUNNELS

From the base of each of the two downstream intake towers a header tunnel extends downstream approximately parallel with and 170 feet above the inner diversion tunnel (closest to the river) to the location of the canyon wall outlet works. These tunnels were excavated 41 feet in diameter to a point past the power penstock branches and then 35 feet in diameter to the end and lined with a 24-inch thickness of concrete. To facilitate the subsequent installation of the steel power penstock headers, construction adits, 26 feet wide and 43 feet high, were driven from the canyon wall to the line of each header tunnel.

From each of the header tunnels, four penstock tunnels on an incline lead to the back wall of the powerhouse. These were excavated 21 feet in diameter and lined with 18 inches of concrete to form a completed tunnel section 18 feet in diameter. The downstream penstock tunnel on the Arizona side of the river branches into two tunnels 12 feet 6 inches in diameter about 45 feet from the face of the canyon wall, to serve two smaller turbines.

At the downstream end of each header tunnel six 11-foot horseshoe-shaped tunnels were excavated leading to the canyon wall outlet works. The horseshoe tunnels were not lined until after the 102-inch outlet pipes were placed, and then the space between the pipes and the tunnel walls was backfilled with concrete. From the bases of the upstream intake towers, 41-foot diameter inclined tunnels connect with the inner diversion tunnels. The inclines are lined with a 2-foot thickness of concrete to give a finished diameter of 37 feet. Four penstock tunnels, almost horizontal, similar to the penstock tunnels from the header tunnels, extend from the inner diversion tunnels to the powerhouse.

Excavation of all penstock and outlet tunnels was performed by methods similar to those used in the excavation of the diversion tunnels.

Concrete for the lining was produced at the high-level mixing plant and delivered by cableway to the portal of each construction adit. The concrete was hauled into the tunnel by train or truck and placed by conveyor-belt system or by pneumatic concrete gun.

TUNNEL PLUGS

The diversion tunnels were designed and constructed to permit the outer tunnels (away from the river) to be used as spillway tunnels and the inner tunnels (closest to the river) as penstock tunnels after they were no longer needed for diversion of the river.

With the dam completed to a height safe from possible overtopping, the inner diversion tunnels were bulkheaded at the upstream ends and the permanent 50- by 35-foot Stoney gates at the downstream ends were lowered, after which the 306-foot-long sealing plugs of concrete were placed. With these permanent seals completed, the connecting tunnels from the intake towers and the power penstock tunnels branching from the diversion tunnels were excavated and the tunnel-plug outlet works toward the downstream ends of the diversion tunnels were constructed.

Next, the Nevada outer diversion tunnel was closed by lowering the 50- by 50-foot steel bulkhead gate at the upstream end and bulkheading the downstream end. The 313-foot-long concrete tunnel plug for the permanent seal contained four temporary conduits, each controlled by a pair of 6-foot by 7-foot 6-inch slide gates to provide for maintaining the required minimum river flow until the rising reservoir reached the openings in the base of the intake towers.

With water flowing through the temporary outlets, the Arizona outer diversion tunnel was closed by the same means employed for the Nevada tunnel and the permanent concrete plug constructed. After the reservoir reached the intake tower openings, the Nevada outer tunnel was again closed and the permanent seal completed by plugging the temporary outlet works.

A total of approximately 88,000 cubic yards of concrete was placed in the tunnel plugs, all of which was produced at the high-level mixing plant. In general, concrete was placed in 5-foot lifts by pumps and transit mixers, and cooling pipes were placed on top of the lifts in a manner similar to that used in the dam. Cooling and grouting of the plugs was performed to effect a tight seal in the tunnels.

PLATE STEEL OUTLET PIPES

Studies and investigations of tunnel rock conditions and requirements for the power plant and outlet works indicated that steel pipes installed in the tunnels would be more practicable than using the concrete-lined tunnels as pressure tunnels. Specifications were issued covering fabrication and installation of the pipes, and a contract was awarded to the Babcock & Wilcox Company of Barberton, Ohio.

For each of the four 30-foot diameter header pipes (one from each intake tower), four 13-foot diameter branch penstocks lead to the turbines, and a 25-foot header beyond the penstock pipes connects with smaller branch pipes leading to the outlet valves. The 30-, 25-, and 13-foot diameter pipes are supported on reinforced concrete piers, two piers being used for each pipe section.

The total weight of the 2¾-inch-thick plate steel required to complete over 14,000 linear feet of the various sizes of pipe was approximately 88,000,000 pounds. The large diameter of the greater part of the penstocks made it impossible to ship completed sections of the pipe by rail to the dam site. Accordingly, a complete field fabrication plant was erected by the contractor about 1½ miles from the rim of the canyon.

After the completion of shop welding, every inch of weld was photographed by X-ray to render visible any internal defects. The completed pipe sections were then hauled from the fabrication plant to the dam site by a specially constructed trailer and lowered to the tunnel adits by the 150-ton permanent cableway constructed by the Government.

The installation of the 30-foot pipe sections was commenced at the upper ends of the tunnels at the base of the intake towers. The installation of the 25-foot headers below the construction adits was commenced at the lower ends of the tunnels, progressing in an upstream direction to the adits. The last pipe section closing the gap left in the 25-foot conduit was located opposite the adits.

OUTLET WORKS

The primary purpose of the outlet works is to provide regulation of the reservoir and to supply water for downstream use in addition to that passing through the power plant. They also provide a means for evacuating flood storage space below the spillway crests. With ideal operation of the outlets works, there has been little occasion for the use of the spillways.

The outlet works, similar in design on both sides of the river, are divided into two systems: first, the canyon-wall outlet system consisting of the valves and structures supported on benches blasted out of the canyon walls; and second, the tunnel-plug system consisting of the valves in the downstream tunnel plugs of the inner diversion tunnels.

The canyon-wall outlet valves are located in reinforced concrete houses approximately 800 feet downstream from the dam. The six 84-inch needle valves in each structure discharge approximately 175 feet above the normal elevation of the river. The valves are pointed at an angle of 60° downstream, to give the issuing water a downstream component, and are located so that the jets from the two sides of the river meet approximately in the stream bed.

The steel outlet conduits enter the house through the back wall and connect with 96-inch paradox emergency gates embedded in concrete blocks which form the operating floor for the gates and valves. Directly in front of the gate blocks are the 84-inch needle valves, supported on reinforced concrete pedestals built integral with the foundation. On the front of the needle valves

are attached conical sheet metal discharge guides which carry the water through discharge apertures in the front wall. Each house is equipped with a 30-ton crane, with 10-ton auxiliary for installation and maintenance of heavy equipment. The 96-inch paradox emergency gates are closed when the needle valves are not in operation to relieve the valves of the pressure head of the reservoir.

Excavation of the benches for the valve houses involved the removal of approximately 56,000 cubic yards of rock. Concrete for the houses was manufactured at the high-level mixing plant and transported by cableway to the point of placement in transit mixers and hopper buckets. A total of 17,000 cubic yards of concrete was placed in both houses.

The downstream tunnel-plug outlet works are located in the inner diversion tunnels, several hundred feet upstream from the outlet portals. They regulate the flow of all stored water released through the upstream intake towers which is not used for the generation of power.

The 25-foot steel outlet pipes leading to the needle valves branch into three pipes, 13 feet in diameter, which in turn branch into six pipes 86 inches in diameter. A block of concrete 100 feet in length, with operating rooms and galleries, encases the outlet pipes for anchorage. The six 72-inch needle valves with 86-inch emergency gates are located in an operating room at the downstream end of the anchor. This room, 35 feet wide, 100 feet long, and 66 feet high, contains equipment similar to that installed in the canyon-wall valve houses for the operation and maintenance of the valves and gates.

Outlet pipes at the Nevada tunnel-plug outlet works.

The emergency gates are housed in blocks of reinforced concrete, the tops of which are used as the operating floor. Needle valves are placed 10 feet downstream from the emergency gates, on reinforced concrete pedestals built integral with the floor of the operating chamber. The distance between the front wall and the needle valves is spanned by discharge guides which confine the water issuing from the valves. An observation platform is provided on the downstream side of the chamber from which to view the action of the water in the tunnels as it issues from the needle valves.

The combined discharge capacity of all outlet works is approximately 91,000 cubic feet per second.

POWER PLANT

The power plant is located at the toe of the dam and consists of two wings, one on either side of the river, with offices, shops, operating and storage rooms located in a connecting structure to form a U-shaped structure 1,650 feet in length. Each wing housing the power plant equipment is 650 feet long, 150 feet high above normal tailrace water surface, and 229 feet above the lowest foundation elevation. Construction of the powerhouse involved 455,000 cubic yards of excavation, 240,000 cubic yards of concrete, 24,000,000 pounds of steel reinforcement, and 12,000,000 pounds of structural steel. The hydraulic and electrical machinery, equipment and wiring were installed by Government forces.

The power plant was designed for an ultimate installation of 15 main generating units of 82,500-kilowatt capacity each and two main generating units of 40,000-kilowatt capacity each, making a total generating capacity of 1,317,500 kilowatts. The initial installation in the power plant was composed of four 82,500-kilowatt units and one 40,000-kilowatt unit.

CONSTRUCTION DATA

Bids for construction of the dam and power plant were opened March 4, 1931, at Denver, Colo. Three regular bids were received, as follows:

Six Companies, Inc., of San Francisco, Calif__ $48,890,995.50
Arundel Corporation, of Baltimore, Md_____ 53,893,878.70
Woods Brothers Construction Co., of Lincoln,
  Nebr., and A. Guthrie & Co., of Portland,
  Oreg _____ 58,653,107.50

Six Companies, Incorporated, of San Francisco, Calif., a company composed of six western contracting firms— Utah Construction Company, Pacific Bridge Company, Kaiser Paving Company, Ltd., MacDonald and Kahn Company, Morrison-Knudsen Company, and J. F. Shea Company—was awarded the contract on April 20, 1931.

The construction of Hoover Dam involved approximately 6,000,000 cubic yards of all classes of excavation

Architectural treatment of both the dam and power plant included the forming of some concrete surfaces with rough lumber, deep V-grooves accentuating vertical lines, and minor variations in surfaces to create shadow patterns.

and placement of 4,400,000 cubic yards of concrete. Additional quantities of other principal work or materials included drilling grout and drainage holes, 410,000 linear feet; pressure grouting, 422,000 cubic feet; earth and rock fill, more than 1,000,000 cubic yards; gates and valves, 21,670,000 pounds; plate-steel outlet pipes, 88,000,000 pounds; pipe and fittings, 6,700,000 pounds; structural steel, 18,000,000 pounds; miscellaneous metalwork, 5,300,000 pounds; reinforcement steel, about 45,000,000 pounds; and cement, 5,000,000 barrels.

Under the terms of the contract 2,565 days (7 years and 8 days) were allowed the contractor for the completion of the project. On March 1, 1936, the dam and power plant were accepted by the Secretary of the Interior, terminating the contract and marking the end of actual construction (exclusive of penstock installation and work by Government forces) on the project in 11 days less than 5 years, or 2 years 1 month and 28 days ahead of schedule.

Some idea of the rapidity with which the work was performed may be gained by a review of the construction program, which is outlined in the accompanying tabulation.

PROGRAM OF CONSTRUCTION AT HOOVER DAM

Feature	Date started	Date completed
Diversion tunnels	June 1931	March 1933.
Upstream cofferdam	September 1932	March 1933.
Downstream cofferdam and rock barrier.	November 1932	March 1933.
Removal of the same.		May 1935.
Excavation for dam	October 1932	June 1933.
Intake towers	February 1932	March 1935.
Spillways.	February 1932	March 1935.
37-foot penstock tunnels.	February 1933	May 1934.
Installation of 30-foot diameter outlet pipes in upper tunnels:		
Arizona	January 1935	September 1935.
Nevada	October 1934	August 1935.
Installation of 30-foot diameter outlet pipes in lower tunnels:		
Arizona	January 1935	June 1936.
Nevada	February 1935	July 1936.
18-foot penstock tunnels.	December 1932	August 1934.
Installation of 13-foot diameter penstock pipes in branch tunnels.	December 1934	January 1936.
Canyon wall outlet works.	November 1932	August 1935.
Tunnel plug outlet works.	July 1935	April 1936.
Stoney gates at downstream portal of inner diversion tunnels.		August 1933.
Tunnel plugs in inner diversion tunnels.	March 1934	November 1935.
Tunnel plugs in outer diversion tunnels.	December 1934	March 1935.
Concrete in dam	June 1933	May 1935.

The unprecedented quantity of concrete involved and the exacting conditions governing the manufacture of it, necessitated the installation of automatic equipment of unusual size for the production of concrete. The contractor built a gravel plant, conveyor systems, a cement blending plant, storage silos, two mixing plants (high level and low level), and construction railroads for transporting the materials. With this equipment the contractor was able to place 10,417 cubic yards of concrete in one 24-hour period, a record at that time.

A system of five cableways, spanning the canyon, was provided by the contractor to handle concrete and other materials used in the construction. All five cableways were designed for manual operation with 20-ton loads and infrequent loadings up to a maximum of 40 tons. The two longest cableways, with spans of 2,575 feet each, were identical units with head and tail towers traveling on parallel tracks and were used to cover the area of the spillways. Two shorter cableways, with spans of 1,405 feet each, operated on concentric radial tracks and were used to cover the downstream portion of the dam and central portion of the powerhouse. The shortest cableway, with a span of 1,374 feet, was a radial-traveling type with the head tower fixed and the tail tower movable, and covered most of the power plant area.

For handling the penstock pipe and power plant equipment at the dam, a permanent 150-ton cableway was constructed by the Government. This 1,200-foot cableway spans the canyon at the downstream end of the powerhouse and is of the fixed type. A 90-foot structural steel tower supports the Nevada end of the cableway, but no tower is required on the Arizona side. The track is composed of six 3½-inch, 6 by 37, parallel cables, spaced at 18½-inch centers. Each of these cables is connected to eyebolts and cross beams anchored in tunnels filled with concrete on either side of the canyon. The hoist house containing all of the operating machinery is located on the Nevada side of the canyon between the anchorage and the tower. Operation of the cableway is by remote control from the main control station, supported on cantilever I-beams over the canyon, or by interlocking change-over switch to any of four stations at the landing platforms.

SUPPLEMENTARY CONSTRUCTION

Since the completion of Hoover Dam and appurtenant works in March of 1936, additional facilities and improvements have been constructed to provide conveniences for visitors and for increased capacity and greater efficiency in operation of the power plant.

A one-story, reinforced concrete exhibit house was opened to the public in October of 1946. This building contains exhibits of Boulder Canyon project and other projects of the Bureau of Reclamation to acquaint visitors with the work being done to develop the West. Additional roads and parking areas, architectural features at the crest of the dam, and a concrete sun shelter have been added for the convenience and interest of hundreds of thousands of visitors annually.

A contract was awarded in 1945 for tunnel and river channel improvement. The work consisted of excavating and deepening the river channel immediately downstream from the dam to increase the net energy head on the turbines. In addition, the downstream portals of all four diversion tunnels were modified to improve hydraulic discharge characteristics.

Ten additional 82,500-kilowatt generating units and another smaller unit have been added to the initial installation in the power plant. Because of the great need for power from Hoover, the smaller unit was designed to deliver 50,000 kilowatts instead of 40,000. Likewise, the remaining unit planned to be installed in 1957 will deliver 104,500 kilowatts instead of 82,500, providing a total plant capacity of 1,349,500 kilowatts as compared with the 1,317,500 originally contemplated.

BENEFITS

With the completion of Hoover Dam the Colorado River was transformed from a destructive menace into a public servant. The major benefits obtained as a direct result of the construction of the Boulder Canyon project may be enumerated as follows:

1. Flood control—Protection for the lives and property of thousands of people.

2. Water for irrigation—Development of productive lands, creation of new homes, and occupations for a great many people.

3. Power—Production of low-cost power for industrial progress, the profits from which make the project financially solvent.

4. Water for domestic and industrial use, stimulating expansion of the southern California coastal region and other areas.

5. Elimination of damaging silt deposits, removal of which was costing tremendous sums annually.

6. Creation of a national playground and recreational area.

7. Creation of a wildlife and bird refuge.

Through these benefits and other indirect returns, the construction of Hoover Dam has amply demonstrated its economic and social value to the Southwest and to the country as a whole.

SHASTA DAM

CENTRAL VALLEY PROJECT, CALIFORNIA

By Jerome M. Raphael

Shasta Dam is the northern outpost of the great Central Valley project of California. It is eclipsed by only two concrete dams in the world—it ranks second in volume to Grand Coulee, and second in height to Hoover. In appearance, however, it is the equal of any other major dam. Its gracefully curved plan and open site, with Mt. Shasta mirrored in Shasta Lake for a background, mark it as one of the more impressive of the works of man.

Shasta Dam is situated on the Sacramento River 12 miles north of Redding, Calif., and 5 miles below what used to be the confluence of the Pit and Sacramento Rivers. Redding stands at the northern end of Central Valley, a basin which is divided into the Sacramento Valley on the north and the San Joaquin Valley on the south, and which contains two-thirds of the agricultural lands of the State.

The San Joaquin basin has one-third of the water supply with two-thirds of the irrigation needs of the entire Central Valley. Extensive development and exploitation of the San Joaquin Valley eventually exhausted surface water supplies and depleted underground storage until large tracts of rich land were on the verge of abandonment for lack of water. The Sacramento Valley, on the other hand, with only one-third of the agricultural lands to be irrigated, has two-thirds of the water inflow of the entire basin. In the short and intensive rainy season, the Sacramento River at one time habitually flooded the lowlands, causing great damage to property and crops, yet summer stream flow was low and inadequate. Shasta Dam now stores the spring run-off of the Sacramento River, releasing it as needed throughout the year for irrigation in the Sacramento Valley and for transfer as supplemental water to the San Joaquin Valley. By increasing summer flow, salt water encroachment on the rich agricultural lands of the delta region is checked. In addition, Shasta Dam develops power for use in the comprehensive Central Valley project scheme.

RESERVOIR

The principal tributaries to the Shasta Reservoir drainage area of 6,600 square miles are the Pit, Sacramento, and McCloud Rivers. The average annual rainfall varies over this area from 15 to 65 inches. Shasta Reservoir is 35 miles long, has a surface area of 46 square miles, and a storage capacity of 4,493,000 acre-feet, of which 104,000 acre-feet is inactive storage and 500,000 acre-feet is reserved for flood storage.

Construction of the reservoir necessitated the relocation of facilities of all types. The main line of the Southern Pacific Railroad, which followed the Sacramento River through the reservoir area and dam site, had to be relocated to the east, involving the construction of 30 miles of railroad, 8 bridges, and 12 tunnels. Twelve miles of United States Highway 99, the main north and south highway, also had to be relocated. Both of these relocated arteries cross the Pit River branch of Shasta Reservoir on the Pit River bridge, the world's highest double-deck bridge. The old mining town of Kennett, 3 miles above the dam site, was inundated, necessitating the removal, with suitable ceremonies, of the town's cemetery. After filling of the reservoir, ferry service had to be provided across the McCloud arm of the lake for access to an operating mine producing barite, an iron ore much used as ballast for ships.

DAM SITE

After consideration of three possible sites—the Baird site on the tributary Pit River 5 miles upstream from Kennett, the Table Mountain site on the Sacramento River about 10 miles upstream from Red Bluff, and the Kennett site about 3 miles downstream from Kennett—the latter site was selected. At this site, a durable rock existed in two ridges forming a very obtuse V with the point upstream. Bedrock suitable to support a concrete dam existed at moderate depth, although very

Shasta Dam, in a setting of mountains dominated by Mt. Shasta on the horizon, stores the waters of the Sacramento and Pit Rivers.

infrequently exposed at the surface. The rock, locally called greenstone, is the Copley meta-andesite, an altered rock of various volcanic origins. The soil and rock outcrops at the surface are predominantly red and brown in color due to weathering. However, as excavated for construction of the dam, the rock had a fairly uniform bluish-green color. While the rock was predominantly hard and strong, there were dikes and shear zones of altered or crushed material which had to be removed and the space refilled with concrete in order to form a competent foundation for the dam.

DESIGN OF DAM

In designing the dam, its height was fixed within fairly close limits by the storage requirements of the project as a whole, which established a gross reservoir storage capacity of 4½ million acre-feet on the Sacramento River. Of this capacity, the top 500,000 acre-feet was to be used primarily for flood control, and a dead storage volume of 500,000 acre-feet was to be maintained to provide a minimum power head of 238 feet. In order to meet these requirements at the Kennett site, the top of the dam would necessarily be 500 feet

above the river bed. This height, while unprecedented for an earth- and rock-fill dam, did not preclude consideration of the type; however, the large capacity required for the spillway, together with the numerous outlets necessary to serve power and other purposes, led to adoption of a type of dam that would include all these facilities in a single structure, both for simplicity and economy.

A concrete gravity dam with a 375-foot-long straight spillway section occupying the old river channel and adjacent nonoverflow sections curved on 2,500-foot radii most economically fitted the configuration of the ridges forming the dam site. The total length of 3,500 feet and the maximum height of 602 feet from lowest foundation to crest were disproportionate for an economical arch dam and the left abutment was not strong enough to support the heavy thrusts from arch action. Accordingly, the dam, while curved in plan to accommodate the configuration of the site, was designed as a gravity structure throughout, resisting the pressure of the reservoir by its sheer mass. At the left end of the dam, weathering affected the competency of the rock to great depth. To obviate removal of great quantities of material unsuited for foundation of a concrete dam,

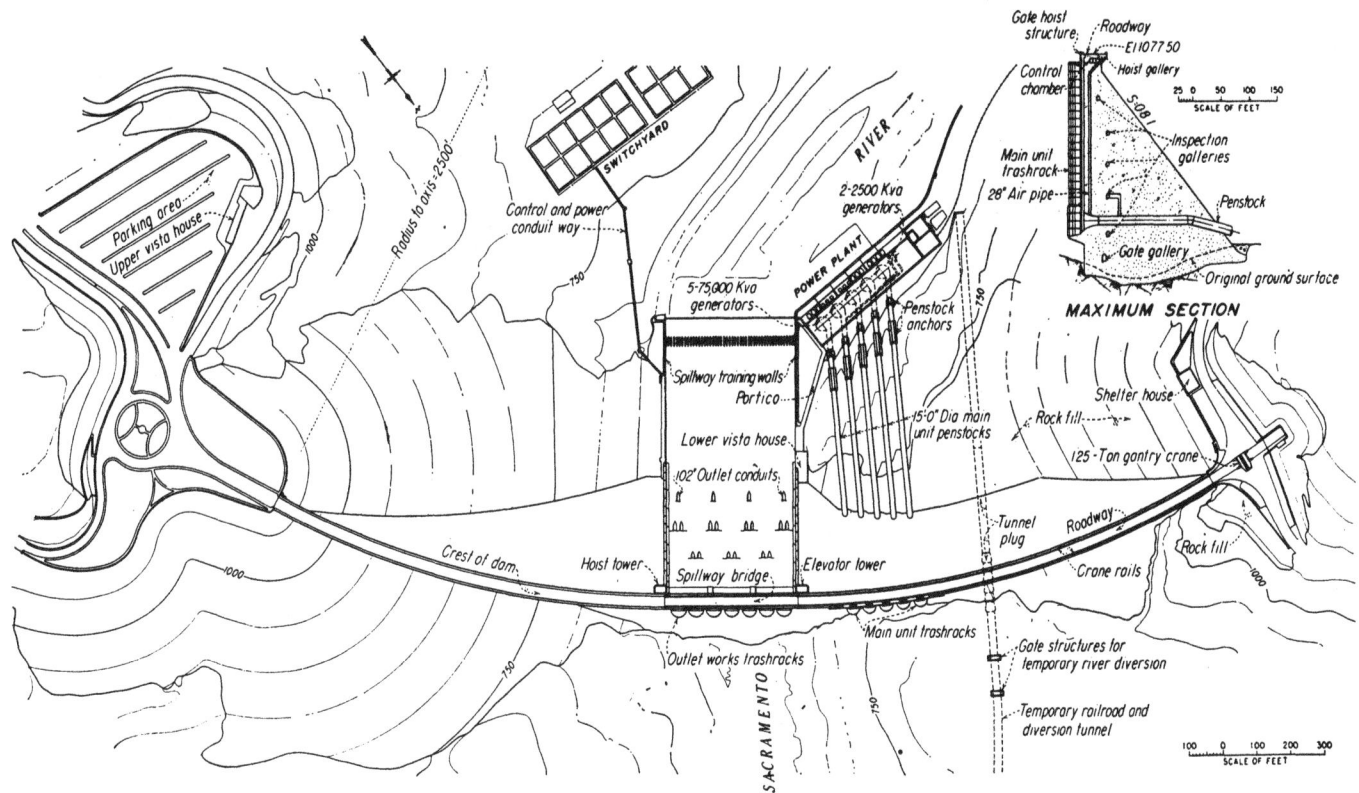

an earth- and rock-fill embankment 525 feet long with a maximum height of 115 feet was constructed in this area.

Necessity of passing a major portion of the flood flow over the downstream face of the dam led to the design of the world's highest man-made waterfall. Hydrologic studies had determined that the total discharge capacity of the spillway and outlet works should be 250,000 cubic feet per second. The spillway was designed for a maximum capacity of 185,000 cubic feet per second at the full reservoir water surface elevation of 1,065 feet. Under maximum discharge conditions, the water will drop 480 feet on the downstream face of the dam, creating dynamic energy at the rate of 7,200,000 horsepower. A sloping apron at the base of the spillway provides for dissipation of this excessive energy in the turbulence of the hydraulic jump formed on the apron, thereby preventing damage to the river bottom downstream from the dam.

Flow over the crest of the spillway and the water surface elevation of the reservoir above the elevation of the spillway crest are controlled by three floating-type drum gates. These gates are each 110 feet long with an effective operating height of 28 feet. The gates are automatically operated so that the power head can be controlled closely within given limits. These gates, in the form of a long triangular prism hinged along one edge, in the lowered position occupy special recesses in the crest of the spillway, with the upper surface of the gate shaped to continue the smooth curved line of the spillway crest. The gates are raised by being floated out of their recesses by water pressure. In normal automatic operation, the drum gates rise as the reservoir rises above the spillway crest until a certain predetermined elevation is reached, at which point the automatic controls cause the gates to lower and discharge excess water from the reservoir. When the reservoir is lowered a sufficient amount, the gates automatically rise, shutting off the flow, after which further lowering of the reservoir causes the gates to lower at the same rate the water surface falls, still retaining the reservoir water. In addition to the automatic operation of the gates, they may be operated by hand wheels in the gate piers or by remote control from the powerhouse.

When outflow from the reservoir is required in amounts greater than the discharge from the powerhouse turbines, the excess is discharged by the outlet works. With a total combined capacity of 65,000 cubic feet per second, 18 outlet conduits are arranged in three tiers, four at centerline elevation 742, eight at elevation 842, and six at elevation 942. All are 102-inch diameter welded steel pipe, passing horizontally through the dam. Within a very few feet of the downstream face they turn downward abruptly until nearly parallel with the face so that the issuing jet will merge with minimum interference to the overflowing sheet from the spillway when it is in operation. A prominent feature of the downstream face of the spillway is the "eyebrow" located above each conduit outlet, which further prevents interference between the overflowing spillway sheet and the jet issuing from the conduit. Location of the outlet conduits in the spillway section made possible the economical dissipation of energy of the outlet discharges on the sloping apron provided for the spillway. Flow through the outlets is controlled in the two upper tiers by regulating gates, and in the lower tier by tube valves. Emergency closure of all conduits can be made by a movable coaster gate lowered in special guides within the trashracks on the upstream face of the dam.

More than 5 miles of galleries in the dam provide for passage of operating personnel to valve chambers, grouting and cooling outlets, cable galleries, instrument terminal boards, and other operating points and also permit visual inspection of the interior of the dam at all times. Power circuits and water lines carried in these galleries are readily inspected, tested, and replaced when necessary.

Cooling pipes were installed throughout the dam on

Transverse contraction joints were formed in this manner. Keyway and gallery opening details are plainly discernible. It is unusual for so large a surface to be exposed in Bureau construction, but this was feasible at Shasta because of the method of river diversion.

top of each 5-foot lift of concrete as it was placed and at an average horizontal spacing of 5½ feet to provide a means of minimizing the temperature rise associated with the heat of hydration of the cement in the concrete. River water for preliminary cooling was circulated through the pipes from pipe headers located on the downstream face of the dam or in galleries in the spillway section. Final cooling to specified temperatures was by water refrigerated in a special plant on the left abutment.

In order to control the development of cracks that would normally be associated with such temperature changes as did actually occur, a system of contraction joints at approximate 50-foot intervals, both parallel and perpendicular to the axis of the dam, was incorporated in the dam. Thus, the dam was built of a series of columns roughly 50 feet square. A system of piping and grout outlets was provided and, after completion of cooling and attainment of final volume of the dam, the contraction joints were filled with cement grout to make the structure monolithic.

Grout pumped into a series of vertical holes drilled into the foundation near the upstream face of the dam forms a continuous curtain, or barrier, against the infiltration of water under the dam. Development of pressure from any slow seepage of water is inhibited by a system of vertical relief drains immediately downstream from the grout curtain. Another system of 5-inch porous tile drains in the concrete near the upstream face of the dam serves to prevent the development of internal water pressure in the mass of concrete itself.

CONSTRUCTION HISTORY

The first major item in the construction of Shasta Dam was the tunnel through the right abutment to provide initially for bypassing the Southern Pacific Railroad around the dam site until the permanent relocation was completed, and later to serve as a diversion tunnel to handle the flow of the Sacramento River past the dam site. The Colonial Construction Company of Spokane, Wash., began excavation for the tunnel in June 1938, and the tunnel was accepted on June 12, 1939. Railroad traffic began on June 22, 1939.

Bids for the construction of Shasta Dam and power plant were opened on June 1, 1938. Bids were received from Pacific Constructors, Incorporated, for $35,939,-450, and from Shasta Construction Company for $36,202,357. The contract was awarded to the low bidder, and Pacific Constructors, Incorporated, received notice to proceed with the work on September 6, 1938. Even before this, on August 21, 1938, the first bulldozer began knocking down thickets of manzanita and poison oak to make way for roads and construction plant and equipment.

By the end of 1939, the major part of the excavation had been completed, and construction facilities erected. During the first 10 months of excavation, material was removed from the foundation of the dam at the average rate of nearly 200,000 cubic yards per month with a record month of 476,700 cubic yards. The first steel for the cableway head tower was erected on October 6, 1939, and the last piece of corrugated sheeting was attached to the completed tower on July 15, 1940. The first concrete was placed in the dam on July 8, 1940, and the first diversion of the river to the east side of the riverbed took place in August 1940.

By the end of 1940, 4,150,000 cubic yards of material had been excavated for the dam and the power plant, and 500,000 cubic yards of concrete had been placed. Concreting proceeded at an average rate of about 2,000,000 cubic yards per year until the 6-millionth cubic yard of concrete was placed on December 23, 1943. The last bucket of concrete was placed on the spillway bridge of the dam on December 22, 1944, marking the production and placement of 6,535,000 cubic yards of concrete in the dam and appurtenant works.

The river was diverted alternately across low rows of blocks in the dam during 1942 and 1943, and on July 10, 1943, the entire flow of the river was passing through the diversion tunnel.

While some construction activities were slowed down by the prosecution of World War II, others were correspondingly speeded up. Construction of the drum gates and three of the penstocks was deferred to save steel. On the other hand, storage of water commenced early in the construction of the dam in order to speed up production of power, and the rising reservoir stopped all activity in the concrete mixing plant and head tower area on the right bank upstream from the dam on February 15, 1944. A new mixing plant on the left bank of the river below the dam commenced in November 1943 and was completed in February 1944. The last 200,000 cubic yards of concrete needed for the dam were produced here. Delivery of commercial power was commenced on June 26, 1944. At the beginning of 1944 there was storage of only 18,000 acre-feet in Shasta Reservoir. One year later the reservoir was at elevation 925 with a storage of 1½ million acre-feet. By December 31, 1945, the reservoir elevation had increased to 1018 and the storage was 3¾ million acre-feet.

In January 1945, Pacific Constructors, Incorporated, commenced cleaning up the area and removing buildings and equipment. This work was completed in June and the last of the work required under the contract was completed at 11 a. m. on June 20, on which date the United States accepted the Shasta Dam and the power plant.

After cessation of World War II hostilities, work recommenced on the fabrication and assembly of the penstocks, and Western Pipe and Steel Company completed the fabrication of the main unit penstocks on June 6, 1947. Erection of the penstocks was completed by the Eichlay Corporation in November 1947. The last generator was placed in service on April 27, 1949. By the end of 1949 all work had been completed on the drum gates, construction of which had been deferred during the war years.

DIVERSION TUNNEL

The diversion tunnel was 1,821 feet from portal to portal and had a horseshoe-shaped cross section with a width of 27 feet and a height of approximately 29 feet, excavated dimensions. In the unsupported section, the lining was 18 to 24 inches thick. Eleven contracting firms submitted bids for construction of the tunnel, and the contract was awarded to the Colonial Construction Company of Spokane, Wash., at an estimated cost of $426,475. The principal items were 36,-000 cubic yards of open-cut excavation for the roadway; 46,000 cubic yards of tunnel excavation; and installation of 35 tons of permanent steel tunnel supports, 13,000 cubic yards of concrete lining, 25 tons of steel cover sections, and 5 tons of reinforcement bars. The contractor was notified to proceed with the work on June 23, 1938, and began excavating for the portals immediately. Tunneling proper began in August 1938; concrete lining began in January 1939. The tunnel was holed through March 4, 1939, and concreting was finished in May. All work was completed and the tunnel was accepted on June 12, 1939.

FOUNDATION PREPARATION

The next major item in the construction of Shasta Dam was the excavation of the foundation area. The excavation and hauling equipment consisted of three 4½-cubic-yard electric, and two 1¾-cubic-yard Diesel shovels, served by eighteen 25-cubic-yard butane dump trucks and twelve 7-cubic-yard butane dump trucks. An impressive fleet of 42 wagon drills, which consumed a total of 12,000 cubic feet per minute of air supplied by four compressors, drilled the necessary holes into the greenstone for blasting. This equipment handled nearly 5,000,000 cubic yards of excavation of all kinds.

Completion of the excavation verified the prediction, made during exploration operations, of the presence of numerous weakened zones in the foundation which took the form of faults, crushed zones, seams, small fractures, and incipient cracks. There were, however, no major faults and there was no indication of recent movements. While some joints were open, others were filled with broken rock and gouge. The seams of weak material were excavated to a depth of twice the width of the seam at the surface, and shafts were excavated at the

Excavation for the left abutment of Shasta Dam was gouged out of the hillside. Concrete placement proceeded at the bottom, and the illustration shows the sequence of placement in alternate blocks. Catenary spans of the cableways reached to the movable tail towers in the distance.

upstream and downstream ends of the seams and backfilled with concrete to provide increased resistance to percolation through the weak zone. Systematic washing of the seams and consolidation of the rock by grouting to a depth of 30 feet was done over the entire foundation area.

CARE OF THE RIVER

A diversion channel was first excavated in the rock on the east bank of the river and this channel carried the river flow while excavation for and construction of the dam proceeded in the natural riverbed. Following this, the river was diverted over a row of blocks in the completed construction which had purposely been left low, while the remainder of the river bottom was excavated and the concrete placed. Diversion through two low rows of blocks was continued until the railroad was permanently relocated. The river was then diverted alternately back and forth between the two low rows in the spillway section in successive raises of 10 feet. Concrete construction proceeded in the row

not in use for diversion, with the area protected by a temporary cofferdam made of a leaf gate structure bearing against the adjacent blocks at the upstream face. When the reservoir reached the elevation of the tunnel through the right abutment, the river flow was carried through it until the lowest tier of outlets was completed. Closure of the diversion tunnel was effected on February 4, 1944, by a set of two coaster gates operating through a shaft from the surface of the ground beneath the contractor's cableway head tower. The diversion tunnel was permanently closed by a concrete plug beneath the right abutment of the dam. After closure of the tunnel, the alternate diversion over low blocks continued until the reservoir reached the lowest outlets.

CONCRETE PRODUCTION AND PLACEMENT

With almost 6½ million cubic yards of concrete of all classes to be placed in the dam and appurtenant works, the plant for the production and handling of this material was necessarily extensive, elaborate, and efficient. Sand and gravel for the concrete came from the Kutras Tract in the Sacramento River bed just north of Redding, Calif. Under a separate contract, the Columbia Construction Company of Oakland, Calif., dug the raw material out of an old river bar, and screened, washed, and delivered it to the dam contractor at the dam site. In gold-conscious California, and with about 12,000,000 tons of material to be processed, it was natural that someone would think of gold recovery. Provision of the necessary riffles in the processing plant enabled the recovery of over one-half million dollars worth of gold from the washed material.

The aggregate plant was capable of producing 22,000 tons of material on peak days and of maintaining a delivery of 500,000 tons per month for several successive months. Material was excavated from the deposit by two walking draglines of 7- and 10-cubic yard capacities and carried by pit conveyors to the processing plant. There was some deficiency in the fine sand sizes in the gravel as dug from the pits, and rod mills having a capacity of between 50 to 100 tons per hour were used in the processing operations to make up this deficiency.

Although the main line of the Southern Pacific Railroad ran between the processing plant near Redding and the stock piles at Coram, just below the dam, study showed that economies could be effected by moving the tremendous quantity of gravels and sands to the dam by conveyor belt. The 26-flight, 9½-mile-long conveyor belt that was constructed by the Columbia Construction Company is still the world's record-sized conveyor belt. The flights of conveyor belts were of various lengths, depending on the grade. The longest flight was two-thirds of a mile long and the steepest grade was 25 percent. The 36-inch-wide belt moved at the rate of approximately 6 miles per hour and its capacity was 1,100 tons per hour. Each flight was driven by a 200-horsepower motor, with those on the downhill flight acting as generators by utilizing the available kinetic energy of the moving material. All flights were remotely controlled from the Coram end of the system. Electrical interlocks assured sequential starting and stopping of the belts, so that there was never any overload at transfer points. The conveyor discharged into hoppers at Coram, where it was picked up by the prime contractor's shuttle belt system discharging onto stock piles just upstream from Coram. From the stock piles, the sand and aggregate were moved as needed to fill the hoppers in the top of the concrete mixing plant.

Production of concrete for Shasta Dam required approximately 6,000,000 barrels of low-heat cement, enough to keep a mill continuously active for a number of years. The Permanente Corporation of Oakland, Calif., was the successful bidder for supplying the cement, at a time when it had no plant, quarry, or cement mill. However, the corporation had explored certain deposits of raw materials in the vicinity of Permanente Creek near San Jose, Calif., and had assured itself that sufficient deposits of suitable materials were present to justify the erection of a completely modern cement plant. When Permanente captured the bid at $1.90 per barrel delivered, or about seven-eighths of the combined bid of six other West Coast cement companies, the corporation had only about 8 or 9 months in which to design and construct the complicated mill and quarry plant and to begin deliveries of cement in bulk to Shasta Dam. Nevertheless, although the contract award was made on June 10, 1939, the first cement was manufactured on December 25, 1939, and deliveries to Shasta started early in 1940 on time and at the required rate.

Cement was delivered to Coram in hopper cars and unloaded pneumatically into a battery of nine blending bins. From here, the bulk cement was pumped to the top of the mixing plant.

The mixing plant itself was conventional. Hoppers for cement, sand, and several sizes of aggregate up to 6 inches maximum were provided at the top of the structure. These fed by gravity to batching hoppers which automatically weighed quantities required for batches of each of several possible mix combinations. Two operating panels controlled all operations of weighing and mixing, and continuously recorded all operations of the mixing plant. On the mixing floor five 4-cubic-yard mixers were arranged around a common hopper. Concrete was discharged directly from the common hopper into self-propelled concrete transfer cars, which

Seven cableways radiated from the gigantic construction cableway head tower, which rose 466 feet above the right abutment of the dam.

ran on a circular track connecting the mixing plant and the bucket-loading dock. Concrete transfer cars carried two buckets, each holding 8 cubic yards of concrete. The transfer cars dumped directly into 8-cubic-yard bottom dumping buckets, which were spotted on the loading dock by the cableway operator. The buckets were then picked up by the cableway and moved to a point directly over the point of placement, under the direction of a bellboy in telephonic communication with the cableway operator. After safety latches on a bucket had been released by workers at the point of placement, dumping was accomplished by the cableway operator's manipulation of the load lines on which the concrete bucket hung. Mechanical vibration was used for final compaction of the concrete.

An indication of the efficiency of this concreting plant was the record placement of 11,790 cubic yards of concrete made on August 9, 1941.

Probably the outstanding feature of the construction of Shasta Dam is the immense cableway head tower projecting 466 feet above ground and anchored over 100 feet deep into the rock of a small promontory just upstream from the dam on the right abutment. This single-cantilever structure was designed to take the surging loads from seven cableways radiating outward from the top of the tower. On the hoist floor, 260 feet above ground, were located seven 3-drum hoists for the operation of the cableways. Arranged around the outside at this level were cubicles for the cableway operators. All cableways had movable tail towers, three of them being as much as one-half mile away from the head tower. The carriages for all cableways ran on 3-inch lock-coil track cable, weighing about 22 pounds per running foot. This cableway system handled over 19,000,000 tons of material during the construction of the dam, with no changes needed in the original lay-out.

Some idea of the amount of construction plant that Pacific Constructors, Incorporated, assembled for the construction of Shasta Dam may be gained from the fact that 7,100 tons of structural steel were used in the plant, of which 4,100 tons were in the head tower alone.

One of the major difficulties in the construction of any concrete dam is to provide a downstream face free from bug holes or air pockets. With an overflow spillway of such an unprecedented height as at Shasta Dam, it was especially important to secure as smooth and strong a facing as possible, to withstand the abrasion of the down-rushing water. Accordingly, special forms were provided for the downstream face of the spillway section, to which vacuum was applied by pumps after concrete deposition and during its consolidation in order to remove all entrained air and water at the surface, forming a much stronger, "case-hardened" skin.

During the construction of Shasta Dam, instruments were embedded and observational points were established from which to obtain data and measurements for checking the assumptions used in the design of the dam and also to record its structural behavior, all in the interest of assurance as to its safety. Electrical strain meters, stress meters, jointmeters, and resistance thermometers were embedded to measure deformations, stresses, and temperatures in the concrete of the dam. At four locations along the length of the dam, plumb lines were hung in special shafts accessible to galleries at 50-foot intervals, so that deflection of the dam might

These 75,000-kilowatt capacity hydroelectric generators produce most of the Central Valley project's power to run pumping plants and serve irrigation needs.

be checked. Three lines of precise traverses and levels were run, starting at undisturbed ground off the dam, to give indications of the movement of the dam as a whole. A seismograph station was set up just outside the Government camp 3 miles distant from the dam and placed in operation before the dam was finished to record normal seismic behavior of the region, and to detect any additional seismic activity due to the increased load upon the earth caused by the reservoir, whose weight amounts to 6¼ billion tons. A system of pipes was embedded in the base of the dam leading to pressure gages in the foundation galleries in order to measure the amount of pressure against the base of the dam due to water percolating through the rock. Other electrical gages were embedded directly in the concrete to see if pressure of the water percolating directly through the concrete could be detected. Routine tests of concrete cylinders made during the construction gave an indication of the quality and strength of the concrete placed. A special block of concrete was placed on the left abutment downstream from the dam from which large cores are drilled periodically and tested, giving long-time measurements of the development of the strength of the concrete in the dam.

POWERHOUSE

The Shasta power plant, situated on the right bank of the river downstream from the dam, is a huge structure although dwarfed by the immensity of the dam. It covers the ground area of a city block and towers to the height of a 15-story building.

In the powerhouse are five 103,000-horsepower turbines driving five 35,000-kv.-a. generators and two 4,250-horsepower turbines driving two 3,000-kilowatt station-service generators. Five 15-foot inside diameter, welded-steel penstocks pass through the dam and convey water from the reservoir to the power plant. Water for running each of the two station-service units is tapped from two of the main penstocks.

It is interesting to note that this is perhaps the only power plant whose generators were placed in service before there was water behind the dam. In the early days of World War II, while the West faced a critical power shortage, parts for two generators and turbines lay idle at Shasta Dam power plant, with no prospect for using them visible. No water had yet been stored in the reservoir, nor would there be for a long time to come. There was ample water at Grand Coulee Dam and space to install the turbines and generators. Hydraulic characteristics were such that the turbines and generators could be used at Grand Coulee. Accordingly, the parts were shipped to Grand Coulee Dam, where they were installed and provided power during the critical war years, after which they were returned to Shasta. First power from Shasta Dam power plant flowed onto the line on July 14, 1944, 6 years after construction began.

Water plunges through these penstocks into the powerhouse turbines. In the background is the spillway, framed by the training wall, topped by the drum gates, and dotted by the discharge openings of the outlet works.

With the reservoir at low level, Roosevelt Dam resembles a battlemented castle of medieval Europe. The bridge in the foreground crosses the south spillway; visible on the canyon wall beyond the end of the bridge are the butterfly valves of the north outlets. In the background can be seen the bridge and gate structure of the north spillway.

ROOSEVELT DAM

SALT RIVER PROJECT, ARIZONA

By Ben R. Elliott

Roosevelt Dam was one of the first large dams built under the Reclamation Act of 1902. Construction started in 1905 and the dam was completed in 1911. President Theodore Roosevelt, in whose honor the dam was named, attended the dedication ceremonies held on March 18, 1911. The dam is located on the Salt River in the south-central part of Arizona, about 30 miles from Globe. It was built to provide adequate storage of irrigation water for the Salt River project, and its power plant was the beginning of a large power system which now includes power plants at Horse Mesa, Mormon Flat, and Stewart Mountain Dams, all located farther down the river in the order named.

The general features of the dam and appurtenant works are shown in the accompanying drawing. Preliminary holes drilled into the foundation at the dam site showed the rock to be tough, fine-grained sandstone, dipping upstream at an angle of about 29° and at approximately right angles to the dam. A cyclopean masonry, arch-gravity type dam was selected for the site. The masonry consisted of massive stones set in portland cement mortar with the vertical joints between the large stones filled with concrete and spalls. The face stones were rough cut and laid in courses with 2-inch mortar joints. The upper 100-foot height of dam was reinforced with steel rails.

A large spillway capacity was required since the flow of the river varied from a few cubic feet per second to an estimated 150,000 cubic feet per second which occurred in February 1891. The design provided for a spillway on each abutment of the dam with a combined discharge capacity of 150,000 cubic feet per second. Each spillway consisted of a masonry overflow weir about 10 feet high across an unlined open channel through the rock. The combined length of the two spillways was 402 feet. Rock from spillway excavation furnished practically all of the stone required for the dam.

A diversion tunnel 12 feet wide, 10 feet high, and about 480 feet long was driven in the canyon wall at the south (left) abutment of the dam. The tunnel was located at river level and had a capacity of about 1,300 cubic feet per second with water at the top of the cofferdams. Flows in excess of this amount were to be carried in a wooden sluiceway over the cofferdams. For later use as part of the outlet works, a reinforced concrete intake structure was built on the upstream end of the tunnel about 125 feet from the dam.

One of the factors which influenced the type of dam selected was the problem of materials transportation. To reach the site in the rugged, mountainous country, some 112 miles of roads had to be built, relocated, or improved from the rail point towns of Mesa and Globe, Ariz., to the dam site. Realizing that the roads would open up undeveloped areas, the people of Phoenix, Mesa, and Tempe contributed a substantial portion of the funds for the road to Mesa. The road from Mesa to Globe by way of Roosevelt is now a part of the State highway system and is known as the Apache Trail. A secondary highway crosses the dam via a three-span reinforced concrete arch bridge over each spillway channel.

Truck transportation was practically unheard of at the time the dam was built and all freight was hauled in by horse- or mule-drawn vehicles. Because of the difficulty in transporting freight and the high resulting cost, all materials possible were obtained locally. The Government built a sawmill in the Sierra Ancha Mountains about 30 miles from the dam site, which supplied practically all of the lumber used on the job. A plentiful supply of dolomitic limestone and clay was found near the dam site and a cement mill was built which furnished all the portland cement required for the dam as well as for other structures on the project. Fuel to operate the cement plant was the greatest problem, which was solved by shipping in oil from California.

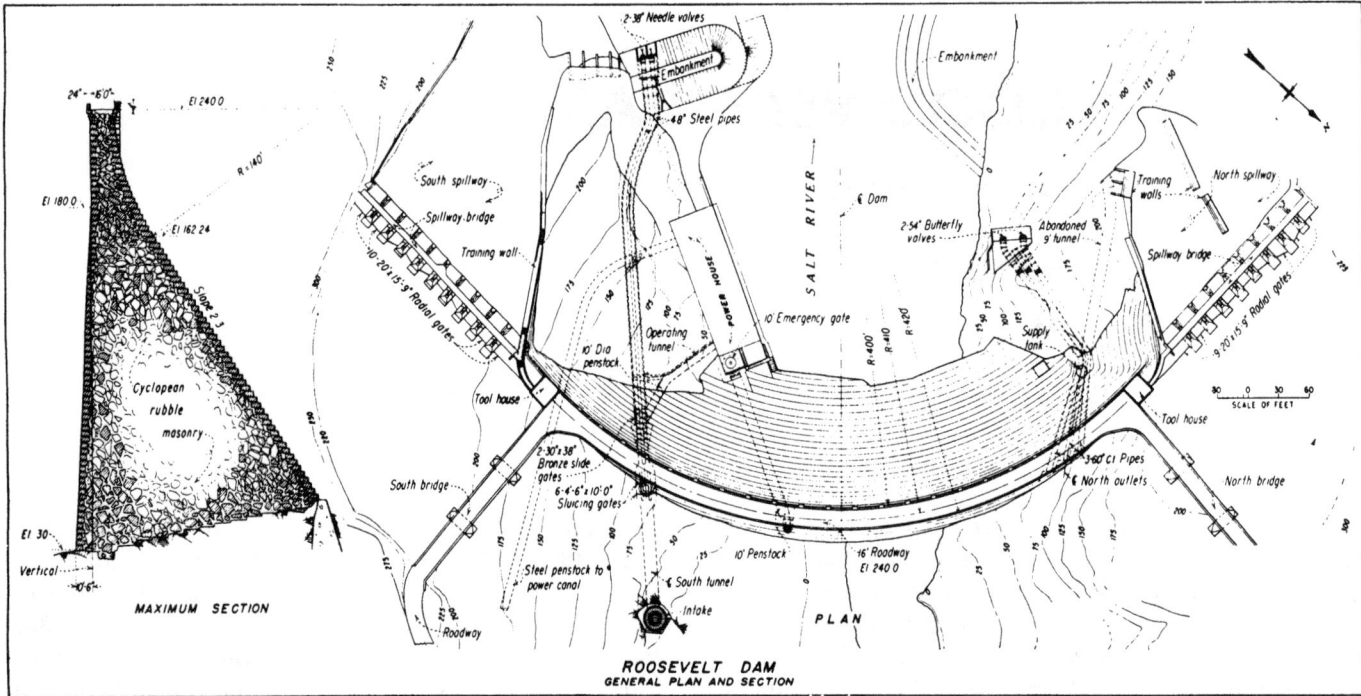

MAXIMUM SECTION

ROOSEVELT DAM
GENERAL PLAN AND SECTION

From a diversion dam on the Salt River about 19 miles above the dam site, a power canal was built which delivered water to a 7-foot diameter penstock leading to a temporary hydroelectric unit installed in a cave excavated behind the powerhouse. The unit operated under a head of 220 feet and supplied the power required during construction. Although built primarily for construction uses, the power canal is a permanent feature of the project. During construction, the canal

Construction methods at the turn of the century are illustrated by these old pictures of Roosevelt Dam. Spaces between the large stones were filled with cement mortar and rock spalls.

also supplied water to operate hydraulic jet pumps for all of the contractor's pumping operations.

A crushing plant was installed to produce sand from the dolomitic limestone; however, the product lacked fine particles and produced a harsh mortar. After trials, it was found that a satisfactory mortar could be produced by using equal parts of crushed sandstone and limestone.

The contract for construction of the dam was awarded in April 1905 to J. M. O'Rourke & Company of Galveston, Tex. The contractor erected two cableways, each about 1,200 feet long, spanning the canyon walls at the dam site, which handled materials to stifflegged derricks on the dam. He also built a 1,700-foot aerial tramway to carry cement and sand to the mixer.

By November 1905, the contractor was driving piles for some of the temporary work when one of the worst floods in the history of the river occurred. The water surface rose about 30 feet in 15 hours and reached an estimated discharge of 130,000 cubic feet per second. All work done in the river channel was destroyed and the contractor lost some of his equipment. High flows made it impossible to work in the channel until the following March. After this experience, it was decided to abandon the wooden sluiceway plan and to work on the lower part of the dam only when the flow of the river could be handled through the diversion tunnel. During construction, the contractor was delayed on several other occasions by floods. As the dam rose, the north end was kept at a lower elevation so that floodwaters passing over the structure were diverted away

Contrast the primitive appearance of the cableway head towers built of timbers and the steam-powered hoisting engine with present day construction machinery.

from the powerhouse area at the south abutment.

The rock in the foundation was sound and required little excavation. Seams carrying river water were sealed with grout. Several warm springs in the base were piped together and furnish warm water for the powerhouse. At the upstream face of the dam a minor fault was uncovered with a displacement of about 8 inches. The curve of the arch missed the fault in the abutment areas, and where the fault lay within the lines of the dam foundation it was cleaned out and a masonry cut-off constructed.

The original design provided for two power penstocks; the 7-foot diameter steel pipe from the power canal, and a 10-foot diameter steel pipe passing through the dam with intake 165 feet below the top of the dam. Also, in the original design, three 60-inch diameter cast-iron pipes located on the north side of the dam converged into a 9-foot diameter concrete-lined outlet tunnel. The pipe intakes, located 123 feet below the top of the dam, were controlled by 58-inch balanced valves. The outlet tunnel discharged from the face of the cliff in the north spillway area. For additional outlet capacity, two 43-inch balanced valves were installed in the powerhouse and connected to the 10-foot diameter penstock.

The powerhouse at the downstream toe of the dam was constructed by Government forces. It is a concrete structure faced with stone. Provisions were made for the ultimate installation of six units. Three 1,200-horsepower units installed in 1909 were served by the power canal penstock. Later, two more units of the same output were installed and connected to the 10-foot diameter penstock. The sixth unit, of about 5,000

horsepower, was installed in 1915 and connected to the larger penstock.

Power and outlet equipment were in the pioneer stages of development at the time the original installations were made at Roosevelt Dam. Since that time, changes and improvements have been made in the various features as originally installed; the power units at Roosevelt and some other dams still generate 25-cycle power which is converted to the normal 60-cycle before use.

As the Stoney gates in the south (diversion) tunnel did not operate satisfactorily under high heads, two 58-inch diameter bronze pipes controlled by 30- by 38-inch bronze slide gates were installed in a concrete plug downstream from the Stoney gates. Steel pipes 48 inches in diameter connect to the bronze pipes and extend to the outlet of the tunnel. By connecting the two 48-inch pipes to the 7-foot penstock with risers and installing 38-inch needle valves on the downstream ends of the 48-inch pipes, the power units can be operated from the reservoir whenever there is a failure in the power canal due to heavy rains. In 1923, the two 43-inch balanced valves in the powerhouse were removed and a 10,000-horsepower generating unit installed in the space they had occupied. To serve this new unit, a 10-foot diameter penstock was installed in a 14-foot diameter tunnel driven from the powerhouse to a chamber in the south tunnel.

The north outlet works was altered by plugging the 9-foot diameter tunnel and relocating it to the left from near its upstream end. Three 54-inch diameter steel pipes were installed within the cast-iron ones and were extended through the relocated tunnel to a canyon-wall outlet structure where they discharged through 54-inch needle valves. These valves were later removed and two 54-inch butterfly valves located on the outlet ends of the

Laboriously, stone upon stone, Roosevelt Dam rose from the river bed. Wooden derricks for handling the stone were used in 1908.

two outside pipes, the middle pipe being bulkheaded.

In 1913, the spillway crests were raised 5 feet to provide additional storage. In 1923 the spillways were remodeled; the channels were partially lined with concrete below the weirs, concrete gate structures added, and radial gates installed to further increase the storage capacity. In 1936, the crests were lowered 6 feet, the gate structures reinforced, the gates altered, and new operating equipment installed. The last changes were made to restore the desired spillway discharge capacity of 150,000 cubic feet per second which had been curtailed with the raising of the crests and installation of the gates.

PERTINENT DATA

Reservoir capacity	1,398,000 acre-feet.
Height of dam	280 feet.
Length of dam (including spillways)	1,125 feet.
Size of radial spillway gates	15 feet 9 inches high by 20 feet long.
South spillway	10 gates.
North spillway	9 gates.
North outlet works	Two 54-inch butterfly valves.
South outlet works	Two 38-inch needle valves.
Power output	24,000 horsepower.
Approximate cost (1913)	$3,650,000.

ARROWROCK DAM

BOISE PROJECT, IDAHO

By Carl J. Hoffman

Shortly after the Reclamation Act was passed in 1902, the Boise project in southwestern Idaho was established to supply irrigation water to project lands and to supplement the supply to irrigated lands then being served through Warren Act contracts. The Arrowrock Reservoir was completed in 1916 as a part of the storage system for the project, impounding surplus winter and flood flow of the Boise River for use on the portion of the project lands situated between the Boise and Payette Rivers. Together with other reservoirs in the Boise and Payette River basins, Arrowrock Reservoir is now a part of a coordinated and integrated multiple-purpose storage system providing irrigation storage, flood control regulation, and power storage regulation for the project and surrounding developments.

Arrowrock Dam is located on the Boise River, about 4 miles below the junction of the north and south forks and about 22 miles upstream from Boise, Idaho. The river above the dam drains about 2,200 square miles on the western slope of the Sawtooth Mountain Range. The average annual flow of the river at the dam is 1,600,000 acre-feet, which, before upstream reservoirs were built, varied from high flood rates in the spring to a few hundred cubic feet per second late in the summer.

The original storage capacity of the reservoir was about 276,500 acre-feet with the water surface at the top of the spillway drum gates, when raised. In 1936, the dam and spillway crest were raised 5 feet, adding about 15,000 acre-feet of storage. Considering the deposition of silt within the reservoir storage space, the estimated usable storage capacity is about 284,000 acre-feet with the water surface at the top of the gates.

The reservoir is about 17 miles long, and covers an area at maximum storage level of about 3,000 acres. In shape it resembles the letter **Y**, as water backs up both the north and south forks of the Boise River.

DAM SITE

In 1903 and 1904, reconnaissance surveys were made at several reservoir sites on the upper waters of the Boise River, including rough surveys of several possible dam sites. Comparative cost estimates of dams at various sites were made, and in 1910 a drilling crew was sent out to explore the foundations at the more promising sites. Through a process of elimination, the Arrowrock site was chosen as most favorable, and testing of the foundation was continued in more detail during the latter part of 1910 and the early part of 1911.

Fifty-nine diamond-drill holes were sunk over the foundation area and along the spillway site, supplemented by numerous test pits, shafts, and tunnels above water line. The foundation rock was found to be a good quality hard granite, whose surface was nearly 90 feet below river level at the deepest place and at a depth of 65 to 70 feet over most of the remaining river area. The river-bed material was mostly sand, gravel, and boulders which were suitable for concrete aggregate.

The canyon at the dam site is fairly narrow, with high, steep, bare granite cliffs on the north side and less precipitous granite slopes on the south side, covered along the lower part for some distance by a cap of lava. The face of the lava cap rises almost vertically to a height of about 50 feet from its contact with the granite, which starts about 20 feet above the river water surface. The top of the lava bench is nearly level and, like the granite, is covered with a thin layer of overburden.

THE DAM

In 1916, when Arrowrock Dam was completed, it was the highest dam in the world. The dam is a curved gravity type, built of sand-cement concrete. The dam as originally built rose about 260 feet above river level, and was about 349 feet high from the deepest point in

With Arrowrock Reservoir at a low stage, irrigation water for the Boise project pours from the upper tier of outlets.

the foundation to the top of the parapet. It was 223 feet thick at the base, 15.5 feet thick at the thinnest point near the top, and carried a 16-foot roadway along its top length of 1,100 feet. In 1936 the dam was raised by adding a 5-foot concrete capping along the crest, placing the crown of the road at the normal high storage level.

An inspection gallery and two operating galleries are situated inside the dam. The inspection gallery is located 25 feet from the upstream face of the dam. The lower, central portion is level and the abutment portions follow the profile of the foundation. The gallery has entrances at each end of the dam near the crest, and an additional entrance at the level of the lava bench on the left side. Two 24-inch square drains lead from the inspection gallery to the downstream face of the dam to discharge drainage water which collects in the gallery. Connecting with the inspection gallery are two horizontal operating galleries, each placed under a set of the regulating outlets which pass through the dam. The balanced valves controlling the flow through the outlets are operated from these galleries. The galleries, located 16 feet from the upstream face of the dam, are 87 feet apart in vertical distance and are con-

nected by a spiral stairway. The lower gallery runs under the lower set of regulating outlets only, while the upper gallery extends along the entire length of the dam, providing access for inspection of the upper portion of the dam.

The foundation of the dam was pressure-grouted through two lines of holes drilled along the entire length of the dam at distances of 5 and 13 feet, respectively, from the upstream face. The holes were drilled at 10-foot centers to an average depth of 26 feet into the bedrock. For draining the foundation downstream from the grouted curtain, drainage holes, 27.5 feet from the upstream face of the dam, were drilled on 10-foot centers 26 feet into the bedrock. These holes emerge in the inspection gallery. For draining any seepage penetrating the concrete from the upstream face of the dam, vertical drainage conduits are spaced at 15-foot centers at a distance of 12 feet from the upstream face of the dam, extending from the inspection gallery to within 10 feet of the top of the dam.

Radial contraction joints were formed in the upper portion of the dam by building alternate sections at different times. The joints are spaced at various intervals dependent on the elevation and thickness of the

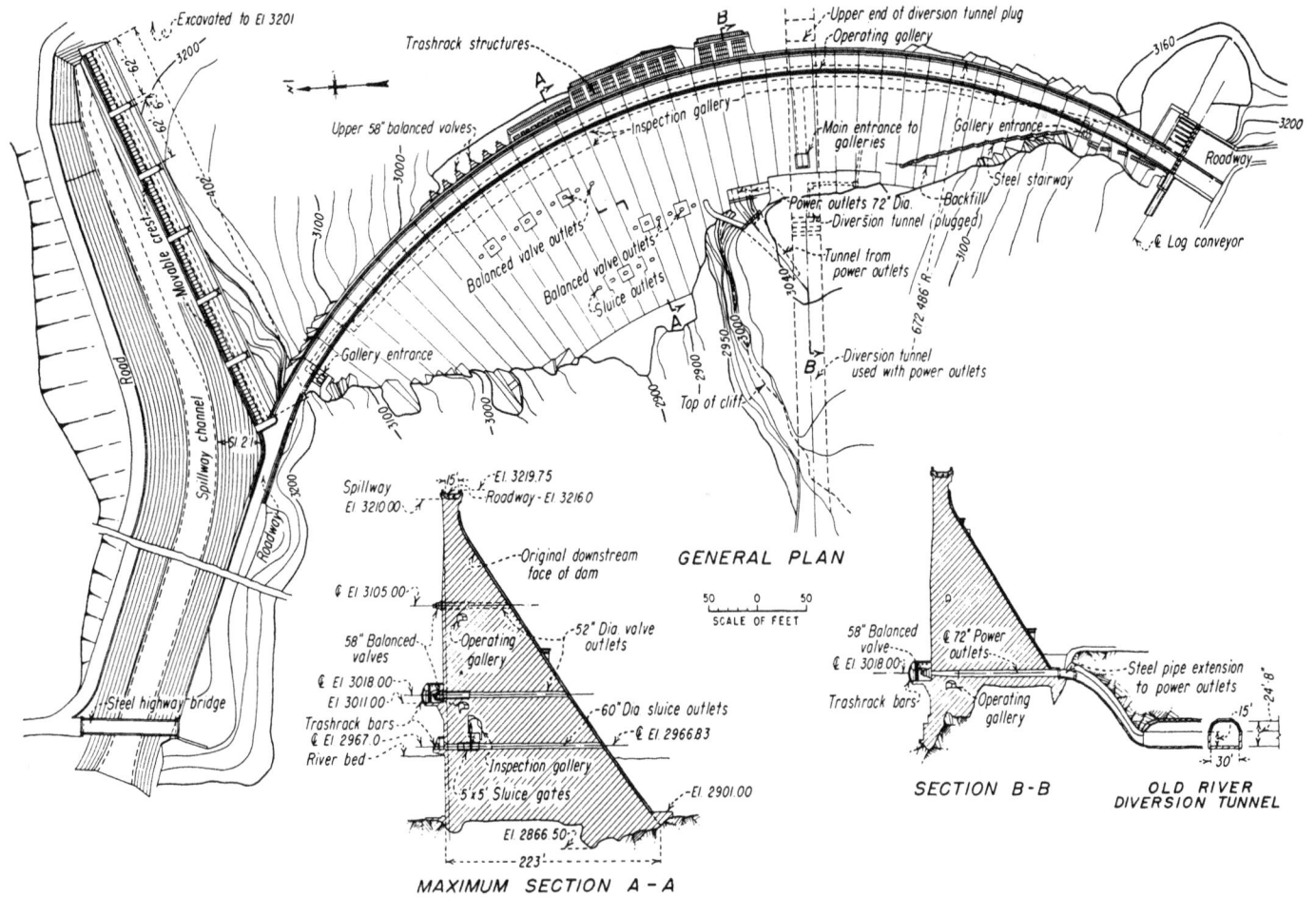

GENERAL PLAN

MAXIMUM SECTION A-A

SECTION B-B

OLD RIVER DIVERSION TUNNEL

dam. Three vertical wells were formed in each joint which were later filled with concrete during cold weather, after concrete in the dam had undergone some contraction. A **Z**-strip, annealed-copper water stop was installed in each joint 5 feet from the upstream face of the dam, and immediately downstream from this strip a triangular drain was formed in the joint. These drains, provided to collect any water which percolates past the water stop, terminate in the inspection or operating galleries.

The principal quantities involved in the original construction of the dam were: 322,390 cubic yards of excavation, 585,165 cubic yards of concrete, 603,020 pounds of reinforcement steel, 10,490 linear feet of grout holes in bedrock, 24,540 linear feet of drainage conduits in concrete, 1,067 linear feet of drains, and 2,182 linear feet of operating galleries.

SPILLWAY

The spillway, a gated side-channel type, is located in a granite cut at the north or right end of the dam. The spillway weir is adjacent to the dam and extends along the north bank of the reservoir a distance of 402 feet.

The spillway channel is approximately parallel to the weir, so that flow below the crest is turned at right angles and carried around the end of the dam to a point where the water discharges into Deer Creek, which flows into the Boise River about 800 feet below the dam. The spillway is designed for a capacity of 40,000 cubic feet per second with a head of 10 feet over the fixed crest of the spillway.

Flow through the spillway is controlled by six structural steel drum gates, each 62 feet long and 6 feet high. These gates, housed in recesses in the crest, are separated by 6-foot piers which contain the gate control mechanisms. The fixed crest of the spillway, when the gates are lowered, is at elevation 3210. The normal high storage level at elevation 3216 is maintained by raising or lowering the gates as necessary, either automatically or by manual manipulation of valves.

The channel cross section adjacent to the spillway weir increases in width and depth along its length and, at the lower end, has a 30-foot bottom width and a 50-foot depth. Beyond the end of the overflow weir, the channel continues for about 400 feet and terminates high on the cliff above the Deer Creek ravine. The channel is lined with reinforced concrete, anchored to

Arrowrock spillway discharges beneath the bridge on the left, and the outlet works discharge into the river channel in the center. The chute at the right is the logway, by which saw timber is passed over the dam.

the rock, and thoroughly drained with vitrified tile and weep holes.

The roadway across the dam passes along the spillway channel to the lower end, where a 96-foot-long steel highway bridge, 16 feet wide, crosses the channel.

The principal quantities contained in the spillway as originally built were: 359,000 cubic yards of excavation; 1,300 cubic yards of backfill; 25,564 cubic yards of concrete; 708,690 pounds of reinforcement steel; 641,770 pounds of gates, machinery, and structural steel; 5,030 linear feet of drains; and 50,200 pounds of structural steel in the highway bridge.

OUTLET WORKS

There are 25 outlet conduits through the dam, in three tiers, arranged so that when the conduits are discharging, the water falls into the river bed. The lower tier of five sluicing outlets is located at the elevation of the river bed, 249 feet below the crest of the dam. These outlets are 60 inches in diameter. They are protected by a trashrack and are controlled by 60- by 60-inch slide gates, operated by oil pressure.

Ten outlet conduits are located 198 feet below the crest of the dam. Three of these are 72 inches in diameter and are reinforced so they may become penstocks in connection with a possible future power development. The remaining seven conduits are 52 inches in diameter. All 10 conduits in this middle tier are controlled by 58-inch balanced valves placed at the conduit entrances and protected by trashracks.

The top tier of 10 outlet conduits is located 111 feet below the crest of the dam. These are similar to the 52-inch diameter conduits in the middle tier, except that the trashracks are omitted.

The sluice gates are not intended to be operated at heads greater than 75 feet, but they are so designed that they can be opened under a head of 125 feet in an emergency. The discharge at the 75-foot head is about 5,000 cubic feet per second, and at the 125-foot head about 6,700 cubic feet per second. The regulating outlets are designed to operate under a maximum head of 100 feet, at which head they will discharge 1,000 cubic feet per second each. The total length of all outlet conduits is 2,821 linear feet, and they involved the placing of 2,672,300 pounds of gates, machinery, and structural steel.

PRELIMINARY CONSTRUCTION WORK

Before beginning actual construction on the dam, it was necessary to perform considerable work of a preparatory nature, including the construction of a telephone system, access railroad, wagon road, power plant, transmission line, sawmill, construction camp, and diversion works for carrying the river past the dam site during early construction. All work was performed by Government forces with the exception of the grading of the railroad.

The entire construction plant at Arrowrock, with the exception of a steam shovel, dragline excavator, and dinkey engines, was operated by electricity. To furnish the necessary power, a 1,500-kilowatt power plant was constructed at the Boise Diversion Dam, 14 miles below Arrowrock. A duplicate transmission line was built from the power plant to Arrowrock, and a single line to Barberton, 3 miles from the plant, where it connected with the line owned by the local power company.

The diversion works consisted of an upper and lower cofferdam and a tunnel constructed through the granite below the lava bench at the south abutment of the dam.

Smooth flow over the spillway crest at Arrowrock breaks into turbulence in the outflow channel.

The tunnel was designed for a flow of 20,000 cubic feet per second and the cofferdams were designed to withstand overtopping in case of a larger flood.

The diversion tunnel was 470 feet long and in cross section was 30 feet wide and 25 feet high, with an arched roof. The bottom and sides of the tunnel were lined with concrete, and the roof was supported by timbers. The entrance and exit of the tunnel were both bell shaped to avoid loss of head and to increase the discharge. The upper cofferdam was about 200 feet long and 40 feet high, and was built of timber cribs filled with rock, gravel, and fine material sluiced into the crevices. The faces of the cribs were of solid timber construction, with the joints calked with oakum. A sheet piling cut-off was provided. The lower cofferdam was 150 feet long and 25 feet high with construction similar to that for the upstream cofferdam. Centrifugal pumps, mounted on cars which could be raised or lowered on an inclined track, served to pump out seepage water from the excavation.

When there was no longer any need for the diversion tunnel, it was plugged with concrete for a length of 190 feet under the dam.

CONSTRUCTION

Dam construction in the river bed was performed in two stages. During the first stage, excavation and construction for an upstream portion of the dam were carried forward to an extent sufficient to protect the work during the flood season. This excavation and concrete work was successfully accomplished between the flood seasons of 1912 and 1913. The completed section of the dam then served as an effective cofferdam to prevent flood damage and seepage into the deeper workings during the remainder of the excavating and concreting.

The material in the river bed was largely gravel and sand, with perhaps 5 to 10 percent large boulders. Inasmuch as concrete materials were not plentiful in the vicinity of the work, all excavated material which was suitable for concrete aggregate was stored for such later use.

On the south abutment, the lava cap was entirely removed in order that the dam might be founded upon the underlying granite. This material, along with other materials not suitable for concrete, was deposited upstream and downstream from the dam site. The excavation for the abutments was carried on just ahead of concreting and was completed in November 1915.

Concrete work on the dam started in November 1912, and was completed in November 1915. The dam was actually built in three stages. The first stage, as indicated previously, served to protect the subsequent work.

It consisted of the lower upstream portion of the dam to a height of about 40 feet above high water in the river, and was of ample thickness to withstand water pressure to that height. The second stage brought the full thickness of the dam to the top of the first section, and the third stage completed the dam.

The best progress was made during the months of April, May, June, and July 1914, when more than 200,-000 cubic yards of concrete were placed, an average of more than 50,000 cubic yards per month. In June 1914, 56,500 cubic yards were placed in 26 working days, an average of 2,170 cubic yards per day of two 8-hour shifts.

The concrete aggregate for the first and second stages of the dam, 186,000 cubic yards, was obtained from the river-bed excavation. The remainder was hauled over the railroad from a gravel pit located near the Boise Diversion Dam, 14 miles below Arrowrock.

The sand-cement for the concrete was composed of standard portland cement to which was added a little less than an equal amount of pulverized granite, reground to such a fineness that 90 percent would pass a No. 200 sieve. Besides the rigid fineness test, the sand-cement was required to pass all the standard physical tests for portland cement. The concrete made of this sand-cement was slower in setting and hardening than that made with straight portland cement. The sand-cement plant was located at the north abutment of the dam and had a capacity of 2,000 barrels of sand-cement in a 24-hour period. The plant manufactured a total of 586,450 barrels of sand-cement at an estimated saving of approximately $300,000.

Two 12-ton cableways, each 1,500 feet long with 60-foot stationary head towers, were installed so as to command the entire area of the dam. These cableways were assisted by four 10-ton stationary derricks, two 6-ton traveling derricks, and several small stiff-leg derricks. A steam-operated dragline excavator with a 2½-cubic-yard bucket was used for excavating part of the foundation of the dam. Also, a 70-ton steam shovel with a 2½-cubic-yard dipper was used for excavating and for loading gravel at the Boise Diversion Dam gravel pit.

The aggregate screening and crushing plant and the concrete mixing plant for constructing the first and second stages of the dam were located on the cliff on the south abutment, directly beneath the main cableways. The mixing plant was a two-unit installation, consisting of 1-cubic-yard mixers electrically operated. Concrete was delivered into movable hoppers and transported by the cableways to any point desired.

Upon completion of the second stage of the dam, the mixing plant was moved to the lava bench just below the dam and another mixing unit added. Aggregates for the second plant were hauled by railroad from the gravel deposit near the Boise Diversion Dam. The concrete was discharged from the mixers into 2-cubic-yard trolley cars and carried to a central distributing tower 150 feet high. From this tower the concrete was distributed via cableways by buckets and hoppers in the same manner as that employed in the first stages of the dam.

For concrete placing in the upper portion of the dam, the cableways could no longer conveniently cover the work on the long sweeping area, and the distributing system was changed. A track and dump-car system was employed, the concrete being conveyed from the mixers by the main cableways to hoppers placed at both ends of the dam. From the hoppers, the concrete was carried to any point by dump cars running on the track. The track and hoppers were raised as the work progressed.

Concrete for the spillway was obtained from the second mixing plant, and was transported from the plant to a dumping hopper at the spillway site by one of the main cableways. The concrete was distributed from the hopper to the various parts of the work by chutes, dump cars, an auxiliary cableway, and a traveling derrick.

Since there are some 3 billion board-feet of timber growing on the Boise River watershed above Arrowrock Dam, it was considered desirable to provide a means for floating this timber past the dam. Therefore, a log conveyor system was built at the south end of the dam, designed to handle 60 million feet of logs during a season lasting from May 1 to July 15. The conveyor consists of a lift to raise the logs from the reservoir to a log deck by cable loops, "live" rolls across the dam, an endless chain chute 390 feet long, and a gravity chute 245 feet long. The structures are built of reinforced concrete and structural steel. All the original construction was performed by Government forces.

1936–37 ALTERATIONS

During the 20 years following the construction of the dam, the downstream face, the spillway channel lining, and other exposed surfaces which were made of the sand-cement concrete, suffered some disintegration. This was due, primarily, to the fact that the sand-cement concrete was very porous and absorbed water freely, resulting in rapid spalling with alternate freezing and thawing. It became apparent as early as 1927 that remedial measures would ultimately be necessary to protect the concrete surfaces from freezing and thawing action.

In 1935, funds were made available for making repairs to the dam and spillway and the contract for construction was awarded to T. E. Connelly, Incorporated, of San Francisco, Calif., in 1936.

The repairs to the dam included an 18-inch reinforced concrete slab on the downstream face below elevation 3197.75, and a gunite covering on the remaining portion of the face above that elevation. The spillway channel floor was covered with a reinforced concrete slab and the side lining covered with a layer of gunite, reinforced with steel mesh. Incident to the repairs, the dam and spillway crest were raised, as described previously.

To unwater the foundation and gain access to the downstream toe of the dam for the face repairs, a tunnel was constructed connecting two of the power outlets and one irrigation outlet in the lower tier of conduits with the original diversion tunnel below the concrete plug. A cofferdam was constructed upstream from the old diversion tunnel outlet, and water for irrigation demands was diverted through this tunnel during the repair period.

The cost of original construction was approximately $4,800,000 and the cost of repairs in the neighborhood of $600,000.

MARSHALL FORD DAM

COLORADO RIVER PROJECT, TEXAS

By Merlin D. Copen

⌄⌄

The Colorado River of Texas rises in the southeastern section of New Mexico and flows 800 miles southeasterly into the Gulf of Mexico at Matagorda, Tex. The upper part of the basin lies in semi-arid plains and produces little run-off. The lower basin, however, is subject to torrential rains which produce high run-off. Since the gaging station was established at Austin in 1898, the flow of the Colorado River of Texas has varied from a minimum of 13 cubic feet per second in August 1918 to a momentary peak of 481,000 cubic feet per second in June 1935. The average annual run-off at Austin is estimated at 1,900,000 acre-feet.

Of the objectives in developing the Colorado River, flood control to minimize damage at Austin and below has always been the most important. The inconstant and inadequate supply of water for rice culture on the fertile plains between Austin and Matagorda Bay, and the production of power to aid the project financially, assist industrial development, and spread the domestic use of electricity were other important factors that demanded consideration.

The Lower Colorado River Authority, a Texas State organization, was formed in 1935 to conserve and utilize the waters of the Colorado River. In the summer of 1937, the Buchanan (Hamilton) Dam, with a storage capacity of 1,000,000 acre-feet, was completed by the Authority. Thus was realized the first part of the flood-control program. The Roy B. Inks Dam, a power development, was also completed by the Authority at about the same time. The Austin Dam, originally built in 1893, destroyed in 1900, rebuilt in 1915 and damaged to an unusable degree in the same year, was rebuilt, enlarged, and put into service in 1940, also by the Lower Colorado River Authority.

Marshall Ford Dam, the principal flood-control feature of the Colorado River project, is located on the south side of Horseshoe Bend at the head of the lake formed by Austin Dam, and is approximately 22 miles by river and 18 miles by road northwest of the city of Austin, Travis County, Tex. The main purpose of this dam is to regulate the flow of the river for prevention of floods, storage for irrigation, and the production of power. The dam was constructed in two stages under contracts supervised by the Bureau of Reclamation. The initial development, known as the low dam, was completed in July 1939. The second stage, or final development, was completed May 17, 1942.

DAM SITE

In order to select the most advantageous location for Marshall Ford Dam, the valley of the Colorado River above Austin was carefully studied, with the result that five possible dam sites were located. The three of these nearest Austin, Bull Creek, Mount Barker, and Mount Bonnell Dam sites, were eliminated after estimates showed that because of the great length of these dams their costs were prohibitive. The final choice was therefore between the two upper sites, Hughes and Maxwell.

A board of engineers, approved by the Lower Colorado River Authority and the Bureau of Reclamation, investigated the Hughes and Maxwell sites. They found the geologic and foundation conditions were similar at both sites, but recommended the Maxwell site, principally because more storage and greater power output would be provided there at lower unit costs. After consideration of the board's recommendations, the Lower Colorado River Authority selected the Hughes site in preference to the Maxwell and the dam was constructed at the Hughes site.

The entire region in which the dam site is located is covered with a heavy growth of trees and brush. The profile of the river valley at the site shows a steep right bank and a sloping left bank which, however, is steep enough to permit the required height of dam without a lengthy abutment section. The topography is quite

Marshall Ford Dam stands athwart the Colorado River of Texas. It fulfills the triple roles of irrigation storage, flood prevention, and power production.

rough and rugged, the river having cut its channel deeply into an old plateau.

The rocks of this region are sedimentary in origin and belong to the Cretaceous period. The dam site is located on the Glen Rose formation which is composed of a series of horizontal, alternating beds of hard and soft limestone, sandy lime, shale, shell limestone, and marls. In spite of these variations, the formation as a whole is sound. The beds are generally uniform, continuous, and tight but with some porous layers. The strata forming the floor of the valley are strong, but have well-developed block pattern joint cracks. The valley sidewalls, while showing considerable weathering on the surface, are generally satisfactory beneath the surface.

THE DAM

Marshall Ford Dam is a composite structure consisting of a concrete section extending across the river channel and flanked on both ends by rolled earth embankments. In addition to the main structure, two earth dikes were constructed across low saddles, one on each side of the river.

The concrete structure is of the straight-gravity type with a central overflow section. The total crest length is 2,422.5 feet and the maximum structural height from foundation to crest is 278 feet. Details of the structure are shown in the accompanying drawing.

The crest of the uncontrolled overflow spillway is 36 feet below the top of the nonoverflow section; and the spillway is divided into five equal bays, each with 140 feet clear length, by concrete piers supporting a steel girder bridge with reinforced concrete deck. Training walls are located on each side of the spillway, and a concrete apron with dentated sills at the lower end extends 215 feet downstream from the dam proper. The capacity of the spillway is 508,000 cubic feet per second.

Twenty-four 102-inch diameter conduits extend through the overflow section of the dam and are used to control irrigation discharges from the reservoir. Trashracks are located at the upstream end of the conduits and the upstream 91 feet of each conduit is lined with semisteel castings. Regulation of the flow of water through the conduits is accomplished by two sets of 102-inch gates, the downstream set or normal

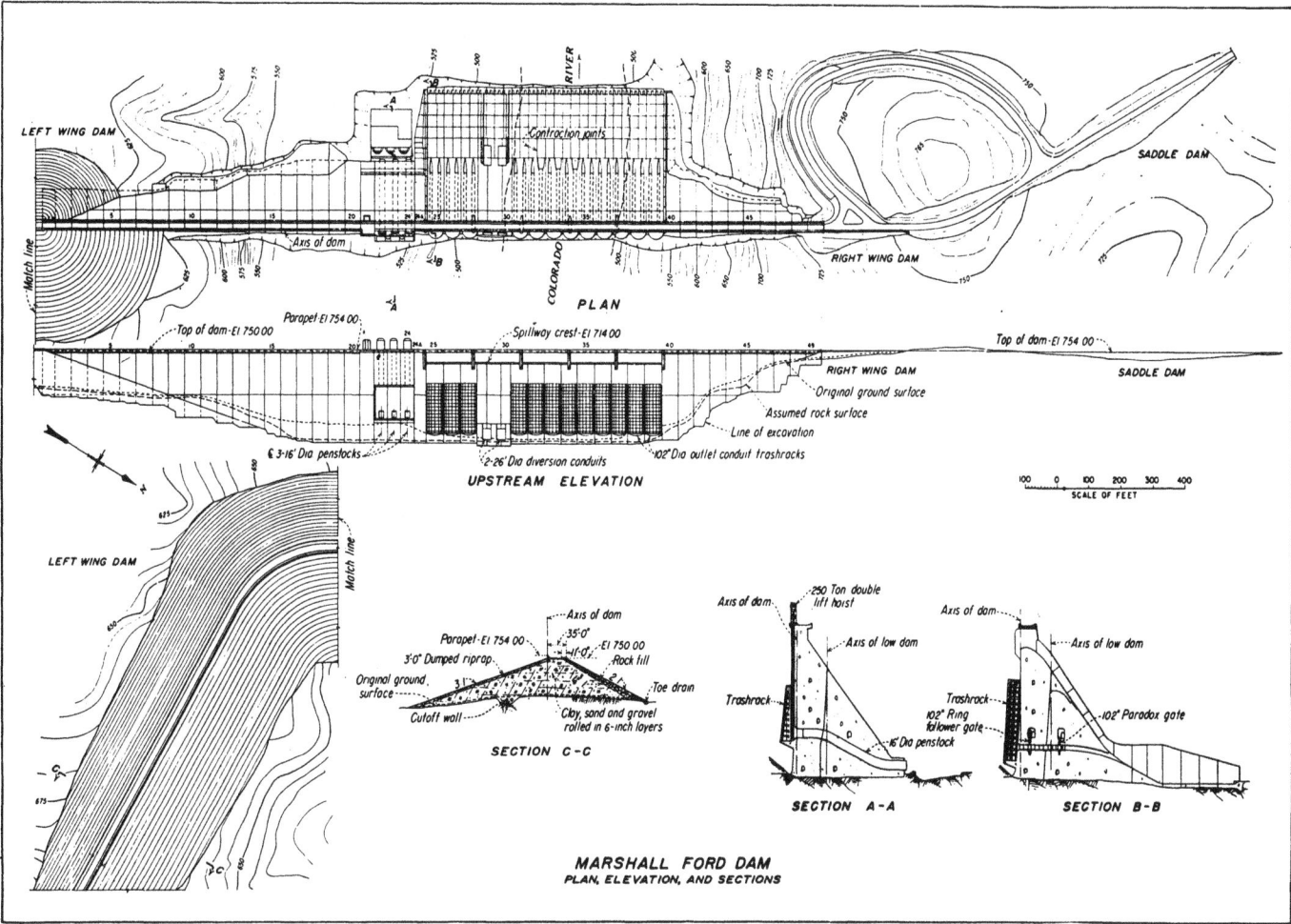

MARSHALL FORD DAM
PLAN, ELEVATION, AND SECTIONS

operating gates being of the electrically operated para-dox type, and the upstream or emergency gates being of the hydraulically operated ring-follower type. In addition, for emergency use, one 12- by 12-foot, crane-operated, steel bulkhead gate is provided to close the entrance to any conduit. The capacity of these conduits is 126,000 cubic feet per second.

Adjacent to and at the left of the spillway section, three 16-foot diameter steel penstocks provide for passage of water from the reservoir to the powerhouse. A concrete trashrack with steel trash bars and Broome-type, hoist-operated steel slide gates is located at the upstream end of the penstocks.

During the construction period, the normal flow of the river was passed through the dam by two 26-foot diameter diversion conduits with low training walls to confine the discharge at their outlets on the spillway apron. The diversion conduits were closed by steel stop logs which were dropped into place on the upstream side when storage of water behind the dam was begun in July 1940. These conduits were later plugged with concrete.

The earth embankment sections, including the two

dikes, have an overall crest length of 4,910 feet and contain 1,635,900 cubic yards of earth, gravel, and rock. The maximum structural height is 120 feet.

A powerhouse, housing three 25,000-kv.-a. units, located on the left bank of the river, downstream from the dam, was designed and constructed by the Lower Colorado River Authority. With this exception, all features of Marshall Ford Dam were constructed by contracts under the supervision of the Bureau of Reclamation.

After the dam was placed in operation, condensation of moisture in the gate-operating galleries damaged the operating mechanisms and control circuits, thus necessitating the installation of dehumidifying equipment to supply dry air. A ventilating house for this purpose was constructed to the right of the spillway on the downstream face of the dam in 1944.

CONSTRUCTION

Marshall Ford Dam was constructed in two stages with the second stage work performed on the upstream side of the first stage. Construction under the contract

The paved roadway atop Marshall Ford Dam curves away into the distance as it follows the crest of the earthen left wing dam.

for the initial stage or low dam was begun in March 1937 by Brown & Root, Incorporated, and the McKenzie Construction Company and was completed in October 1940. Orders for changes under this contract provided for construction of the base for the enlarged dam to about elevation 620, 50 feet below the crest of the low dam.

Completion of the dam to its full height was divided into two schedules. The concrete section, schedule No. 1, was constructed by Brown & Root, Incorporated, and the McKenzie Construction Company. The earth embankments were constructed by Cage Brothers and W. W. Van & Company under schedule No. 2.

Construction of this dam in the sparsely settled hills northwest of Austin necessitated the construction of a number of necessary facilities. The area was not easily accessible over the unimproved county roads, and the nearest railroad was 12 miles from the dam site. Consequently, soon after the contract for construction of the dam was awarded, the contractor began construction of a 12-mile, standard gauge access railroad. The line extended from the dam site to Rutledge, on the Austin and Northwestern Branch of the Southern Pacific Railroad. The existing county road leading from Austin to the dam site on the left side of the river was improved by widening, relocating, and oil surfacing. The work on the county road was done by Travis County. Later, by construction of a reinforced concrete, low-water bridge some 1,500 feet below the dam site, the access road was extended and connected with other county roads on the right side of the river, providing a scenic loop from Austin to the dam site.

A low-heat cement was specified for use in the concrete for the dam to reduce the heat of hydration and subsequent cracking in the concrete. The first cement contracts were proportioned between two companies in

order to assure a sufficient quantity of cement to keep pace with the construction program. Concrete was placed at a slower rate during the final-stage construction and all cement was supplied by one company. Cement was received from the mills in bulk in standard railroad boxcars, each containing about 300 barrels.

The principal sources of concrete aggregates for Marshall Ford Dam were the Horseshoe Bend river deposit, 1½ miles upstream from the dam site on the right bank of the river, and the similar Butre Bend deposit on the left bank of the river, one-half mile upstream from the Horseshoe Bend deposit. This natural aggregate was supplemented as needed by the use of crushed rock from the dolomitic-limestone quarry in the Sudduth area, 75 miles upstream from the dam site. The source of an aggregate supply for the second-stage development presented a problem in that the aggregate pits were in the reservoir area and a rise in water elevation would flood the deposits and roads leading to them. It therefore became necessary to store sand and fine gravel from the Horseshoe and Butre Bend deposits and use crushed dolomitic-limestone for coarse aggregate.

The contractor located an aggregate processing plant

First phase construction at Marshall Ford Dam was near completion at the time of this photograph. Later the crest of the dam and spillway was raised by additional construction on the upstream face.

with a capacity of 300 tons per hour on a terrace near the end of the Horseshoe Bend deposit. The aggregate material was delivered by truck from the pits to the processing area. The processed aggregates were transported from storage piles at the processing plant approximately 2 miles to the dam site by an aerial tramway. Normally, 85 buckets of 36-cubic-foot capacity were used on the system. The buckets, spaced 237 feet apart, were made of steel and traveled at the rate of 550 feet per minute. At the loading terminal, the buckets were detached from an endless cable and spotted under the appropriate aggregate bin where they were loaded. The buckets then rolled by gravity to the dispatching point where automatic equipment engaged them to the traction cable at the proper spacing. At the unloading structure, the buckets were automatically tripped while in motion, depositing the aggregate into one of five 940-cubic-yard capacity storage tanks.

The concrete mixing plant was located downstream from the dam site on the left abutment. The plant had a capacity of 200 cubic yards per hour, but the normal output during the peak construction period was 130 cubic yards per hour.

Most of the concrete was transported from the mixing plant to the various sections of the dam in 8-cubic-yard capacity buckets. The buckets were carried on a 3-inch lock-coiled cableway of 25-ton capacity with a 2,108-foot span. The topography of the site was such that a fixed tail tower 185 feet in height was required for the left abutment. A 68-foot movable head tower, containing all the controls, operated on an 840-foot curved runway on the right abutment.

Originally, the tail tower rested on natural ground near the downstream toe of the dam, but to provide sufficient clearance between the cable and the structure during second-stage construction, it was necessary to move the tail tower to the top of the low dam. For the first-stage work, stiff-leg derricks were used to place much of the concrete which was out of reach of the cableway, and a Whirley derrick equipped with a 110-foot boom was used in placing concrete not within reach of the cableway in the second-stage construction. The latter was operated on two sets of rails mounted on the downstream face of the dam.

The first excavation for the foundation of the dam was begun in March 1937. Most of this excavation was accomplished with a fleet of ten 6-cubic-yard capacity dump trucks and three 1½-cubic-yard shovels. The work proceeded rapidly, reaching a peak in May 1937. As the construction continued, it became necessary to excavate a number of drifts and trenches to intercept and confine soft seams of material in the abutment and foundation rock. These trenches were later backfilled with concrete.

Because of the nature of the foundation material, an extensive grouting program was necessary to prepare the foundation for the dam. The general grouting plan provided for preliminary low-pressure grouting to be followed by final high-pressure deep grouting. The shallow holes for the low-pressure grouting were drilled to a depth of about 25 feet and grouted prior to the placing of concrete. Holes for this low-pressure grouting were usually placed at 20-foot centers over the entire area. Some grouting was done in the spillway apron and many additional holes were required in the abutments and foundation along cracks and faults. After several lifts of concrete were placed, a line of high-pressure holes was drilled and grouted from the grouting and drainage gallery. These holes, spaced from 5 to 10 feet on centers, were in general 75 to 100 feet deep, although holes to greater depths were required in some areas. After completion of the grouting, drainage holes were drilled from the gallery to 50-foot depths on approximate 10-foot centers.

In July 1937, the contractor built two cofferdams for the care and diversion of the river during construction. One cofferdam was built on the left side of the river surrounding the 26-foot diversion conduit section of the dam. It was composed of earth and gravel with a 6-foot concrete core wall. The second cofferdam was built on the right side of the river and enclosed that section of the dam from block 37 to the right abutment. This cofferdam was also of earth and gravel construction, but was more heavily riprapped than the other and did not have a concrete core wall. In May 1938, a third cofferdam of similar construction was built across the river section and the river diverted through the diversion conduits. When work was begun on the enlargement of the dam, the upstream cofferdam was removed and built further upstream to permit this construction.

Heavy rains in the upper basin resulted in the river surpassing flood stage by many feet in July 1938. The river reached a peak flow of 276,000 cubic feet per second and discharged approximately 2,000,000 acre-feet during this flood. Construction of the dam was delayed about 1 month as the result of the flood, the principal delay occurring at the aggregate plant on Horseshoe Bend, which was flooded out. Large quantities of debris passed over the low concrete blocks in the dam proper and through the diversion conduits, but very little damage occurred to the structure.

The first concrete in the dam was placed in the left abutment October 30, 1937. This work progressed steadily, with a maximum of 89,998 cubic yards for 1 month being placed in February 1938. In May 1942 the last concrete work was completed, bringing the total of all concrete placed to approximately 1,413,000 cubic yards.

Artificial cooling to reduce volume change of the mass

Flood waters swirled over low blocks of Marshall Ford Dam, but construction continued on the abutment sections.

concrete in the dam enlargement was provided by circulating river water through a system of thin-wall tubing placed on the surface of 5-foot lifts. The concrete was cooled to about the mean annual temperature. To shorten the period of cooling, a refrigeration plant was constructed and the water pumped through it before being circulated in the cooling system.

Following cooling of the concrete, the contraction joints, which were fitted with copper sealing strips, were thoroughly grouted through a system of thin-wall tubes installed for this purpose.

The construction of the embankments involved placing four types of material into four structures: the left and right wing dams, and the left and right saddle dams. The left wing dam is the largest of the embankments and was the only one affected by the two-stage construction. A cut-off wall for the left wing dam, extending from the concrete dam approximately 845 feet up the left abutment, was completed in September 1938. This cut-off consisted of a trench with sloping sides, a 3-foot-wide concrete footing varying in depth from 3 to 20 feet placed in the trench, and a thin reinforced concrete wall extending 6 to 10 feet into the embankment and resting on the footing. Two

more cut-off walls of this type were required in the enlarged left wing dam, and one was constructed under the right wing dam. The cut-offs under the saddle dams consisted of trenches backfilled with compacted impervious material.

Construction of the embankment proper began in April 1939. Before the first-stage work on the left wing dam had been completed, imminent construction of the enlarged embankment was decided upon. Accordingly, only the impervious section was built and riprap was placed on the upstream face as a precaution against erosion from possible floods. Prior to beginning the construction of the enlarged dam the riprap was removed.

The enlarged embankment was constructed in four zones, the major portion of the yardage involved being the impervious zone 1. It was composed of a sandy clay taken from the left bank of the river downstream from the dam site. The material was hauled to the embankment in trucks and spread by dozers in approximately 6-inch layers. These layers were in turn compacted by sheepsfoot rollers (or pneumatic hand tampers in confined areas) to the required density as determined by penetration tests. At optimum moisture content, these layers required 12 passes of the rollers.

Zone 2 material, a pit-run sand and gravel, was placed on the upstream side of the wing dams and at the downstream toe of the left wing dam. It was rolled in 12-inch layers and well bonded to the impervious sections.

Zone 3, a local limestone, was obtained from a quarry on the left abutment upstream from the dam. This material was used on the downstream slopes of all embankments, and also on the upstream slopes of the saddles. The rock was spread roughly in 3- to 5-foot layers with a bulldozer supplemented by some hand shaping and leveling.

A 3-foot layer of riprap, zone 4, was placed on the upstream slopes of the wing dams. This material, a hard dense limestone, was secured from the Spicewood quarry, 35 miles upstream from the dam, and from the Sudduth area, 75 miles upstream from the dam.

An oil-surfaced roadway built across and connecting the tops of all embankments, supplemented by parking areas and sidewalks, completed the construction of the embankments.

SEMINOE DAM

KENDRICK PROJECT, WYOMING

By Loyd R. Scrivner

❯❯

Seminoe is the only dam of its size for which, under rigorous climatic conditions, the placing of concrete was completed in one construction season. This was accomplished by starting concrete placing on January 19, 1938, with the entire job covered with canvas and heated by means of steam and salamanders, and completing the parapets of the dam on November 28 with the work again steam-heated and under canvas.

Seminoe Dam is a thin concrete arch with vertical upstream face and has the following principal dimensions:

Height	295 feet.
Crest length	530 feet.
Minimum thickness (elevation 6353)	17 feet.
Base thickness (theoretical)	88 feet.
Radius to upstream face	290 feet.

Construction involved the placing of 173,127 cubic yards of mass concrete in the dam and 27,519 cubic yards of concrete in spillway and powerhouse. The required excavation amounted to 527,673 cubic yards, which amount included excavation in the diversion and spillway tunnels, the tailrace channel, and the foundation and abutments of the dam.

This dam has the dual purpose of providing irrigation storage for the Kendrick project and producing power. Of the total storage capacity, 60,000 acre-feet are set aside to maintain a minimum power head of 93 feet and the remaining 960,000 acre-feet are used to store water which is released for power production and irrigation as needed. No flood storage is provided. A total of 6,874,600 acre-feet of water was released at Seminoe between August 3, 1939, when the power plant was put into operation, and the end of 1948. Of this amount, 98.6 percent was used for the production of power. During periods when the quantity of water released for the production of power is greater than the amount required for irrigation or stream flow, the surplus water is stored in Pathfinder Reservoir, the next storage reservoir downstream.

DAM SITE

Seminoe Dam is located in the narrowest point of Seminoe Canyon on the North Platte River, 37 miles north of Sinclair, Wyo. Two sites for the proposed dam were thoroughly investigated. Site No. 1 was at the upstream end of a granite gorge, a point where the river, after following a soft sandstone in a westerly direction, turned sharply north into the granite. The granite is fresh, hard, and strong but has extensive jointing.

Site No. 2, located about 1½ miles upstream from site No. 1, had greater accessibility, being nearer the mouth of the canyon and in somewhat less rugged country. The geological formations at this site consist primarily of a hard, dense, cherty limestone which would have formed the right abutment and the lower part of the left abutment. The upper part of the left abutment would have been anchored in the basal red sandstone beds of the Amsden formation.

There was little room for choice between the two sites with respect to storage possibilities since they were so close together as to have almost the same drainage and reservoir areas. Economically, each site had a major point in its favor. Site No. 1 was narrower and would therefore require less concrete; site No. 2, being more readily accessible and being nearer the aggregate areas, would have required less expenditure to provide access facilities and transportation.

The granite gorge site (site No. 1) was selected primarily because its geological formation was more favorable for the thin arch dam which design studies indicated would be the most economical type. By the time that the contract for construction of the dam was awarded, investigation of the site involved 1,800 feet of diamond core drilling, 175 feet of vertical shafts, and 235 feet of horizontal drifts. Additional investigational drilling and tunneling was carried out after the contractor had begun excavation of the diversion tunnel.

Seminoe Dam and power plant. The power take-off structure, serving also as the switchyard, rises from the powerhouse roof. The spillway tunnel is at the left.

DESIGN

The preliminary designs of the dam were of a concrete gravity structure with overflow spillway in the center of the river channel and a power plant at the base of the right canyon wall. Water would have been delivered to the turbines through penstocks leading from the plugged diversion tunnel.

In 1934, after the granite site had been investigated, it was decided that the abutment rock formation would be adequate for an arch dam. Specifications and invitations to bid were issued on a constant-arch dam design which included a gravity tangent at the right abutment. Bids received from this proposal were, however, all rejected.

Further investigation in the field indicated that redesign of the dam, together with a slight shift in position, would eliminate the undesirable gravity tangent and also the possibility of overhang at the upstream face if there should be need to extend the excavation further into the abutments than assumed in the design. The final design provided a uniform 290-foot radius to the upstream face and variable radii to the downstream face to give horizontal arches thicker at the abutments than at the crown. Additional bearing surface was obtained at the abutments by adding fillets of 50-foot radius to the upstream and downstream faces. Extra bearing area and percolation distance was provided at the base of the dam by a spread footing on the upstream side of the dam.

The trial-load method of analyzing arch dams was used to compute stresses in the designs. The analysis of the final design included effects of reservoir water pressure combined with a horizonal earthquake shock having an acceleration of 0.1 of that due to gravity.

Photoelastic studies were made to get experimental evaluation of arch stresses with particular reference to the effects of the fillets on stress magnitude and distribution.

CONSTRUCTION

Funds for construction of Seminoe Dam and the Casper-Alcova irrigation project (now known as the

The reservoir is full. From the face of the dam protrude the outlet works trashrack and the three penstock intake trashracks. At the right is the spillway gate structure.

Kendrick project) came from an allocation of $22,700,000 announced by the Public Works Administration on August 1, 1933.

Access roads and Government camp facilities were constructed by contract and by Government forces prior to awarding a contract for construction of the dam and powerhouse. Winston Brothers and Associates, the successful bidders on the latter contract, were given notice to proceed on January 15, 1936, and their contract specified that the work should be completed in 950 calendar days after receipt of the notice to proceed.

The 30-foot diameter spillway and diversion tunnel was started in the spring of 1936 by excavating the rock in the cut and cover section at the outlet to the elevation of the midheight of the tunnel section. The top half of the tunnel section was then removed to the junction of the inclined spillway tunnel and the diversion tunnel. After the top half of the horizontal spillway tunnel was excavated, three crews were used; one crew started from the upstream portal of the diversion tunnel, a second crew drifted a pilot tunnel up the centerline of the inclined spillway tunnel, and a third crew excavated the lower half of the horizontal spillway tunnel. Fairly sound rock was encountered in the spillway tunnels except at the downstream portal where 16 sets of steel supporting ribs were used. However, even in sound rock, considerable overbreak occurred due to the blocky, jointed nature of the granite.

The contractor started construction of the diversion tunnel as a 26-foot unlined horseshoe section, excavating from a full heading, but had to install heavy timber supports on reaching disintegrated material. About one-half of the total length of the diversion tunnel had to be supported. The following spring, when the spillway tunnel was lined, the diversion tunnel was also lined as a 24-foot horseshoe section to assure stability of

the right abutment and spillway gate structure which could have been endangered by scouring and caving of the portion of the tunnel in the disintegrated material.

Lining of the horizontal spillway tunnel and the diversion tunnel, using movable forms, was completed in July 1937. The river was diverted in August and the inclined spillway tunnel was then lined using built-in-place forms because the cross section changes shape and size for its entire length. The spillway inlet and gate structure was built during 1938.

Excavation for the dam abutments and the removal of large accumulations of talus material from the dam site was begun early in 1936 and continued through 1937, except for a few weeks when additional exploratory drifts were being driven into the right abutment to prove that satisfactory rock existed in this area despite the disintegrated material found in the diversion tunnel. The abutments above river level were excavated to design grade before the river was diverted, after which the overburden was removed from the foundation area of dam and power plant.

When the floor of the canyon was excavated to the designed base elevation of the dam, one-third of the area was still covered with disintegrated material. Excavation to an additional depth of 30 feet revealed two fault zones. Since it was unsafe to excavate the seams using open-cut methods, due to a large mass of unstable talus material on the upstream side of the excavation, it was decided to form galleries and shafts in the base of the dam and to mine the seams after the base of the dam was placed.

Control of excavation in the field was novel in that it was done entirely by triangulation rather than by chaining. The field party consisted of two transitmen with telephone headsets, a recorder, and a rodman who traversed the precipitous slopes by means of ropes with a life belt. The transitmen, in constant communication with each other, spotted the rodman at a predetermined point by horizontal angles. Then, one transitman, knowing the height of his instrument, read the vertical angle, computed the elevation, and signaled the required cut to the rodman.

The river channel was deepened for 2,200 feet downstream from the dam in order to lower the tailwater elevation, thereby increasing the head available for power production. A total of 179,636 cubic yards of material was excavated in forming a channel with 70-foot bottom width and 1 to 1 side slopes. A dragline with an 80-foot boom and a 5-cubic-yard bucket loaded the material into trucks which hauled it to a side canyon waste area.

All aggregates for concrete were obtained from pits located below stream flood line and 2½ to 4 miles upstream from the dam site. A washing and screening plant was located in the reservoir area about 2½ miles

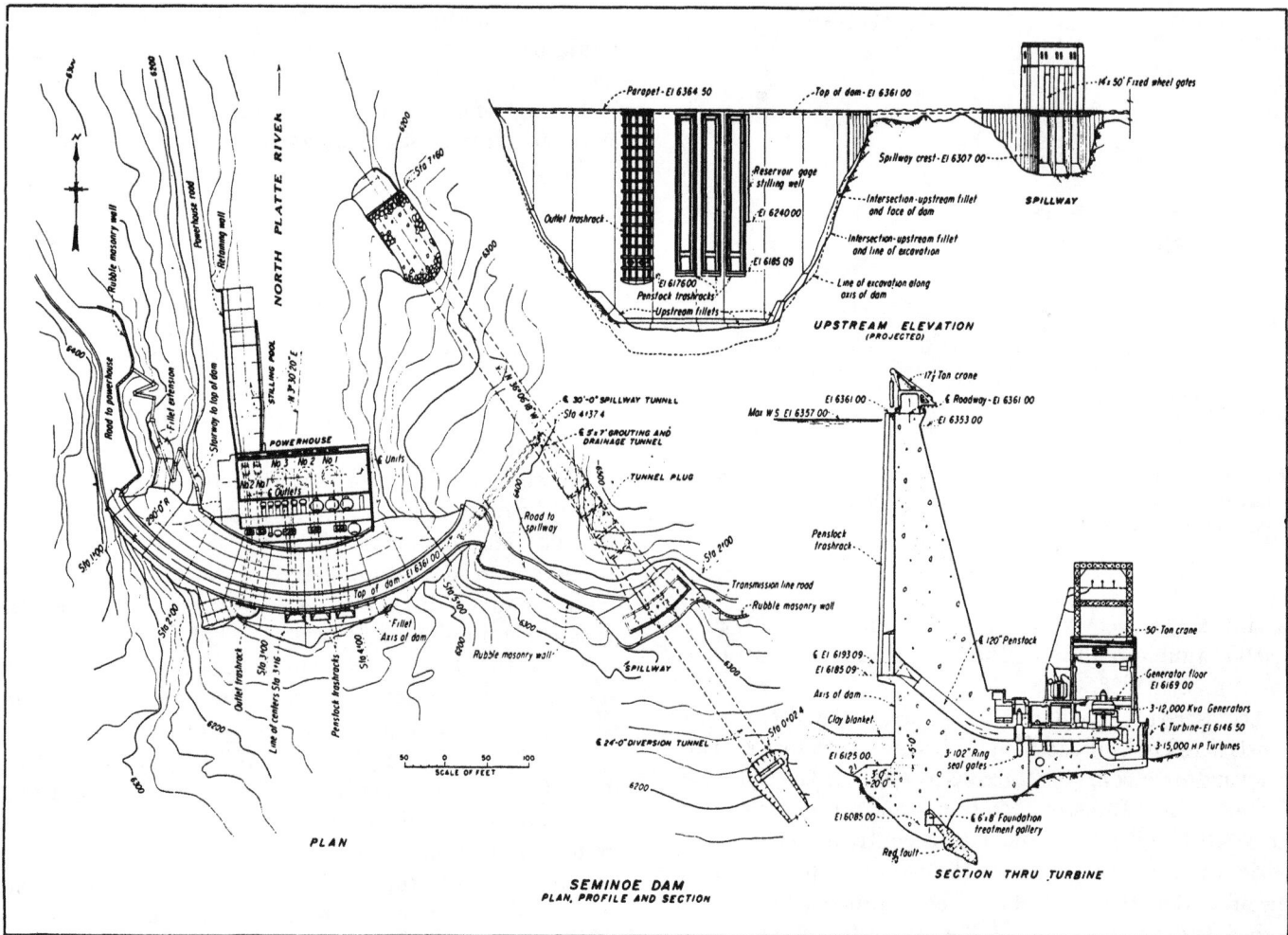

SEMINOE DAM
PLAN, PROFILE AND SECTION

from the dam. The raw aggregates were removed from the river during low-water months of 1936, 1937, and 1938. After being processed, the aggregates were stockpiled near the screening plant until required at the mixing plant. The various sizes of aggregate were transported by aerial tramline to the mixing plant located on the left canyon wall upstream from the dam.

The concrete for the tunnel lining and spillway gate structure was placed using a pumpcrete machine located on the right side of the river. Maximum distance pumped was 800 feet and maximum size aggregate pumped through the 8-inch discharge pipe was 3 inches. While working the pump to its maximum duty, the aggregate was held to 1½-inch size, but 3-inch maximum aggregate was used where it could be handled.

For the dam and powerhouse, the concrete was discharged from the mixer into 4-cubic-yard bottom-dump buckets on flat cars which were pushed by shuttle train to a position under the 12-ton traveling cableway. The cableway then transported the concrete to any point in the dam or power plant, where it was dumped and vibrated into place.

The dam was built in radial blocks, approximately 50 feet wide at the axis, with variations from this width as necessary in order that no penstock or outlet pipe would cross a radial joint. Each radial contraction joint had vertical keyways and was provided with grout stops at upstream and downstream faces and at the top of each 50-foot lift. Evenly spaced grout outlet boxes in the radial contraction joints were connected to a system of small pipe through which grout could be pumped into the joints under pressure after the concrete had been cooled. Alternate blocks were built from two to three 5-foot lifts higher than the intervening ones, and no more than one 5-foot lift could be added to a block during any 72-hour period.

Flexible panel forms 56 feet long and 5 feet wide were used on both faces of the dam. They were constructed of 1¼-inch vertical grain tongue-and-groove flooring as lagging and double 3- by 8-inch timbers as studs. No walers were used. With the studs placed on 3-foot centers, the lagging gave the panel as much strength and rigidity as was required. The forms were not lined and whenever the surface of a panel became

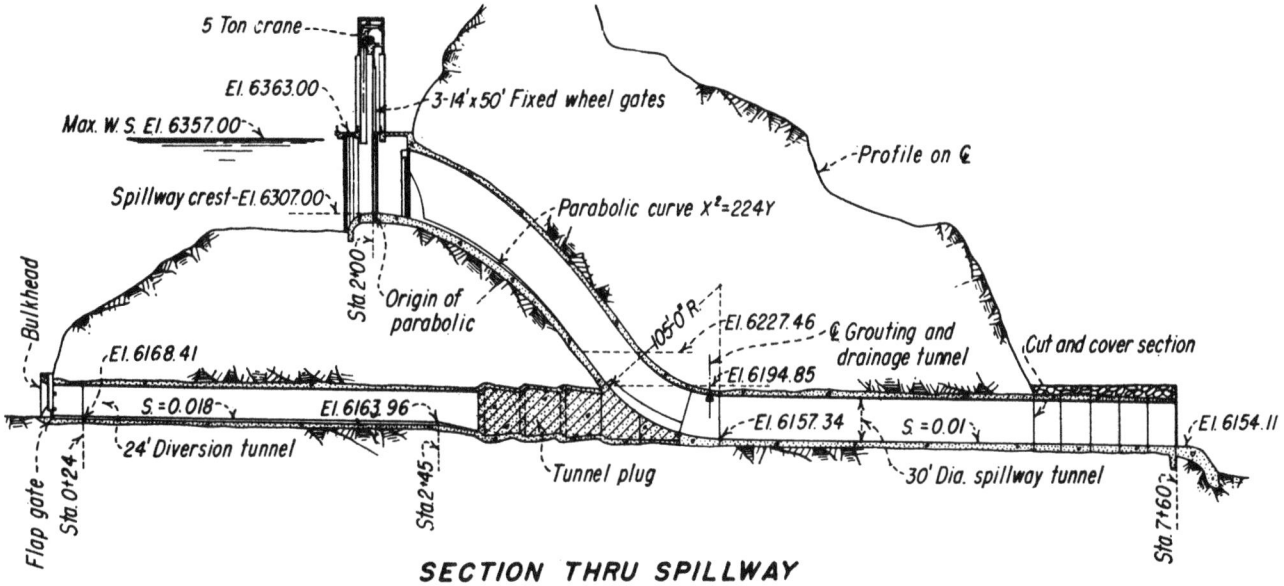

SECTION THRU SPILLWAY

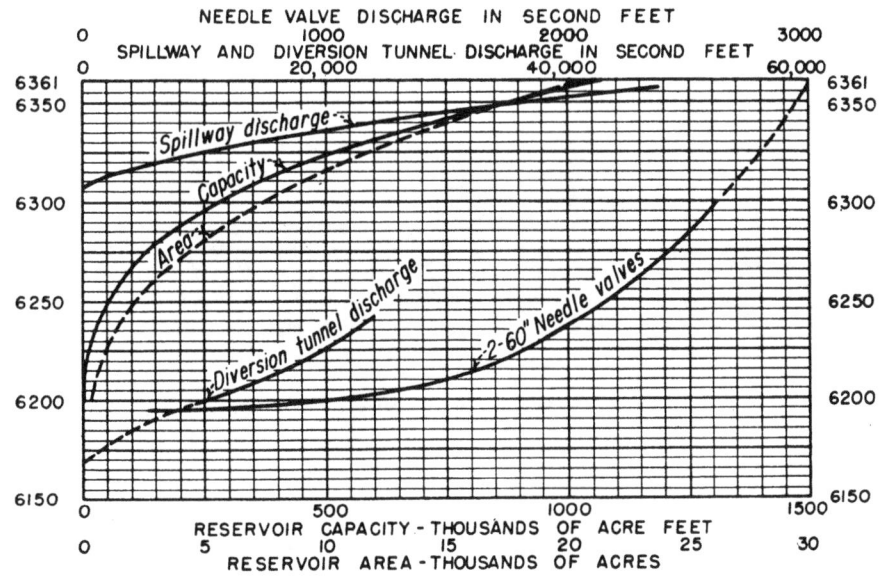

5 Ton crane

El. 6363.00

Max. W.S. El. 6357.00

Spillway crest-El. 6307.00

3-14'x50' Fixed wheel gates

Profile on ₵

Parabolic curve $X^2=224Y$

Bulkhead

Origin of parabolic

Sta. 2+00

105'-0" R.

El. 6227.46

₵ Grouting and drainage tunnel

Cut and cover section

El. 6168.41

S=0.018

El. 6163.96

El. 6194.85

El. 6157.34

S=0.01

El. 6154.11

Flap gate

Sta. 0+24

24' Diversion tunnel

Sta. 2+45

Tunnel plug

30' Dia. spillway tunnel

Sta. 7+60

AREA-CAPACITY AND DISCHARGE CURVES

NEEDLE VALVE DISCHARGE IN SECOND FEET

0   1000   2000   3000

SPILLWAY AND DIVERSION TUNNEL DISCHARGE IN SECOND FEET

0   20,000   40,000   60,000

Spillway discharge

Capacity

Area

Diversion tunnel discharge

2-60" Needle valves

RESERVOIR CAPACITY - THOUSANDS OF ACRE FEET

0   500   1000   1500

RESERVOIR AREA - THOUSANDS OF ACRES

0   5   10   15   20   25   30

---

rough, that panel was taken out, thoroughly dried, and sanded smooth with a mechanical sander. This method kept the surfaces in good shape and the same panels were used from base to top of dam. All forms were raised by manually operated jacks suspended from light timber tripods. The form-moving crew raised the forms and transported their equipment without having to interfere with the work of the cableway.

The modified portland cement used for concrete was shipped in bulk, by rail, to Parco (now Sinclair), where it was pumped into tank trucks for the haul to the mixing plant.

The mass concrete in the dam was cooled for the following reasons: (1) to reduce temperature cracks by averting a high temperature gradient between the center of the concrete mass and the exposed faces, (2) to

Concrete was placed in the 50-foot-wide blocks in 5-foot lifts, with alternate blocks two or three lifts higher than the remainder.

reduce the temperature early and uniformly to a point below mean annual in order to obtain a maximum joint opening for better contraction joint grouting, and (3) to improve the stress distribution in the structure by providing a temperature rise in the concrete after the radial joints were grouted.

To provide for cooling the concrete, 1-inch diameter tubing was assembled on top of each 5-foot lift in a horizontal coil with 5½-foot spacing between adjacent lines of the coil. River water was pumped through the cooling coils from the beginning of concrete placement until the concrete temperature was reduced to within 5° F. of the river water temperature. The average of all the highest temperatures read was 88.5° and the primary cooling reduced this average to 58.3°. Cooling was then suspended until the fall of 1938 when secondary cooling was carried on until the structure

had reached a temperature of 35° to 40°. The cooling water was circulated through the coils at a velocity of not less than 2 feet per second, requiring the pumping of 3.7 gallons per minute per coil. A maximum of 105 coils were in use at one time. The secondary cooling was completed on November 29, 1938.

Concrete temperatures were obtained by means of resistance thermometers inserted into 1-inch diameter tubing placed for that purpose at 25-foot elevations so that one end of each tube was near the middle of a block and the other protruded from the downstream face.

The contraction joint grouting program was arrived at by using trial-load analysis methods to determine the effect of the grout pressure in the radial joints on the deflections and stresses in the structure.

The contraction joints were grouted in three stages:

Concreting operations at Seminoe began in January 1938 with the working spaces protected from cold by heaters and canvas covering.

the first in May 1938, when the joints in the central part of the dam were grouted from bedrock to elevation 6125 with the reservoir empty and the average concrete temperature about 55° F.; the second, during the last week of November 1938, when the dam between elevation 6125 and 6200 was grouted with the interior of the dam at a temperature of about 40° F. and the reservoir empty; and the third, in the spring of 1939, with the reservoir water surface rising from elevation 6250 to elevation 6281.5 and the concrete temperature varying from 32° to 35° inside the section.

The average thickness of the grout film in all joints, based on the 802 net sacks of cement used, was 0.101 inch, or one sack of cement for each 119 square feet of joint area.

The foundation grout curtain at Seminoe is similar to those at many other Bureau of Reclamation structures. Alternate 100- and 50-foot deep holes at 5-foot centers were diamond-drilled into the floor of the canyon through vertical pipes set in the concrete footing on the upstream side of the dam. Holes were also drilled into the abutments perpendicular to the rock surface through pipes left protruding from the upstream face of the dam. Pressures of 50 to 250 pounds per square inch and 20,135 sacks of cement were used in grouting 26,406 linear feet of hole—an average of 0.763 sacks per linear foot.

One of the most unusual features in connection with the construction of Seminoe Dam was the excavation of the faulted areas in the foundation by mining methods after the placing of concrete in the dam was well under way. The accompanying drawing shows the red and

black seams relative to the structure and how the excavated material was removed through adits to the foundation treatment galleries. The same equipment was used to transport the backfill concrete which replaced the excavated material.

The main hoisting shaft was formed in the concrete near the toe of the dam at the left abutment and was 100 feet deep. The muck had to be hoisted up this shaft and then transported across a portion of the powerhouse area. Therefore, the mining operations were delayed until concrete placement was well along in the powerhouse in July 1938. By this time mass concrete of the dam had been placed 150 feet above the foundation. For each seam, a 6- by 8-foot horizontal gallery 10 to 15 feet above the foundation rock was located so that access adits to the seams would have slopes coinciding with the dip of the seams. The adits, 5 feet high by 7½ feet wide, were spaced at 25-foot centers. Shafts were driven from alternate adits down both red and black seams and drifts were driven halfway to the adjoining shafts. In the red seam, the first 8-foot-high drift was driven from the base of the shafts. This space was backfilled with concrete and the process repeated upward to the base of the dam. The material in the black seam was wet and crumbly; therefore, the first drift was driven immediately below the base of the dam. The sides of the shaft were bulkheaded and the drift backfilled with concrete. This was repeated on down to the bottom of the seam, after which the shaft was filled.

Air tools and equipment were used exclusively for excavating the seams, and any material that could not be removed with paving breakers was assumed to be a satisfactory foundation material.

Grouting systems were installed on the hanging walls of both seams before backfill concrete was placed. After the concrete had cooled, 3,586 sacks of cement were used in backfill grouting in the two seams.

After all foundation grouting was completed, 4½-inch drain holes were drilled 30 feet deep at 10-foot centers along the foundation treatment and drainage galleries. Pipes were installed to measure uplift pressures at various points in the dam and measurements have been taken since April 1939. The original drains were reamed and some additional ones were drilled in 1950 to relieve the higher than expected uplift pressures which developed.

The diversion tunnel was closed in December 1938 by means of a flap gate suspended from the bottom of a fixed timber bulkhead occupying the top 17 feet of the tunnel portal and consisting of vertical 12- by 12-inch timbers supported by two heavy steel beams. The timbers and beams fitted into recesses formed in the portal concrete. Attached to the lower end of the fixed bulk-

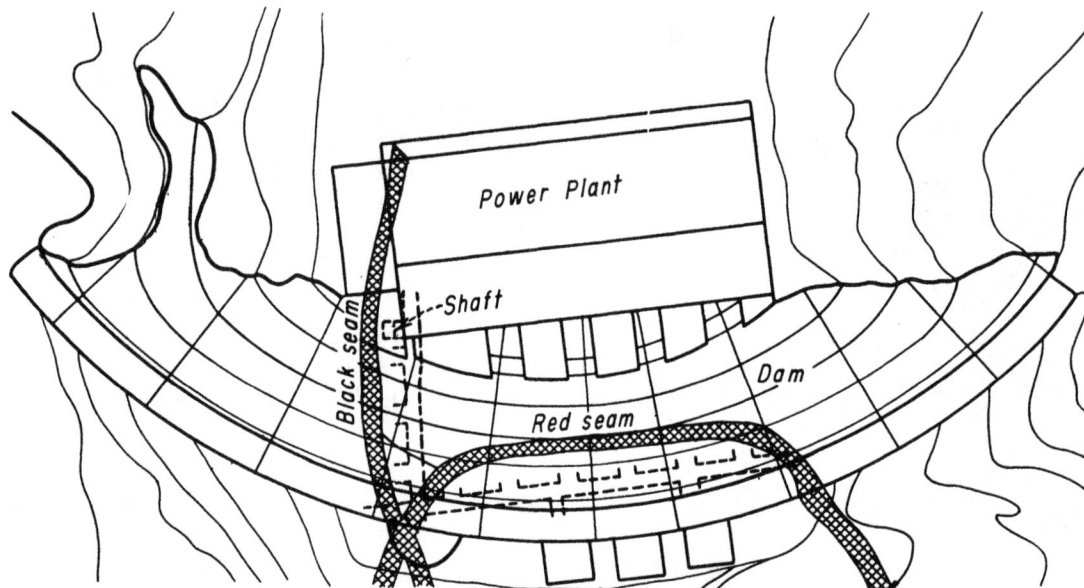

PLAN SHOWING LOCATION OF SEAMS

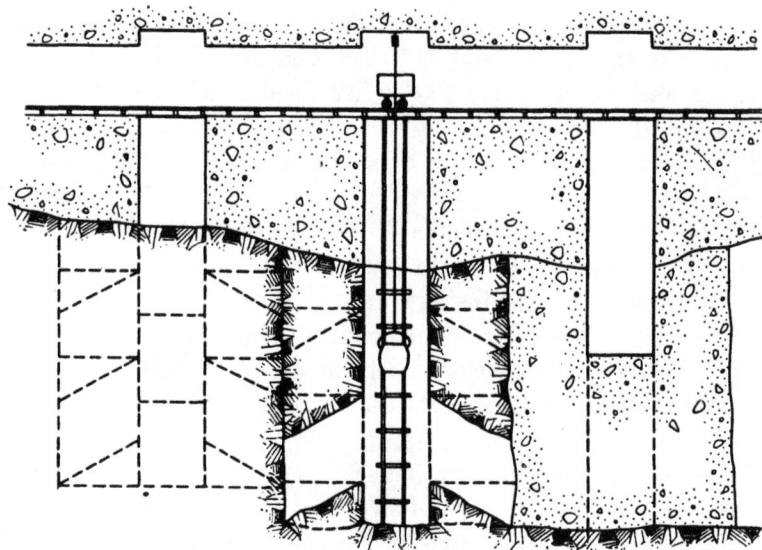

SECTION A-A

SECTIONS THROUGH MINING GALLERIES

head with five heavy steel hinges was the 7-foot-high gate also made of 12- by 12-inch timbers. The gate was held open by a single cable which was cut to effect the closure. After closure, the bulkhead and gate were sealed and calked from inside the tunnel, limiting the leakage to 200 gallons per minute with water 30 feet over the invert. The permanent tunnel closure was made by an 81-foot concrete plug, as shown on the accompanying plan.

The powerhouse is a reinforced concrete structure with steel framed roof and is separated from the toe of the dam by a ¾-inch cork joint. The three vertical-shaft hydraulic turbines have a rating of 15,000 horsepower under full gate opening and a net effective head of 171 feet. The maximum head available at Seminoe is 215 feet. The three generators each have 12,000-kv.-a. capacity. Both the turbines and the electrical equipment were installed by Government forces.

The discharges of water from Seminoe Dam are controlled as follows:

Spillway (capacity 48,500 second-feet)—Three 14-by 50-foot fixed-wheel gates operated by electric motors, manually controlled.

Power outlets—Three outlets with 102-inch diameter ring-seal gates.

Irrigation outlets (capacity 3,000 second-feet)—Two outlet pipes, each with a 72-inch diameter ring-follower gate and a 60-inch diameter needle valve.

The total cost of the dam and power plant including roads, camp facilities, channel improvement below the dam, and other miscellaneous items was $6,900,000.

OPERATION

The dam was completed September 9, 1939, and Government forces completed the power plant installations in July 1939. Water for irrigation was released beginning in June prior to completion of the power installation. The first power was produced August 3, 1939.

In general, the power plant is operated for peak loads only during the winter months, and all three units are used continuously during heavy irrigation in the summer months. The average head on the power plant has been 194.1 feet. The maximum peak load of 43,500 kilowatts is 134 percent of rated plant capacity.

ANGOSTURA DAM

ANGOSTURA UNIT, CHEYENNE DIVISION, MISSOURI RIVER BASIN PROJECT, SOUTH DAKOTA

By Ernest R. Schultz

The irrigation possibilities of the Cheyenne River in the vicinity of Hot Springs, S. Dak., attracted the attention of State and Federal authorities for many years. As far back as 1913 the State investigated the possibilities of constructing a dam and a canal system for developing the area. In 1928, the United States Army Corps of Engineers investigated the area as part of their studies of the Missouri River drainage area. The Bureau of Reclamation also conducted investigations and published reports of its studies in 1917, 1938, and 1939. A project was finally incorporated into a unit of the Missouri River Basin project and authorized for construction under the Flood Control Act of 1944. The unit, as authorized, provided for the construction of Angostura Dam, located on the Cheyenne River 6 miles southeast of Hot Springs, S. Dak., a canal distribution system, and a power plant.

RESERVOIR

Angostura Reservoir is a multiple-purpose reservoir, storing water for irrigation, power generation, recreation, and fish and wildlife conservation, and providing space for temporary storage of floods and deposition of silt. The reservoir has a capacity of 160,000 acre-feet with a surface area of 4,830 acres at normal water surface elevation of 3187.2. Of this capacity, 68,000 acre-feet are allocated to dead storage and silt control and 92,000 acre-feet to irrigation and power. For temporary flood storage, the reservoir provides 60,000 acre-feet of volume above the normal water surface elevation. The Cheyenne River, the source of water for the reservoir, is formed from tributaries in the southwestern part of South Dakota and flows generally in a northeasterly direction to the Missouri River. The drainage basin covers an area of about 9,000 square miles, located in eastern Wyoming, northwestern Nebraska, and southwestern South Dakota. From this area, the average

So recently completed that cableway tail tower had not yet been removed, Angostura Dam began its task of storing irrigation water and stood ready to receive flood flow in the Cheyenne River.

annual run-off of the river at Angostura Dam is about 295,000 acre-feet.

DAM SITE

Five different sites for the dam were investigated in the general area of the present location. The site finally selected is known as the Sheps Canyon site. The rock in this region is a part of the formation in the Black Hills area. Doming of the earth's crust has elevated the sedimentary formations which are resting on very ancient crystalline rocks. The sedimentary formations, sandstones and shales, range in geological age through the Paleozoic into the Cretaceous horizons of the outer plains area. Bedrock at the dam site is composed of sandstone beds having thickness varying from a few inches to as much as 65 feet. These beds are separated by shale (locally called mudstone) and limestone seams of varying thickness. The formation has

Downstream face of the spillway at Angostura Dam shows the hinge anchorages for the five 50- by 30-foot spillway radial gates.   Beyond the spillway, the crest of the gravity section rises as it approaches the earth embankment.

a component of dip of approximately 5° from left to right along the axis of the dam and a downstream dip component of about 1° normal to the axis.   Overburden depth varies from 0 to 10 feet on the left abutment, from 5 to 10 feet in the river channel and, on the right abutment, from 0 to 70 feet.   Originally it was proposed to construct an earth-fill dam across the channel with a spillway located in one of the abutments.   Because of the large spillway capacity required, no suitable economical abutment spillway could be located at either of the five sites being considered.   When the decision was made to construct the spillway structure in the river channel, it was determined that a minimum of 25 feet of sandstone was necessary as a foundation for the dam.   After exploration and studies of all five sites, the Sheps Canyon site was selected due to the presence of a greater thickness of sandstone and the best relative distribution of the sandstone and mudstone beds.

THE DAM

Angostura Dam is a composite type dam consisting of a concrete gravity structure from the left abutment across the river channel, and an earth embankment extending to the right abutment.   The dam has a total crest length of 2,030 feet, the concrete section being 970 feet long and the earth embankment 1,060 feet long. The maximum height above the river bed is 135 feet.

The concrete portion of the dam consists of the spillway section located in the river channel and two nonoverflow sections, one extending to the left abutment and one abutting the earth embankment.   The spillway section of the dam is thicker than the nonoverflow sections in order to accommodate the spillway crest. Also, because of the low modulus of elasticity of the foundation rock (600,000 pounds per square inch), the concrete sections of the dam are somewhat thicker than usual for a gravity dam in order to limit the toe pressure on the foundation to 167 pounds per square inch under normal loading conditions.   The dam was constructed in 5-foot lifts with vertical contraction joints in the transverse direction 60 feet apart to provide for convenience of construction and to allow for contraction of the concrete during the period of setting and cooling. Vertical keyways were built into the contraction joints

PLAN

UPSTREAM ELEVATION

PROFILE OF RIVER OUTLET

PROFILE OF CANAL OUTLET

MAXIMUM EMBANKMENT SECTION

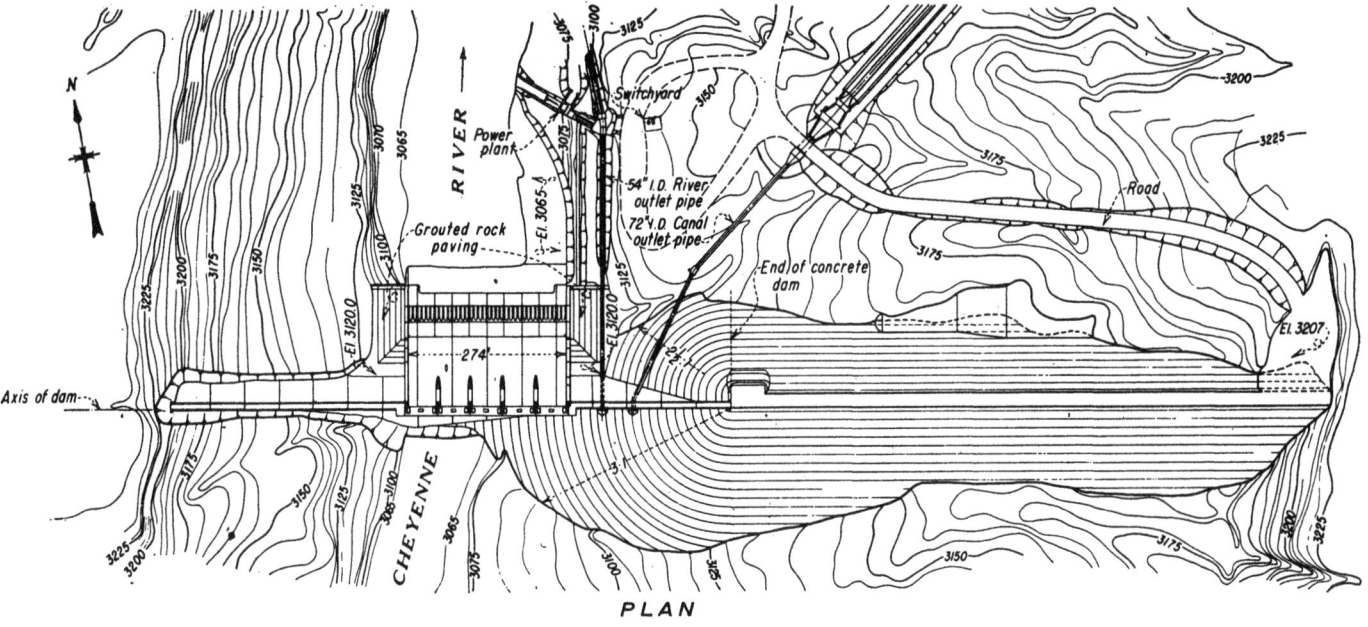

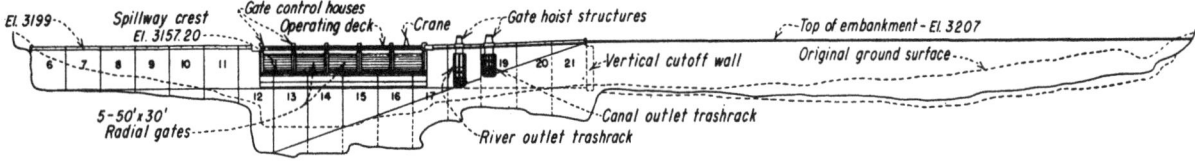

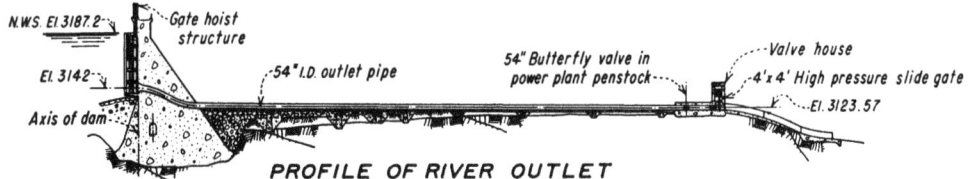

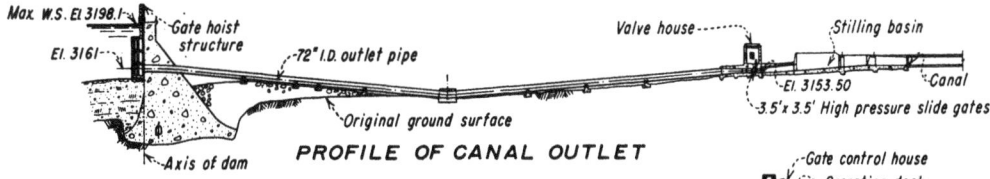

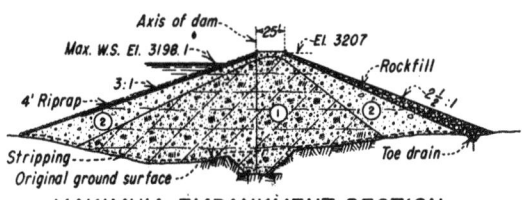

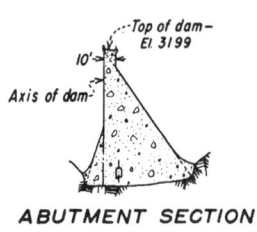

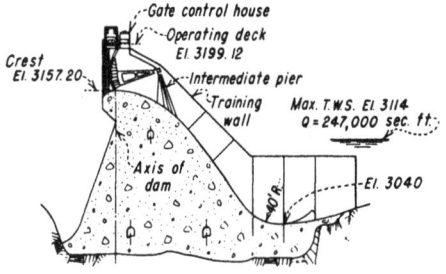

① Impervious material of clay, sand and gravel
rolled in 6" layers.
② Pit run sand and gravel rolled in 12" layers.

ABUTMENT SECTION

MAXIMUM SPILLWAY SECTION

to insure that the dam would act as a unit. Systems of thin-wall steel tubing, fittings, and outlets were placed in the contraction joints to provide a means for grouting them. Because of the very low air temperature during the winter, the considerable upstream and downstream length of the blocks (up to 150 feet), and the probable maximum setting heat of the concrete (115° F.), it was

considered necessary to prevent cracking across the blocks parallel to the axis of the dam by cooling the concrete throughout the spillway section and in the abutment sections where the thickness of the dam exceeds 60 feet. This was accomplished by circulating river water through embedded pipe coils.

A 5- by 7-foot foundation gallery near the upstream

face of the dam collects seepage from the foundation drains. Any seepage through the concrete section of the dam is collected in 5-inch diameter holes formed vertically in the mass concrete on 10-foot centers 6 feet from the upstream face in the nonoverflow sections and 12 feet from the upstream face in the spillway section. The vertical drains are connected to the foundation gallery. A 5- by 7-foot inspection gallery is located at elevation 3115. This gallery contains a pump chamber for pumping the water collected in the foundation gallery to the downstream face of the spillway and a compressor chamber for supplying air to a de-icer system on the trashracks and the spillway crest. In order to increase the path of percolation between the concrete section and the earth embankment section, a reinforced concrete cut-off wall at the right end of the concrete dam extends 12 feet into the earth embankment.

The earth embankment portion of the dam is composed of a moistened and rolled fill of clay, sand, and gravel placed on pervious talus overburden varying in depth from 10 feet to 70 feet. A cut-off trench excavated to bedrock and 30 feet wide at bedrock contact provides a cut-off through the overburden. By the time the embankment had been placed to about elevation 3140 (lower limit of the operational pool) it became apparent that the sand and gravel beneath the riprap was deficient in plus ½-inch sizes and might wash out through the riprap under the influence of wave action. In order to eliminate this possibility, the thickness of the riprap was increased from the planned 3-foot layer to a 4-foot layer above this elevation. The downstream slope is protected from erosion by a rock fill tapering in width from 25 feet at the base to 5 feet at the crest. A toe drain, consisting of an 8-inch sewer pipe laid with open joints in a 4-foot-deep trench backfilled with sand and gravel, extends along the toe of the downstream slope.

In recognition of its greater vulnerability to overtopping by waves, the crest elevation for the embankment is 8 feet higher than the concrete portion of the dam to provide more freeboard. The crest of the embankment is 25 feet wide with a 12-inch unreinforced concrete curb at the upstream face. Since the crest of the concrete dam is only 10 feet wide, a 40- by 60-foot turnaround area for maintenance vehicles is located on the earth embankment where the two dam sections meet. To provide for settlement of the embankment, the crest is cambered to a maximum of 0.70 of a foot.

The general plan for grouting the foundation rock under the concrete dam provided for preliminary low-pressure grouting, followed by a final high-pressure, deep grout curtain near the upstream toe of the dam. The holes for low-pressure grouting were drilled on about 20-foot centers, approximately 25 feet deep, and covered the entire excavated foundation. These holes were grouted prior to placement of concrete in the dam.

After sufficient concrete had been placed in the dam the high-pressure holes were drilled at 5-foot centers from the foundation gallery, through pipes placed in the concrete, to a depth of approximately 100 feet and grouted. The grouting plan under the earth embankment section consisted of a line of holes at 10-foot centers drilled from the bottom of the embankment cut-off trench at depths varying from 15 to 100 feet. These holes formed a continuation of the grout curtain under the concrete dam and were grouted at low pressure.

Drainage of the foundation under the concrete portion of the dam is provided for by means of a series of holes drilled at 10-foot centers approximately 50 feet deep into the foundation rock through pipes extending from the floor of the foundation gallery to the rock surface. A similar set of drain pipes extends to foundation rock from the drainage gallery located downstream from the foundation gallery; however, these drain holes have not been drilled into the foundation. If future conditions require additional foundation drainage, these holes will be drilled as needed. In order to determine the effectiveness of the foundation grouting and drainage system, a system of pipes is installed at four different places in the foundation of the concrete dam, by means of which it is possible to measure any hydraulic uplift pressures that may be present at the base of the dam due to percolation or seepage of reservoir water along underlying foundation strata.

The mudstone seams of varying thickness underlying the foundation of the nonoverflow sections of the concrete dam required special treatment in order to confine them and increase the path of percolation through them. The two seams in the right abutment were treated by excavating cut-off drifts into the seams 100 feet at the upstream toe of the dam and 50 feet at the

Excavation for Angostura Dam was virtually complete when this photograph was taken, and placement of concrete in the spillway section and earth fill in the right abutment embankment section was underway.

downstream toe of the dam and backfilling the drifts with concrete. The large mudstone seam in the left abutment was handled in a similar manner by drifting and backfilling with concrete a distance of about 230 feet at the upstream toe of the dam and about 115 feet at the downstream toe of the dam. To insure an effective seal between the concrete backfill and the overlying sandstone at the top of the drifts, a system of thin-wall tubing, fitting, and outlets was installed for grouting these contact areas.

The principal dimensions and quantities for the dam are:

Maximum height above foundation	189 feet.
Height above stream bed	135 feet.
Length of concrete dam section	970 feet.
Length of earth embankment section	1,060 feet.
Excavation, common, for dam foundation	32,770 cubic yards.
Excavation, rock, for dam foundation	181,750 cubic yards.
Concrete in dam	238,640 cubic yards.
Excavation, common, stripping earth embankment foundation.	34,360 cubic yards.
Excavation, common, for embankment toe drain and cut-off trench.	69,240 cubic yards.
Impervious fill in embankment	320,860 cubic yards.
Semipervious fill in embankment	175,690 cubic yards.
Rock fill in downstream face of embankment.	43,350 cubic yards.
Riprap in upstream face of embankment.	36,110 cubic yards.

SPILLWAY

The spillway is designed to handle a discharge of 247,000 cubic feet per second with the reservoir water surface at maximum flood elevation 3198.1. The discharge over the crest is regulated by five 50- by 30-foot radial gates which are operated by motor-driven hoists from a structural steel operating bridge. The spillway gates are provided with both manual and automatic float controls. The automatic controls are designed to operate the gates only when the rising reservoir water surface reaches elevation 3187.2 (top of spillway gates) in order to prevent overtopping of the dam. The spillway crest is shaped to follow the path of a jet through a 1-foot gate opening with water at normal water surface elevation 3187.2. In order to prevent the formation of ice on the gates during the winter season, a system of pipes discharging compressed air is installed in the upstream face of the spillway section of the dam just below the crest. The energy of the overflowing water is dissipated by a special, slotted type, large-radius spillway bucket set within retaining walls at the downstream toe of the dam. This type of bucket is more economical than the conventional type spillway apron for the large discharge required to be handled.

OUTLET WORKS

The river outlet, located in the right abutment nonoverflow concrete section of the dam, is designed to

Details of the dentated sill of the spillway bucket are illustrated here. The 8-foot diameter pipe used for final river diversion was still in place.

discharge 500 cubic feet per second with normal water surface elevation 3187.2. The river outlet is not located in the spillway section of the dam, as in many dams, since its discharge would create nonuniform flow conditions in the spillway bucket, causing loose river material to be deposited in the bucket with resultant erosive action when the spillway is operating. The river outlet consists of a 54-inch inside diameter steel conduit extending through the dam and 490 feet downstream from the axis to a valve house structure. The intake is protected by a semicircular reinforced concrete trashrack structure with structural steel trash bars. In order to prevent the formation of ice on the trashrack during the winter, a system of pipes discharging compressed air is installed on the trashrack structure.

A 5.81- by 5.81-foot fixed-wheel gate operated by a wire rope hoist located at the top of the dam operates on the upstream face of the dam for use in unwatering the river outlet conduit and for emergency closure. Regulation of discharge from the river outlet conduit is controlled by a 4- by 4-foot high-pressure slide gate with a hydraulic hoist located in the valve house. Discharge from the valve house is guided back into the river channel by means of a trajectory-type reinforced concrete flume. The shape of the bottom of the flume is designed to follow the path of a 1-foot gate opening with water at normal water surface elevation.

The entrance to the canal outlet is located in the right abutment of the nonoverflow concrete section of the dam. The outlet consists of a 72-inch inside diameter steel conduit which extends through the dam and downstream to a valve house and stilling basin or headworks. The intake is protected by a semicircular reinforced concrete trashrack structure with structural steel trash bars. This trashrack also has a system of pipes discharging compressed air to prevent the formation of ice on the racks during the winter.

A 7.74- by 7.74-foot fixed-wheel gate operated by a wire rope hoist located at the top of the dam operates on the upstream face of the dam for use in unwatering the canal outlet conduit and for emergency closure. At the valve house, the conduit branches into two 50-inch inside diameter steel conduits. Discharge into the reinforced concrete stilling basin is controlled by two 3.5- by 3.5-foot high-pressure slide gates with hydraulic hoists located in the valve house. Flow through the stilling basin is controlled by four 60- by 60-inch slide gates installed in the structure. The stilling basin also contains a device for measuring the quantity of water delivered to the canal. The canal outlet will deliver 290 cubic feet per second with reservoir water surface at elevation 3169.4. Only the canal outlet conduit and the canal headworks were included in the contract for construction of the dam.

POWER PLANT

A power plant having one unit with a capacity of 1,200 kilowatts is located in the vicinity of the river outlet works valve house structure. Water for supplying this unit is provided by a penstock connected to the river outlet pipe at the valve house structure. The power plant was to be located at a turnout on the canal system. However, in order to avoid winter water deliveries in the canal structure, it was decided to locate the power plant at the dam and supply water to it from the river outlet conduit.

CONSTRUCTION

The dam was built by the Utah Construction Company of Ogden, Utah, under a contract awarded on June 28, 1946. The contract was completed and the dam accepted by the Bureau on December 7, 1949. Excavation for the foundation of the concrete dam was started in August of 1946. The first concrete in the dam was placed in July of 1947 and concrete in the dam was completed on November 7, 1949. Due to the severe climatic conditions at the site, no provision was made for placing concrete during the winter months. Concreting operations usually ended about the middle of November each year and did not begin until April of the following year. Stripping and excavation for the earth embankment was started in September 1946, the first impervious material was placed in the embankment in August 1947, and the embankment was completed during August of 1949. Placing of the earth embankment during the winter months was also suspended because of the severe climatic conditions.

Diversion of the river during construction of the dam was accomplished in three stages. During the first stage the river was diverted along the right bank while a concrete diversion wall was constructed on the left bank of the river and the bluff to the left of the wall excavated to form a channel. When earth-fill cofferdams at the upstream and downstream ends of the concrete wall and extending to the right abutment were completed, the river was diverted to the left bank for the second stage. For the third stage, an 8-foot diameter diversion conduit was constructed in the block adjacent to the concrete wall used in the second stage. At the downstream face of the dam and in the spillway bucket area, a structural steel sheet-piling cofferdam was built which terminated in an earth-fill cofferdam extending to the left abutment. At the upstream face, a concrete wall was constructed, which extended upstream from the dam and turned into the left abutment. Water was diverted through the diversion conduit for the final stage on January 14, 1948.

Considerable trouble was experienced during the final stage of diversion. In a sudden flash flood on March 16, 1948, the sheet-piling wall downstream from the dam gave way and the excavated area to the left of the diversion conduit was completely inundated. Three men were trapped and lost their lives in the excavated area at the time of the break. Instead of replacing the sheet-pile cofferdam, the contractor connected the 8-foot diameter diversion conduit with a steel conduit of the same size, which extended across the spillway bucket and over the downstream cofferdam. High water again flooded this area by overtopping the contractor's upstream diversion wall on June 18, July 15, July 19, and August 16, 1948, and January 25, 1949. The diversion outlet was finally closed and water storage in the reservoir commenced on October 3, 1949.

The coarse aggregate for concrete was produced by crushing limestone which was quarried from a site located about 6 miles upstream from the dam near the head of Sheps Canyon. No suitable natural gravels of desired sizes were available. After quarrying, the material was run through a primary crusher located in the quarry area and having a capacity of 200 to 400 tons per hour. After separation into two sizes, the aggregates were hauled by truck to the mixing plant area located on the left abutment of the dam site and stock-piled. From the stock piles the aggregate was processed by further crushing, screening, and washing before entering the bins supplying the mixing plant.

Sand for concrete was obtained from a deposit located along the left side of the Cheyenne River, approximately 1 mile downstream from the dam. Sand from the pit was hauled by truck to the processing setup at the mixing plant area where it was washed, screened, stock-piled, and drained before being deposited in the sand bunker at the mixing plant.

Modified (type II) low-alkali cement was used throughout the dam. Bulk cement for the job was received and stored in a 2,600-barrel silo located at the railroad siding at Oral, S. Dak. Cement was hauled by truck to the mixing plant storage bin.

Three 2-cubic-yard tilting mixers dumped into a common hopper which in turn dumped into a 6-cubic-yard bucket carried on a specially built railroad car. The car was pulled by a small Diesel locomotive to the placing cableway which had a fixed tower on the right abutment and a movable tower on the left abutment.

The impervious material for the core of the earth embankment was obtained from a borrow area located at the mouth of Sheps Canyon where stratified layers of clay, silt, sand, and gravel as mixed in shovel cuts resulted in a composite material suitable for the core. Haul distance to the dam was approximately 4,000 feet. The borrow area was irrigated by sprinklers for a minimum period of 100 hours prior to excavating in

order to obtain the proper moisture content. The material was placed in horizontal layers not exceeding 6 inches in thickness after being rolled. Compaction was accomplished by 12 passes of a tamping roller. The sand and gravel material placed on the upstream and downstream face of the impervious core was obtained from the cut-off trench excavation and from an area located on the left side of the river approximately 2,500 feet upstream from the dam. This material was placed in horizontal layers not exceeding 12 inches after being sprinkled and rolled 12 times with a sheepsfoot roller. Rock fill for the downstream face of the embankment was obtained from stock-piled foundation excavation and from a quarry located 1,400 feet upstream from the dam. The rock obtained from the quarry was a fine-grained massive limestone. Material for the upstream face riprap blanket was obtained from a limestone rock outcrop located on the left abutment about 1,000 feet from the dam and from the quarry which produced the rock fill.

BARTLETT DAM

SALT RIVER PROJECT, ARIZONA

By Edward R. Dexter

Bartlett Dam, a multiple-arch type dam, is located in central Arizona about 35 miles northeast of Phoenix on the Verde River, the chief tributary of the Salt River. It was constructed for the Salt River Valley Water Users Association by the Bureau of Reclamation to serve as a regulative means for conserving seasonal run-off. It provides supplementary storage for the Salt River project and for irrigation of Indian lands along the Verde River.

The contract for construction was awarded to the Barrett and Hilp and Macco Corporation of Clearwater, Calif., and notice to proceed with construction was given August 11, 1936. Construction of the dam was completed May 9, 1939.

RESERVOIR

Following are the significant items related to the reservoir and the Verde River above Bartlett Dam:

Drainage area (square miles)	6,160
Annual discharge (acre-feet):	
Maximum (1941)	1,040,000
Minimum (1934)	165,000
Average (32 years)	521,000
Daily discharge (cubic feet per second):	
Maximum (1893)	144,000
Minimum (1939)	56
Average (32 years)	720
Reservoir:	
Normal capacity (acre-feet)	179,480
Normal surface area (acres)	2,783
Normal water surface elevation	1,798

DAM SITE

The Verde River, flowing in a general southerly direction to its junction with the Salt River, flows almost directly west at the dam site. The bottom of the valley at the dam site is approximately 275 feet wide, nearly half of which is occupied by the stream. The confining slopes are fairly steep, joining less abrupt surfaces on top of the abutments. There is a relatively large area a short distance above the adopted normal water surface elevation of the reservoir, but to utilize this area for storage would have required the construction of several small dams between promontories, and also would have resulted in a considerable increase in quantities of materials for the main dam. Economic considerations did not warrant the construction of a larger reservoir.

The reservoir and the dam site lie wholly in a granite formation. Both fine- and coarse-grained granite make up the rock formations at the dam site. However, the dam is located entirely on the fine-grained granite, which is of excellent quality. On the left abutment the granite is fresh and durable fairly close to the surface, while on the right abutment it is considerably weathered. The spillway channel on the right abutment is located for the greater part of its length in the coarse-grained granite which is protected against erosion by concrete lining.

According to drilling records, nearly 70 feet of fill in the river channel covered the fine-grained granite in the dam foundation. All exploratory drilling was done in the river bed, with investigation work above river level being done by test pits, trenches, and tunnels.

Drilling records confirmed the existence of a fault in the river bed formations indicated by rock conditions at the site. The log of an inclined drill hole showed that the fault, approximately 2 feet wide, and filled with gravel, traversed the foundation of the dam. No concern was expressed by geologists over any danger of future movement, as all surface evidence indicated that a considerable period of time had elapsed since the fault was active.

THE DAM

Multiple-arch construction for Bartlett Dam was selected after investigations of several types; namely

Interesting patterns of shadow are formed by the arch barrels of Bartlett Dam in this view of the upstream face.

earth and rock fills, straight concrete gravity, concrete arch, diamond-head buttress, Ambursen slab and buttress, massive multiple arch, and lightweight reinforced multiple arch. The dam has a maximum height of 286.5 feet, measured from the lowest point in the foundation to the top of the parapet, elevation 1803. At present, it is the highest multiple-arch dam in the world. The dam is slightly curved in plan so as to most economically fit the topography, the radius to the axis being 1,379.7 feet. The crest length, including the 10 arch-barrel sections, the gravity blocks at each end, and the spillway gate structure at the right abutment, is 970 feet.

The dam was designed for a normal reservoir water surface at elevation 1798, with a 5-foot freeboard below the top of the parapet. Stresses and stability factors were checked for a maximum water surface 4 feet above the top of the parapet, with water flowing over the dam as well as through the spillway. Effects of earthquake disturbances on the stability of the dam were included in the design by assuming that the dam would move horizontally upstream and downstream with a period of vibration of 1 second, and that the acceleration would equal one-tenth gravity. The effect of air and water temperature variations on the behavior of the arch barrels was carefully analyzed.

Buttresses are of the hollow type, consisting of (1) two main walls each directly supporting an abutment of an arch barrel, (2) an upstream face slab, (3) a downstream face slab, and (4) vertical stiffener walls between the main walls to provide greater lateral stability for the buttress. Each main wall has a vertical plane surface on the inside and an inclined plane surface on the outside, and the traces of planes through points of uniform thickness of the buttress walls are shown on the maximum section on the accompanying drawing.

Buttresses were constructed in vertical column units by leaving 18-inch openings at 41-foot 6-inch centers in the main walls. These were filled with concrete after the initial shrinkage had taken place. The faces of the vertical column units were formed as sloping steps at 5-foot vertical centers, the slope of the steps being determined from the computed directions of the resultant forces acting through the completed buttress. Openings were provided through the top and bottom of the stiffener walls and through the main walls to permit ventilation of the inside of the buttress and to aid in securing a more uniform distribution of temperature in the main walls.

Reinforcement steel, consisting of vertical bars and inclined bars parallel to the upstream edge of the buttress, was placed near the outer surfaces of the main walls to provide a means of counteracting the effects of the variation of the outside air temperatures. In placing, the inclined bars were lapped in the vertical openings so that bond for the laps was provided when the openings were filled. The upstream face slab of each

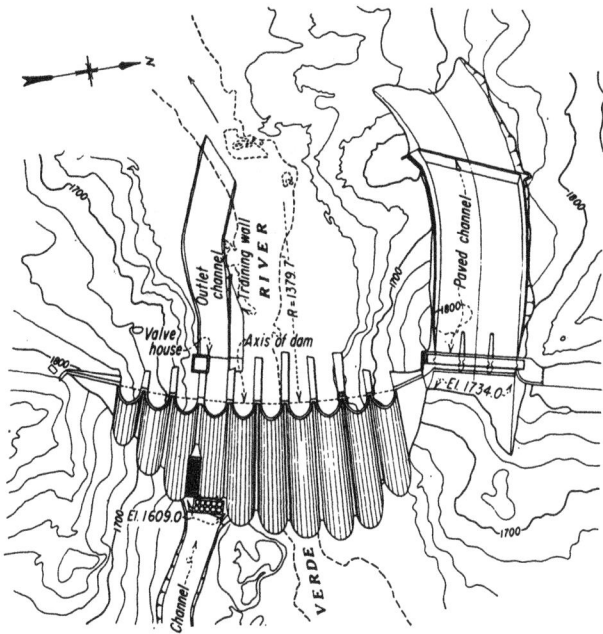

PLAN

SCALE OF FEET

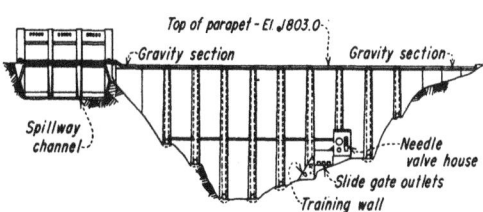

SPILLWAY
PROFILE ON ℄

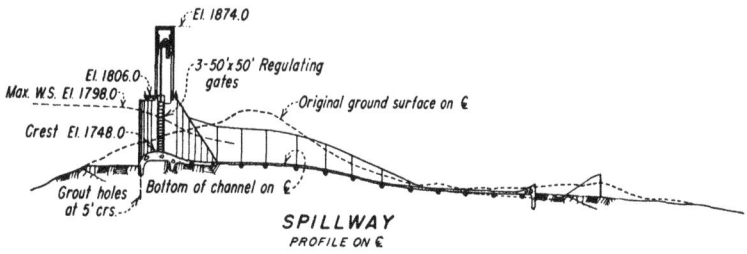

SECTION THRU OUTLETS

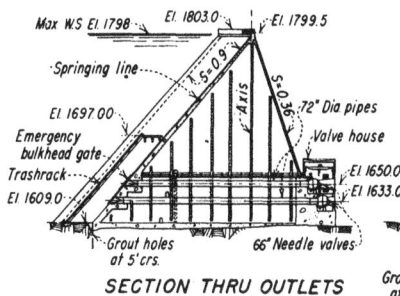

SECTION THRU SLUICE GATE

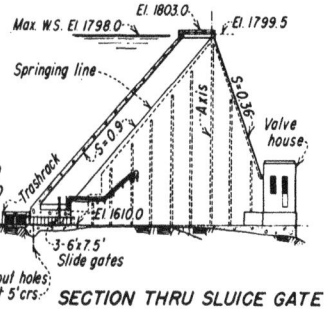

DOWNSTREAM ELEVATION

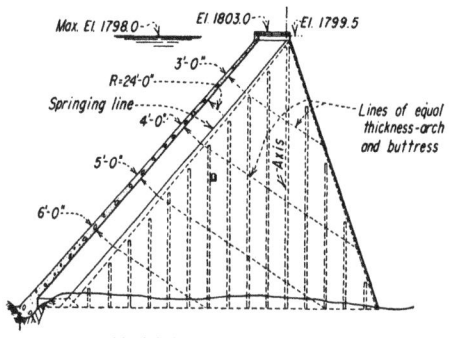

MAXIMUM SECTION

buttress was reinforced to carry the loads due to water pressure, to resist the forces resulting from temperature changes and reactions of the arches at their abutments, and also to assure better structural action for the buttress. Stiffener walls and the downstream face slab were reinforced by steel, which ties the reinforcement in the outer surfaces of the main walls and aids in making a more rigid buttress.

All arch elements of the inclined barrels are circular, have a total central angle of 180° (except at footings), and have constant thicknesses in a plane normal to the springing line. The radius to the inside or downstream face of each barrel is constant and is equal to 24 feet. The thickness of an arch barrel is shown on the maximum section on the drawing. At any elevation on the springing line, the thicknesses of the arch element, the buttress upstream face slab, and the main walls for the buttresses are equal.

Arch barrel footings have a minimum depth of 1½ times the normal thickness of the arch barrel at the point of support and a base width of not less than the normal thickness.

Surmounting the arch barrels are horizontal ribs having a vertical dimension of 3 feet and a horizontal thickness of 4 feet. These ribs serve three purposes: first, to give support and stiffness to the top of the barrel; second, to support a parapet; and third, to provide a walkway across the dam. A 2-inch standard pipe railing flanks the downstream edge of the walkway. A trussed walkway through the buttresses and a stairway in a buttress near the spillway provide access from the crest of the dam to the needle valve house for operating purposes.

Two relatively small gravity sections, one at each end of the multiple-arch portion of the dam, complete the structure exclusive of the spillway. Each gravity section may be visualized as a solid buttress section constructed in lieu of a low and almost completely nonsymmetrical terminal arch barrel with a complex structural behavior. The gravity block at the right abutment was also designed to provide optimum hydraulic conditions for the approach channel to the spillway.

Grace and strength combined are epitomized in Bartlett Dam. The spillway is at the left; the needle valves discharge into the river channel at right center.

SPILLWAY

The spillway gates are the Stoney gate type, except that fixed wheels, instead of roller trains, provide the bearing system which allows the gates to be raised and lowered while subjected to reservoir water pressure. The three 50- by 50-foot gates rest on a sill having a crest at elevation 1748. The discharge channel, lined with concrete for approximately three-fourths of its excavated length, discharges into a natural draw which empties into the river about 900 feet downstream from the dam. The location of the channel was determined partly by the topography and partly by the desire to return the flood water to the river as far from the dam as feasible in the interest of security. An interesting feature of the discharge channel is that it is curved in plan and superelevated.

The spillway is designed to discharge 175,000 cubic feet per second with the reservoir water surface at the normal storage level, elevation 1798.

OUTLET WORKS

Two types of outlets are used for passing water through the dam to meet irrigation demands or for draining the reservoir; namely, high-pressure slide gates and needle valves.

The three high-pressure slide gates, each 6 feet by 7 feet 6 inches, are hydraulically operated and are used only under heads less than 100 feet; however, they are designed to withstand full reservoir water pressure. They are placed in the base of an arch barrel spanning a bench on the left bank of the river and discharge into a channel formed by the two buttresses that support this arch barrel. The bottom of the channel is paved to provide better flow conditions and to prevent possible erosion of the rock.

Two 66-inch needle valves, one directly above the other, are located in the downstream end of the buttress on the left of the high-pressure slide gates. These needle valves are connected to the reservoir by two horizontal 72-inch diameter, $\frac{3}{8}$-inch plate-steel outlet pipes. The center line of the lower needle valve is 23 feet higher than the sills of the high-pressure slide gates, and the upper valve is 15 feet above the lower valve. For emergency and repair closure purposes, an 8- by 16-foot steel bulkhead gate operated by gravity and a hand hoist is situated at the upstream end of the outlet pipes, where it slides along the upstream face of the buttress between the abutments of adjacent arch barrels.

The discharge channel from both sets of outlets is formed by the left canyon wall and a concrete cantilever wall.

The entire outlet works system is designed to provide a discharge of 4,000 cubic feet per second with the water surface in the reservoir at elevation 1700, or 90 feet above the sills of the slide gates.

CONSTRUCTION

Diversion of the river during excavation and construction operations in the river channel was costly because of damage to diversion works and loss of equipment caused by floods. The first cofferdam constructed enclosed an area in the south side of the channel which, when unwatered, permitted excavation of foundations and construction of a part of buttresses 8 and 9. A 63,000-second-foot flood in February 1937 overtopped the cofferdams and flooded the area.

Cofferdams were then constructed across the river channel above and below the site. The river was diverted through a 72-inch pipe and, subsequently, through the permanent outlet channel between buttresses 8 and 9. Construction of the dam to above, or in some portions just below, river level had been completed when a 100,000-second-foot flood occurred in March 1938. This flood caused no damage to the structure but equipment was lost, including a concrete mixer (never recovered), some steel forming, and various other items.

The river was diverted later over a low arch at about midstream to permit construction of the slide-gate and trashrack structures and installation of the slide gates.

Practically all of the common excavation for the entire job was in the stream bed between buttresses 3 and 8. Maximum depth of overburden removed was 70 feet. Rock was excavated to obtain a gradually sloping longitudinal profile with arch footings sloped either uniformly from one springing line to the other springing line or, where impossible without excessive sound rock excavation, from springing line to arch midpoint. At first, buttress footings were excavated with a center ridge of rock remaining between side walls, but close jointing resulted in an insecure ridge of large broken fragments of rock, and a single wide trench was later employed.

Where buttress and arch footings cross fault zones, the fault zone material was removed to a depth of 3 to 5 feet under buttress footings, and to depths where the fault zone was 6 inches wide or to minimum depths of 3 to 5 feet under arch footings.

Excavation for the spillway required cuts up to 75 feet deep. Cut-off trenches were excavated under the spillway crest 10 feet deep, and at the downstream end of the spillway 10 to 39 feet deep.

Power shovels and draglines were used for excavating and 8-cubic-yard dump trucks for hauling excavated material to disposal areas. Mucking was generally by hand into skips which were handled by dragline to trucks. In the spillway channel, a bulldozer pushed loose rock into piles, where it was picked up by dragline and loaded into trucks.

Grout holes in foundations included primary holes at 5-foot centers along the cut-off trench at the upstream face of the dam, and secondary holes in arch groins, faults, and areas of seams and close sheeting of the rock. Alternate primary holes were drilled 100 feet deep across the bottom of the canyon and to lesser depths in the canyon walls with intermediate holes 25 feet shallower. Secondary holes were generally 25 feet deep.

Maximum grouting pressures ranged from 350 pounds per square inch in the river bed to 60 pounds per square inch beneath the spillway. The 351 primary grout holes, with a total length of 25,680 feet, consumed 20,230 sacks of cement and the 302 secondary holes, with a total length of 8,010 feet, took 4,220 sacks of cement.

Foundation drainage for the gravity sections was provided by drilling holes 25 to 35 feet deep at 10-foot centers both ways in the foundation rock. A system of metal piping embedded in the concrete provided outlets for these drainage holes.

Most of the concrete was placed by means of a pumpcrete machine and a maximum length of 1,200 feet of 8-inch discharge pipe. Concrete for some of the footings was placed by bottom-dump buckets hauled by trucks and handled by a crane. During the extremely hot summer months concrete placing temperatures were held to a maximum of 88° F. by spraying aggregate stock piles, by using spray-cooled water for mixing, and by spraying the burlap-wrapped concrete delivery pipe. This very effective method of cooling by evaporation was possible because of relatively low humidity. Cooling of newly placed concrete was effected by fog spraying with water from a system of nozzles placed 30 inches from the surface.

QUANTITIES AND COST

The principal construction quantities for the dam and its appurtenances are as follows:

Common excavation	cubic yards	195,000
Rock excavation	do	233,000
Grout holes, total depth	linear feet	34,000
Cement for grouting	sacks	24,500
Concrete	cubic yards	182,040

Reinforcement steel_____pounds__ 6, 700, 000
Grout and drainage pipe_____do____ 63, 000
Spillway gates, frames and hoists_____do____ 2, 033, 000
Trashrack metalwork_____do____ 271, 700
High-pressure gates and conduit linings_____do____ 564, 500
Outlet pipes_____do____ 157, 100
Needle valves and mechanisms_____do____ 137, 600

Bulkhead gate and mechanisms_____do____ 22, 600
Cranes and rails_____do____ 37, 900
Metal stairways and walkways_____do____ 97, 300
Pipe handrails _____do____ 34, 300
Miscellaneous metalwork _____do____ 23, 800

The total cost of Bartlett Dam was $4,731,032.29.

STONY GORGE DAM

ORLAND PROJECT, CALIFORNIA

By Benjamin R. McGrath

Stony Gorge Dam is one of the two storage dams of the Orland project in north-central California. The dam is located on Stony Creek, about 25 miles northwest of Willows, Calif., the nearest railroad point on the main line of the Southern Pacific Railroad, and about 8 miles west of Fruto on a branch line of the same railroad.

RESERVOIR

At the normal water surface elevation the reservoir has a storage capacity of 50,200 acre-feet and a surface area of 1,280 acres. It extends upstream from the dam for 5½ miles and stores the run-off from a drainage area of 275 square miles.

GEOLOGY

The reservoir basin and dam site are situated within the general region embracing the Knoxville formation, which consists of a thick and well-indurated series of clay shales, hard sandstone, and pebble or boulder conglomerates. The character and tilted position of the underlying shale beds are such that seepage loss from the reservoir is negligible.

The dam site is situated at the north end of the reservoir basin where Stony Creek turns westward through a low range of hills. The rock formation at the dam site is largely conglomerate with some sandstones and lesser amounts of shale. The predominating conglomerate rock is massive, sound, and of excellent quality. A tight and relatively minor fault crosses the dam site approximately along the north bank of the creek channel.

THE DAM

Stony Gorge Dam is an Ambursen type dam, 868 feet long at the crest, and is comprised of 46 bays of slab-and-buttress construction, with buttresses spaced 18 feet on centers. At each abutment, the dam terminates in short, massive gravity sections. The Ambursen type dam was selected principally for economy and also because the noncontinuous slabs, acting as simple beams between adjacent buttresses, were considered best adapted to accommodate possible minor foundation movements. All concrete of the dam is reinforced except in the cut-off trenches, the massive portion of the spillway bucket, and the spillway apron.

The dam, which rises 125 feet above stream bed, is so oriented as to utilize the most suitable rock for dam foundations and the most resistant rock to withstand erosion below the spillway. Also, the spillway and outlet works are located to avoid the main fault line.

Buttress thicknesses are constant in each lift and are offset from 2 to 5 inches at lift elevations. The thickness is 18 inches at the top lift of all buttresses and 3 feet 9 inches at the bottom lift of the tallest buttress. Footings are from 4 to 6 feet in height and 1½ times thicker than the base thickness of the corresponding buttress. Thicknesses of nonoverflow deck slabs vary from 15 inches at the top of dam to 4 feet 2 inches at a height of 120 feet, the variation being 3½ inches in a vertical distance of 12 feet. The thickness of the spillway face slabs between crest and bucket is uniformly 18 inches. Horizontal struts, 18 by 24 inches in cross-sectional dimension, are spaced 24 feet on centers both horizontally and vertically between buttresses.

A walkway and parapet cross the top of the dam, and another walkway between and through buttresses near the base of the dam provides access by means of connecting walkways to the gatehouse and valve house for the outlet works.

SPILLWAY

The spillway, which has a discharge capacity of 30,000 cubic feet per second, occupies six bays of the dam and is

The quiet reservoir lies in the California hills, retained by Stony Gorge Dam. The design of the dam makes it appear both light and sturdy.

divided by piers into three equal openings. From the crest, the downstream face slabs descend in a reverse curve to the mass-concrete spillway bucket and apron resting on bedrock of the creek channel and extending downstream for a distance of approximately 50 feet from the bucket. Spillway discharge is controlled by three 30- by 30-foot structural steel caterpillar gates of the overflow type, which move down the upstream face of the dam to open. This type of gate was decided on to accommodate the passage of the large pieces of driftwood during floods. A gatehouse at the top of the dam contains gate-hoisting machinery and a traveling crane. The gates are operated by screw-stem hoists driven by electric motors, power for which is supplied by a 15-kilowatt generator located in the valve house and driven by a turbine served by a branch from the needle-valve outlet pipes.

OUTLET WORKS

Each of the two main outlets occupies a bay between buttresses and consists of a structural steel trashrack supported on the upstream face of the dam, a 3½- by 3½-foot emergency high-pressure gate, a 50-inch riveted steel pipe, and a 42-inch balanced needle valve. The capacity of the outlet works is 1,050 cubic feet per second. A gatehouse between buttresses at the underside of the slab deck houses the high-pressure emergency gates and operating equipment. A valve house between buttresses at the downstream toe of the dam contains the needle valves and operating equipment, the turbine generator, and a standby gasoline-driven generator.

Provision for diversion of 10 cubic feet per second of water to the privately owned Angle-Troxel ditch was required because of destruction of the diversion dam for this ditch when Stony Gorge Dam was built. The trashrack for the private outlet is within the main trashrack structure of the south main outlet. A 12-inch gate valve, 12- by 7-inch Venturi meter, and 10-inch needle valve are installed in the gatehouse of the main outlets. From the 10-inch needle valve, the water is carried through an 18-inch steel pipe and thence through a 24-inch concrete pipe at the south side of the spillway to the Angle-Troxel ditch.

CONSTRUCTION

Bids for construction of Stony Gorge Dam were opened on August 18, 1926, and the contract was awarded to the low bidder, the Ambursen Construction Company of New York, N. Y. This company sublet the hauling, excavation, backfill, and the production of sand and gravel for concrete to the A. Haidlen Company of San Francisco.

Diversion of the stream during construction was relatively simple since the creek flow branched into two channels at the dam site, separated by a small rocky island. With an area of the north channel enclosed by a cofferdam and stream flow diverted to the south channel, excavation was accomplished for the dam between buttresses 30 and 34. This portion of the dam was then built to an elevation well above probable high water and 8-foot by 10-foot 3-inch rectangular openings were left in the deck in the two bays between buttresses 32 and 34 to allow for diversion. A special structure was built around these openings to permit, at a later date, installation of stop logs and closure of the deck-slab openings. With the stream passing through the diversion openings, the south channel was unwatered by means of additional cofferdams. This permitted excavation for the portion of the dam within the south stream channel

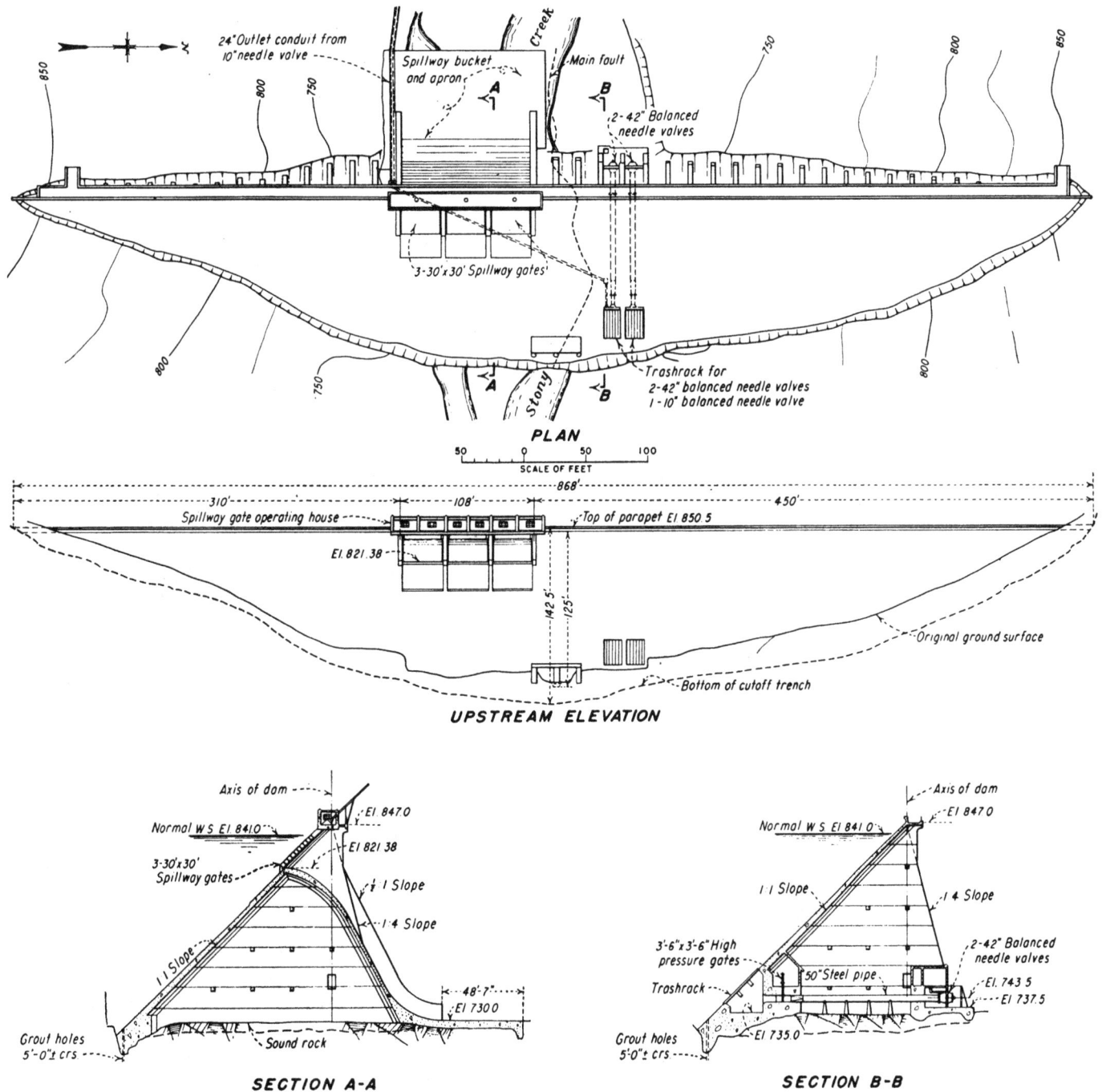

PLAN

SCALE OF FEET

UPSTREAM ELEVATION

SECTION A-A

SECTION B-B

Water pours over the spillway and jets from the needle valves at Stony Gorge Dam. In this view of the downstream face, construction details of the flatslab and buttress (Ambursen type) dam are readily apparent. The pipelike protuberances above the gatehouse surmounting the spillway bridge are the gate stems of the spillway gates.

and construction of the dam to an elevation above any expected flood.

Excavation disclosed good rock for foundations. Depths of excavation for buttresses varied from 4 to 27 feet below original surface, with the greater portion running from 15 to 25 feet. The upstream cut-off trench was excavated from 7 to 29 feet below original surface, the average being 20 feet. The cut-off trench at the downstream end of the spillway apron was excavated 10 feet below the top of the apron.

The main fault line crosses the foundation of the dam as shown in the accompanying drawing. In addition to the main fault, there are a number of secondary ones which branch off and are generally parallel to the main fault. So far as exposed by the excavation, the fault seams were well sealed and tightly filled with clay gouge of about 1-inch thickness.

Earth excavation was done by hand on steep slopes and by horse-drawn fresno scrapers on flatter slopes. Careful blasting was necessary in order to excavate the narrow trenches for buttress foundations. In certain portions of the foundation where rock was close-jointed and thin-bedded, the contractor preferred to take off part of the ridges between trenches, without payment, to permit the use of a gasoline shovel to greater depths. Excavation was largely removed by power shovels and trucks, some being handled by steam derrick, skips, and dump push cars.

Pressure grouting of the foundation involved the drilling of 160 holes at spacings ranging from 4 to 7 feet in a single line in the bottom of the cut-off trench.

The depth of holes below the bottom of the trench varied from 18 to 40 feet, the deepest holes being in the vicinity of the main fault. Grout pipes, 2½ inches in diameter, were set over the drilled holes and, after the concrete of the cut-off trench had set, the holes were grouted at pressures of from 90 to 100 pounds per square inch. Only 5 holes took more than 3 sacks of cement each. The other 155 holes averaged less than 1 sack each— little more than enough to fill the hole and pipe.

Concrete aggregates were obtained from a gravel bar in Stony Creek, one-half mile upstream from the dam site. The bar covered about 10 acres and had a depth of 8 to 10 feet of gravel and sand lying over ledge rock. There was a small excess of sand and the natural gradation was satisfactory. Cobbles up to 6-inch diameter were used in the massive portions of the concrete. For other concrete, all cobbles were crushed to 3 inches and combined with the gravel. The crushing, washing, and screening plant was set up at one side of the gravel bar and, at maximum capacity, furnished aggregate for about 24 cubic yards of concrete per hour.

The concrete mixing plant discharged into bottom-dump buckets of 1½-cubic-yard capacity, which were successively transported by hoist, cableway, and small trucks running on a light track supported on the buttress forms. Chutes were used to convey the concrete from the buckets to final position. About 1,000 cubic yards of concrete in the spillway apron and the Troxel conduit were placed by a small tower and chuting system.

Maximum capacity of the mixing plant was 24 batches per hour, but this was seldom attained. A day's run of concrete averaged 150 cubic yards in an 8-hour shift. The greatest amount placed in 1 month was 4,881 cubic yards in January 1928.

Cableways were used for handling reinforcement steel and forms as well as concrete. The installations included 3 cableways of 4-ton capacity and 1,100-foot span, erected over the dam site on lines 35 feet apart. The upstream cableway was later moved to a location between the other two for use on the upper portion of the dam.

Forms for the cut-off blocks, buttress footings, and the first part of the lift above the footings were built in place. Above the first full horizontal joint, the buttresses were built in 12-foot lifts using panel forms 14 feet in height and from 8 to 24 feet in length. The top forms of the face slabs were panels, 15 by 18 feet; the underside forms were built in place.

Selected clay backfill material was placed in the upstream cut-off trench to a minimum depth of 4 feet after the concrete was placed and the pressure grouting was done. The spaces between buttresses were backfilled with any convenient material to a depth of at least 3 feet above the bottom of the buttress footings to prevent weathering and to give a finished appearance to the structure.

QUANTITIES AND COST

The principal items involved in the construction of the dam were:

Excavation, earth and loose rock_____cubic yards__	11,000
Excavation, solid rock_____do____	27,600
Grout holes, total depth_____linear feet__	3,800
Concrete in cut-offs and buttress footings cubic yards__	7,400
Concrete in apron_____do____	1,440
Concrete in buttresses_____do____	23,940
Concrete in deck slab_____do____	6,090
Reinforcing steel_____pounds__	2,233,000
High-pressure emergency gates_____do____	23,700
50-inch outlet pipes_____do____	49,900
42-inch needle valves_____do____	62,000
Trashrack metalwork_____do____	37,200
Spillway gates_____do____	525,600
Traveling crane_____do____	8,050
Pipe handrail_____do____	25,000

The total cost of Stony Gorge Dam, exclusive of pre-liminary investigations, rights-of-way, relocation of roads, permanent buildings, etc., was $1,069,310.

PERFORMANCE AND BEHAVIOR

In the period since Stony Gorge Dam was completed in October of 1928, the concrete throughout has remained in exceptionally fine condition with only minor leakage at joints. There is no indication whatever of either settlement or sliding.

During construction, after several of the tallest buttresses were started, vertical shrinkage cracks began to appear, and it was decided to add horizontal reinforcement in all buttresses. This extra reinforcement appears to have stopped all vertical cracking. However, a careful check has been kept on these cracks and in 1949 strain-gage stations were established at cracks in five of the buttresses from which readings are taken periodically. Also, two permanent triangulation stations downstream from the dam, one at each side of the creek, and permanent points in three of the buttresses permit observation of deflections under various loading conditions.

ANDERSON RANCH DAM

BOISE PROJECT, IDAHO

By Fred J. Davis

⌄⌄

When the last cubic yard of earth material was compacted in place at Anderson Ranch Dam on October 10, 1947, contractors, under the supervision of Bureau of Reclamation engineers, had successfully completed construction of the highest earth embankment in the world.

To complete such an undertaking, the contractors were required to excavate about 11,300,000 cubic yards of earth and rock materials. This is roughly equivalent to excavating 960 acres, or an area 1 by 1½ miles, to a depth of 10 feet.

Anderson Ranch Dam is located in a relatively narrow rugged canyon of the South Fork of the Boise River, about 30 miles north of Mountain Home, Idaho. Bids for its construction were opened on July 7, 1941, and the contract was awarded to a combine of Morrison-Knudsen Company, Incorporated; J. F. Shea Company, Incorporated; Ford J. Twaits Company; and Winston Brothers Company, Construction Contractors. Active construction work was begun in August 1941, and the embankment, including rock fill and riprap, was completed in December 1947. Labor and material shortages, and the uncertain status of the project during the war years, 1941 to 1946, seriously hampered the progress of construction.

In December 1942, the War Production Board issued an order to cease all activity on the project, except that considered necessary to protect the work already accomplished. This order was later modified to permit construction of the outlet works intake structure and the installation of an emergency discharge regulating device. This work allowed the storage and release of 80,000 acre-feet of water in 1946, and 130,000 acre-feet in 1947, to the Boise Valley, materially aiding the production of foodstuffs in the Boise area.

PURPOSE

Plans for development of the Boise project call for Anderson Ranch Dam initially to provide supplemental storage to prevent a shortage of irrigation water in those project lands lying for the most part between the Boise and Snake Rivers. Later development contemplates diversion of water from the Payette River to that area and Anderson Ranch Reservoir will then serve as a source of supply for irrigation of the 400,000-acre Mountain Home area.

In addition to serving these irrigated lands, Anderson Ranch Dam provides flood control benefits to the Boise Valley and its power plant adds to the power potential of the area. The lower 30,000 acre-feet of the reservoir are reserved for silt control.

THE DAM

The design and construction of an earth-fill dam considered, at the time of its construction, to be the highest in the world, posed unprecedented problems for both the designers and the constructors. The depth of overburden in the stream channel, the relatively weak and broken bedrock, and the absence of nearby sources of large quantities of suitable concrete aggregates were important factors which influenced the decision to construct the embankment of earth materials. Quantitative considerations called for extensive explorations in the field to locate approximately 10,000,000 cubic yards of embankment materials. In addition to the ordinary laboratory tests performed to determine the physical properties of embankment materials, special testing was necessary to establish placement control procedures which would assure a stable structure. The variety of earth materials available permitted the design of a zoned embankment, particularly desirable for a structure of the height contemplated. The problem of securing a stable structure while utilizing the available materials from the most economical locations was frequently reanalyzed as additional information became available, and the zoning was modified several times between the preliminary design and completion of construction.

Water in the reservoir rises nearly to the spillway level and pours through the penstock into the turbines; and Anderson Ranch Dam is in service.

The 40-foot-wide crest of the embankment is approximately 330 feet above stream-bed elevation and 456 feet above the lowest point in the cut-off trench. The crest length is 1,400 feet and the base width is about 3,000 feet in the original river channel section. The central portion of the embankment is constructed of impervious material and functions as the water barrier. A pervious zone is provided downstream to drain the inevitable seepage through the impervious core, and between this pervious zone and the core a transition zone of semipervious material is provided to preclude any possibility of removal of the fine core material by the percolating water. Providing additional safety against failure of the downstream slope is a zone of rock fill. A 6-foot filter blanket of sand and gravel beneath the downstream pervious and rock-fill zones prevents possible loss of fine material from the foundation through these zones. Upstream from the central core, a semipervious transition zone and a pervious zone lend stability to the upstream slope during reservoir drawdown and prevent removal of fines from the impervious zone by wave wash through the 3-foot blanket of protective riprap.

A cut-off trench, 200 feet wide at the maximum section, extends through the river bottom materials and the abutment overburden to bedrock. Two concrete cut-off walls, with a maximum height of 16 feet and with footings extending 3 feet or more into bedrock, are located in the cut-off trench across the bottom of the canyon and up the abutments. These serve to prevent seepage along the contact of the cut-off trench backfill and the bedrock. Cement grout pumped under pressure through vertical holes on 10-foot centers forms curtains under each concrete cut-off wall. These grout curtains extend as much as 150 feet into the foundation rock and,

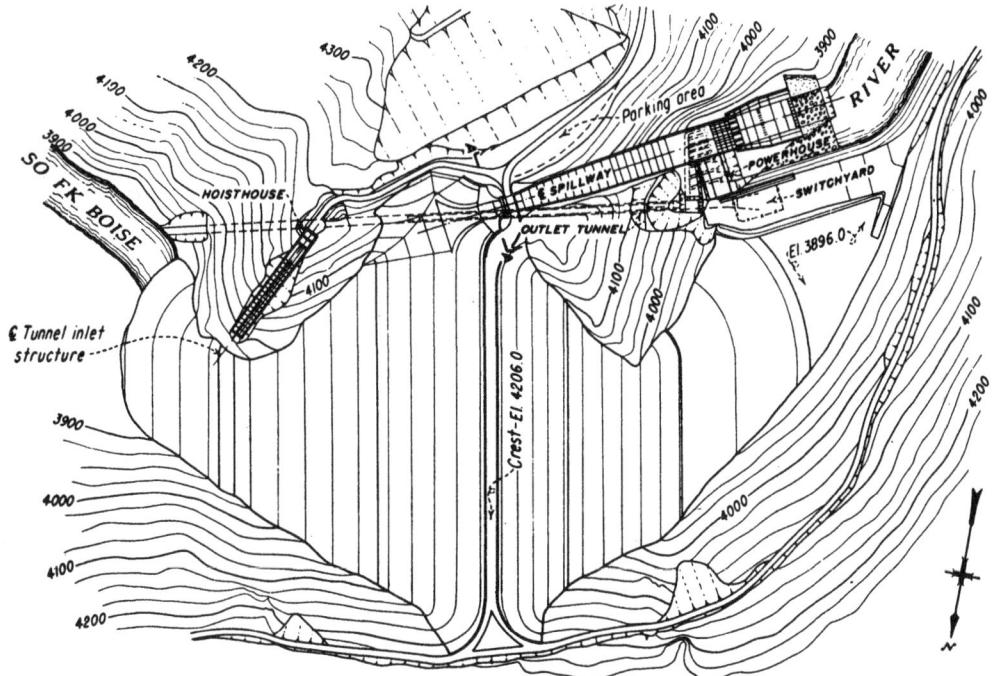

GENERAL PLAN

SCALE OF FEET
100 0   500

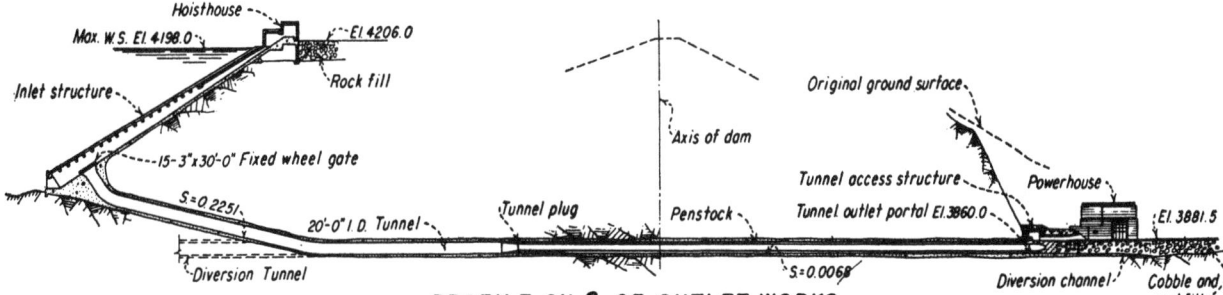

PROFILE ON ₵ OF OUTLET WORKS

SCALE OF FEET
100 0   100   200

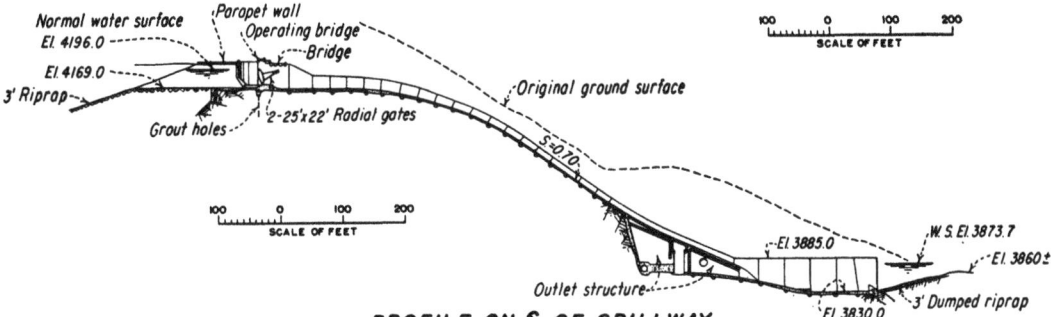

PROFILE ON ₵ OF SPILLWAY

SCALE OF FEET
100 0   100   200

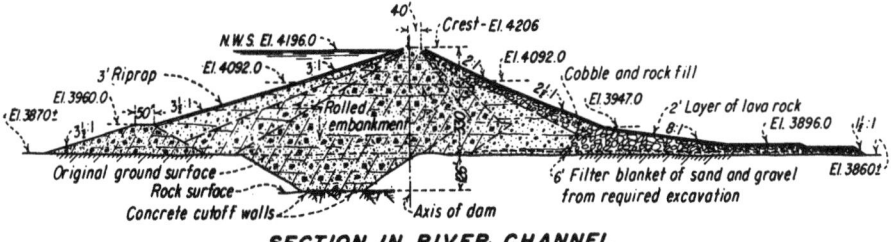

SECTION IN RIVER CHANNEL

SCALE OF FEET
100 0   500

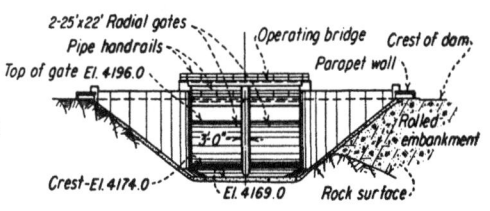

SECTION A-A

together with blanket grouting through holes 30 feet deep on 20-foot centers between the two walls, provide insurance against excessive seepage underneath the dam.

The upstream slope of the embankment is protected from wave action by a 3-foot layer of hard, durable, lava rock riprap. A 2-foot layer of similar rock placed on the face of the downstream rock-fill zone protects the friable granite in this zone from erosional and weathering damage and provides a uniform appearance of the slope.

RESERVOIR

When filled to capacity at elevation 4196, the reservoir is about 13 miles in length and contains 500,000 acre-feet of water, or more than 163 billion gallons. A county road, extending up the river valley, and several Forest Service roads were inundated. To provide access to the ranches, timber areas, mining properties, and recreational areas above the reservoir, more than 30 miles of gravel-surfaced roads were constructed. Much of this was difficult and expensive construction along the steep side slopes forming the reservoir. Nearly 5,000 acres were cleared of all timber and brush. All merchantable timber was salvaged and the remaining trees and brush were stacked and burned. The small village of Pine, Idaho, was purchased, and all usable properties salvaged before storage of water commenced. Three bridges across the river in the reservoir area were salvaged and used on the relocated road.

CONSTRUCTION

An effective means of diverting the river from its channel through the embankment area was provided by driving the 1,500-foot long, 24-foot diameter outlet tunnel through the left abutment. Working from both portals on a three-shift basis, the contractor excavated the tunnel in 80 days. Construction operations were planned so that concreting of the lining could be started while excavation of the tunnel was progressing. Practically all materials from tunnel excavation were placed directly in appropriate zones of the embankment. The original design called for construction of an intake structure at the upstream end of the diversion tunnel. As construction progressed, however, evidences of instability of the left abutment overburden began to appear. After several slides had developed in this area, the location of the intake structure was moved to a ridge of hard rock downstream and to the right of the original location. Construction of the intake structure on the steep slope, and excavation of the inclined tunnel which connects it with the outlet tunnel presented difficult construction problems, which required unique modification of mucking machines and maximum use of cable hoists.

As soon as the diversion tunnel had been lined with

concrete, excavation of the overburden from the nearly vertical abutments was started. Top soil, vegetation, and other organic material were stripped and wasted. The remainder of the material was placed in the cofferdam or stock-piled for later use in the main embankment. With the cofferdam in place, the initial diversion of the river was successfully accomplished in spite of the high flows prevalent in the river at that time, and excavation of the river bottom material was begun immediately.

After completion of the intake structure, including the installation of a fixed-wheel bulkhead gate, the diversion tunnel was blocked by a heavy timber bulkhead supported against the upstream end of the tunnel lining and reinforced by three 36-inch steel beams embedded in the concrete lining. Eighty feet of the diversion inlet was then backfilled with concrete and the river was permanently diverted through the outlet tunnel.

It was anticipated that large quantities of water seeping into the cut-off trench would cause much difficulty during excavation of the cut-off trench, preparation of the cut-off wall foundation, foundation grouting, and backfilling. To handle seepage, the contractor dug two vertical shafts—one upstream and one downstream from the axis near the left abutment—to depths into the rock well below the elevation of the bottom of the cut-off trench, with drift tunnels reaching across the valley to a point near the right abutment. Several plans of draining the foundation water into the tunnels were tried, the most successful being a series of 3-inch diameter perforated pipes extending through the overburden into the roofs of the tunnels. A cave-in of a portion of the upstream tunnel filled it with sand and rendered this system largely inoperative. Pumps were maintained at each shaft, however, and a small amount of water was removed by this system. The greater portion of the seeping water was pumped from sumps in the bottom of the trench as the excavation progressed. Well points installed on the trench slopes, and extended downward as the excavation progressed, removed additional water and prevented sloughing of the steep sloping sides of the trench. In general, the anticipated difficulties from seeping water did not materialize, a maximum of about 2,700 gallons per minute being encountered.

Material from the cut-off trench was classified as it was excavated and directed to appropriate sections of the embankment, to waste piles, or to stock piles. About 630,000 cubic yards of material were removed in 110 days of around-the-clock operation.

The two concrete cut-off walls were constructed across the river bottom and partially up the abutments as soon as possible after the foundation had been excavated to bedrock. Drilling and grouting the cut-off curtains

In the borrow pits, powerful shovels excavated selected material and placed it within reach of a pendulum feeder which straddled the conveyor belt leading to the dam embankment. The pendulum feeder could be swung in any direction to pick up material.

through pipes placed in the cut-off wall footings followed immediately.

When the grouting and cut-off walls had been completed, backfilling of the trench was begun. Most of the material for the central zone 1 portion of the embankment, of which the cut-off trench is a part, was obtained from the Dixie borrow pit, located on a high flat about 2 miles downstream from the left abutment and 1,200 feet above river bottom. Because of the cost and difficulty of constructing haul roads to this area, the contractor decided to install a conveyor system to transport the material to the embankment. The material was excavated by two 5-cubic-yard shovels and dumped into a hopper at the end of a "pendulum" belt feeder. From this feeder belt, material was transferred to the first of the six flights of main conveyor belt. The 2-mile-long system of belts moved the material at a speed of 550 feet per minute—a little more than 6 miles per hour—and had a maximum capacity of 900 cubic yards per hour. The material was discharged from the end of the system over the left abutment and allowed to roll down the steep slope to the embankment level. It was found that the most practical method of adding moisture to the material, when needed to insure the desired placement condition, was by jets spraying the soil as it was dumped from the conveyor belt.

Materials for the semipervious and pervious zones were obtained mainly from the Whipple borrow area (a delta deposit upstream from the dam), from the slide area excavation on the left abutment, from required excavation, and a small amount from the Dixie borrow area. Approximately 2,200,000 cubic yards of these materials had to be processed through a separation plant. The plant had an operating capacity of 900 cubic yards per hour and produced three types of material; namely, rock fill, pervious, and semipervious.

The impervious and semipervious zones were placed in 8-inch loose layers, moistened additionally if neces-

sary to meet the pre-established limits, and rolled 12 passes with ballasted tamping rollers having two 5-foot sections and weighing 20 tons. Lifts that became smooth from truck travel, or that dried out, were scarified and moistened before a new lift was placed. All rocks over 5 inches in maximum dimension were removed from the fill before rolling was allowed. Along the cut-off walls, at abutment contacts, and in other areas inaccessible for the large rollers, compaction was obtained with pneumatic hand tampers and with a job-constructed narrow-drum frameless roller.

In the bottom of the cut-off trench and up the abutments in the area beneath the impervious zone, the bedrock was carefully cleaned by hand and with air and water jets before embankment material was placed. The foundation area under the rock-fill zone was leveled and rolled before the filter blanket consisting of free-draining, coarsely graded sand and gravel was placed. The material was placed in 12-inch layers, saturated, and compacted by travel of construction equipment.

The pervious zone material was placed in 12-inch layers and compacted by sluicing and travel of the construction equipment. The rock fill was placed in 3-foot layers and covered with a 2-foot layer of lava rock. The riprap rock, ranging in size from one-half cubic foot to one-half cubic yard and larger, was placed on the upstream slope, keeping pace with the embankment placing. Some hand work was necessary to obtain uniform distribution of the larger riprap rock and to dress up the slope.

To maintain control of the placement moisture and

Embankment material from the borrow pits was delivered at the dam by belt conveyor, and spread and compacted by the usual construction equipment. In this photograph, the material being delivered and placed was in the impervious zone of the dam. In the foreground, the upstream slope of the embankment was being surfaced with rock for wave protection.

Placing the earth fill at Anderson Ranch Dam required the use of hand-operated pneumatic tampers near the cut-off walls and against steep rock surfaces where the tractor-drawn tamping rollers could not reach.

density values within acceptable limits, Bureau personnel performed over 3,000 field density tests on material for the compacted zones of the dam. A test pit excavated through a portion of the constructed embankment showed that although inspection forces were hampered throughout the war years by personnel shortage, satisfactory placement of the materials had been achieved.

APPURTENANT STRUCTURES

The concrete-lined outlet tunnel, through which water releases are made for power production and irrigation requirements, has an inside diameter of 20 feet for the full length. Average thickness of the lining is about 2 feet. The intake structure is furnished with trashracks and a mechanical rake for removing debris. A hoist house, located at the top of the intake structure, houses a hydraulic winch which, by means of a 6¾-inch diameter steel shaft 380 feet long, operates the bulkhead gate for emergency closure of the tunnel, and also the trash rake.

A concrete plug in the outlet tunnel located a short distance upstream from the dam axis encases the upstream end of a 15-foot diameter, plate-steel pipe. Near the downstream toe of the dam, the pipe is curved downward and to the left and tapers to 96 inches in diameter. Each of the five 72-inch outlet pipes leading from the tapering section are provided with regulating valves to control the irrigation discharges. Owing to the limited space available in the canyon bottom, the outlet works are located directly beneath the end of the spillway chute. Heavily reinforced concrete construction adequately supports the spillway over the control house. The outlet works is designed for a maximum discharge of 10,000 cubic feet per second.

The two-story, reinforced concrete powerhouse is located adjacent to the spillway at the downstream toe of the dam. Three 90-inch diameter steel penstocks branching from the tapering section of the outlet pipe provide water releases to the plant. Flow to the turbines is controlled by 100-inch butterfly valves, and the turbines in turn drive the 15,000-kv.-a., vertical-shaft generators. Space is provided for installation of a third unit in the future. The plant is equipped with a 60-ton traveling crane for use in handling heavy equipment.

The spillway, a concrete-lined open channel cut through a prominence of the left canyon wall and down the abutment to the river channel, is designed for a maximum discharge of 20,000 cubic feet per second. The flow is regulated at the crest by two radial gates operated by automatic, float-controlled power hoists. An ice prevention air system consisting of air compressors, distributing pipes, and strategically located dis-

The intake structure of the outlet works was built on a spur of the mountainside on a long slope. Spidery derricks placed concrete delivered to the structure by truck mixer.

Erection of the powerhouse and construction of the downstream portion of the spillway, outlet works, and the stilling basin were carried out simultaneously.

charge nozzles is installed in the gate structure to prevent ice from rendering the automatic controls inoperative. The broken, shattered condition of the rock at the gate structure location required the construction of heavy, counterforted walls in this section, in lieu of lighter walls placed directly against the rock.

The spillway channel chute is unusually steep, having a maximum slope of 0.70 to 1. Steel bars, 1¼-inch square, spaced on 7-foot centers, anchor the floor of the chute to the rock. The channel width flares from 53 feet near the top to 100 feet at the stilling basin in order to spread the sheet of water as it cascades down the incline into the stilling basin. The lower portion of the channel chute extends over, and is supported by, the outlet works control structure described previously. Dentated sills are placed at the end of the stilling basin to further dissipate the destructive energy of the discharging water.

INSTRUMENT INSTALLATION

To determine the behavior of the completed embankment, piezometers and cross arms were embedded in the dam as construction progressed.

CONCLUSION

Anderson Ranch Dam is rendering invaluable service to the Boise region. For the first four years after storage began in 1946 precipitation and river flow were considerably below normal, yet supplemental irrigation water was released to the developed project lands each year and substantial gains in crop production were realized. Precipitation was slightly above normal in 1950 and 1951 and, on April 10, 1951, the spillway operated for the first time. The first of two 13,500-kilowatt capacity generating units (with provision for a third) began operation in December 1950 and the second in July 1951. In 1952, a dry year, the power plant delivered 141,000,000 kilowatt-hours of electrical energy to the area.

DAVIS DAM

DAVIS DAM PROJECT, ARIZONA-NEVADA

By Carl J. Hoffman

The Davis Dam project was authorized in April 1941 under provisions of the Reclamation Project Act of 1939. The principal purposes of the project are to furnish supplemental power for the Southwest, to afford reregulation of the river flows in coordination with the fluctuating releases from the Hoover power plant, and to service the terms of the United States-Mexican Water Treaty of 1944, which provides for a metered delivery of certain waters beyond the boundary of the United States. Also, the project will contribute to flood control, navigation improvement, irrigation and municipal water supplies, reduction in silt pollution, recreation, wild-life protection, and related conservation purposes.

The project, originally authorized as the Bulls Head Dam project, was later renamed Davis Dam project in honor of Arthur Powell Davis. It was Mr. Davis who, as one of the early directors of the old Reclamation Service, laid the foundation for the planned development of the Colorado River.

The construction of Davis Dam practically completes the development of the Colorado River from Lake Mead (Hoover Dam) to the Mexican border.

RESERVOIR

Davis Dam is located on the Colorado River in Pyramid Canyon, 67 miles downstream from Hoover Dam, and 88 miles above Parker Dam. It is about 10 miles north of the point where the boundaries of Arizona, Nevada, and California meet. Its location is about 34 miles west of Kingman, Ariz., and about the same distance southeast of Searchlight, Nev.

The lake formed by the dam submerges Cottonwood Valley upstream from Pyramid Canyon, traverses Painted, Eldorado, and Black Canyons and, when full, backs up to the tailrace pool at Hoover Dam. At normal water surface elevation 647 the reservoir, named "Lake Mohave," is about 28,500 acres in area. It is nearly 4 miles wide at its widest and 67 miles long, and contains 1,820,000 acre-feet of storage. The active storage capacity between the minimum power head level, elevation 570, and elevation 647 is 1,600,000 acre-feet. All major inflows into the reservoir are releases from Hoover Dam, since no streams enter the Colorado River between Hoover and Davis Dams.

DAM SITE

The dam site is located in a unique position in Pyramid Canyon, where the present river gorge is crossed by an ancient river channel. This old buried channel lies east of the present channel upstream from the site and west of the river downstream from the dam. Bulls Head Rock, just upstream from the dam, and from which the original project name was derived, divides the present river channel from this ancient stream bed. The old and new channels in plan take on the shape of an X, the dam being situated where the two cross, and where a single gorge exists in the canyon.

The canyon at the dam site is about 800 feet wide at river level, from which elevation the depth to bedrock in the river channel is in excess of 200 feet. The river material consists mainly of irregular lenses of silt, sand, and gravel. The abutments above the river are steep and bare, composed of porphyritic granite gneiss, severely broken by persistent jointing, and containing minor, clay-filled faults and soft andesite dikes. The Arizona abutment rises to a level slightly higher than the dam and then continues as a ridge for some distance back from the canyon. The Nevada abutment rises to a knoll about 100 feet higher than the dam.

Alternative sites were explored in Pyramid Canyon above and below the present dam, and alternative schemes and estimates were prepared for all the sites. Preliminary designs were undertaken as early as 1920. Studies showed that construction of a concrete dam at any of the sites would prove costly, because of the great

111

White water pours over the spillway at Davis Dam. The great fixed-wheel gates are open only slightly to regulate the river flow. The square openings on either side of the spillway are the outlet works, controlled by radial gates. The powerhouse on the left now contains the five generators.

depths to bedrock compared to the total height of the dam. Further, at the adopted site, the advantageous use of materials excavated from the appurtenant structures to construct an earth dam placed it in economic favor over a concrete dam.

THE DAM

The arrangement of the Davis Dam structures is considered unusual in that the plan does not follow the layouts ordinarily employed in dam construction. The scheme at Davis Dam entailed, in effect, the digging of a new river channel around a stretch of the existing river, the use of the material excavated from this new channel to dam up the existing canyon, and the construction of concrete control structures across the newly created channel.

The dam is a zoned earth- and rock-fill embankment rising approximately 140 feet above the normal river bed and 200 feet above the lowest point in its foundation. It is approximately 1,400 feet thick at river level

and 50 feet wide at the crest. The top of the dam is bounded by a concrete parapet and sidewalk along the upstream side and a concrete curb along the downstream side, and the 1,600-foot crest length is surfaced to accommodate an Arizona-Nevada State highway.

Of the several storage dams built on the Colorado River, Davis Dam is notable in that it is the only one that rests on original sand and gravel river fill, without the conventional cut-off extending to the rock foundation. However, a cut-off trench about 50 feet deep and 120 feet wide at the bottom, with 2 to 1 side slopes, was excavated under the dam to intercept permeable surface beds of the river fill and to minimize differential settlement of the embankment. The trench was backfilled with impervious materials and the center core of the dam was continued upward from this backfilled trench. As compensation for the omission of the deep cut-off, a 10-foot layer of the impervious material was extended over the entire foundation of the dam upstream from the trench, and a layer of sand and gravel was placed over the foundation downstream from the trench. The upstream impervious layer lengthens the percolation path and reduces seepage, while the pervious downstream layer acts as an inverted filter to dissipate water pressures which would normally build up in a foundation composed of lenses of material of varying permeability.

The dam embankment consists principally of materials produced in excavating the diversion channel and structure foundations, supplemented by minimum quantities of selected materials not available from required excavations. Sand and gravel taken from the cut-off trench was placed in the filter zones of the dam. The material obtained from the channel and structure excavations, composed mainly of soft to hard gneiss which broke up during blasting and handling, provided material for the semipervious and pervious zones of the dam. Material for the impervious zone was taken from borrow-pits about 3 miles upstream from the dam. Riprap was obtained from a quarry about 7½ miles south of the dam.

The impervious materials for the center core of the dam, the backfilled cut-off trench, and the upstream foundation blanket were a selected mixture of clay, silt, sand and gravel, moistened and rolled in 6-inch compacted layers. The semipervious zones adjacent to the impervious core consist of rock screenings ranging from dust to 4-inch fragments compacted in 12-inch layers. The pervious zones which provide the stabilizing bulk of the dam are unscreened rock materials deposited in 2-foot layers without compaction. Layers of screened rock 8 and 6 feet thick, respectively, complete the upstream and downstream faces of the dam. The upstream face above minimum water level for power production is covered with a 3-foot layer of large rock riprap. Two berms, at elevations 580 and 540 on the downstream face of the embankment, accommodate 24-foot, bituminous-surfaced access roads to the upper deck and machine shop deck of the power plant.

On the rock abutments for the embankment, concrete cut-off walls are trenched 3 feet into the foundation rock and protrude 10 feet into the impervious zone material. These walls extend from the elevation of the bottom of the cut-off trench to the top of the dam. Holes were drilled under the cut-off walls and grout was forced into the underlying rock, with the line of holes on the left abutment tying into a grouted cut-off curtain encircling the forebay.

The principal quantities involved in the construction of the dam were 500,000 cubic yards of cut-off trench ex-

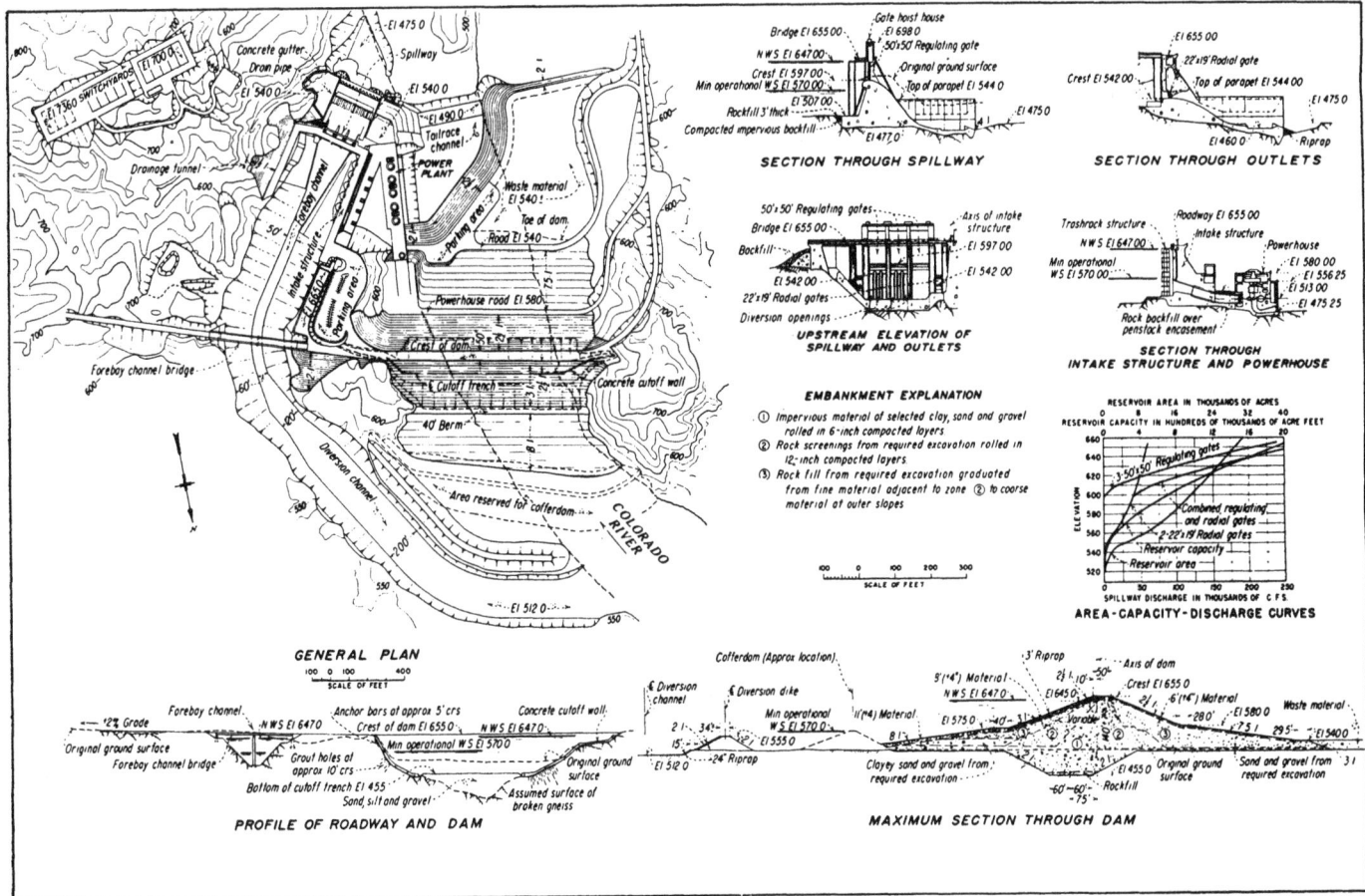

GENERAL PLAN

PROFILE OF ROADWAY AND DAM

SECTION THROUGH SPILLWAY

SECTION THROUGH OUTLETS

UPSTREAM ELEVATION OF SPILLWAY AND OUTLETS

SECTION THROUGH INTAKE STRUCTURE AND POWERHOUSE

EMBANKMENT EXPLANATION

AREA-CAPACITY-DISCHARGE CURVES

MAXIMUM SECTION THROUGH DAM

An Arizona-Nevada highway will cross the crest of Davis Dam on this pavement. The power plant, the power plant transformer deck and take-off structure, the penstock intake structure, and the spillway gate structure are successive steps up to the switchyard on the Arizona abutment.

cavation, 1,000,000 cubic yards of impervious fill, and 2,800,000 cubic yards of rock fill and riprap.

DIVERSION AND FOREBAY STRUCTURES

The channel excavated around the Arizona (left) abutment of the dam served initially as a diversion channel to bypass the flows of the river during construction of the dam embankment and finally became the forebay channel for the spillway, outlet works, and power plant. The channel is more than three-fourths of a mile long, and at its downstream end is enclosed by a U-shaped arrangement of gravity-type concrete structures. The right side of the combination is the power plant intake structure or penstock dam; the center section of the U, located directly across the end of the channel, is the spillway and outlet structure; and the third side of the system, a plain gravity wall, serves to complete the enclosure.

The channel at the upper end begins at right angles to the river just below Bulls Head Rock and follows Bulls Head Wash for about 2,000 feet, crossing the area occupied by the ancient river. The channel in this portion is comparatively shallow, is excavated to river-bed level in silt and sand, and is 200 feet wide at the bottom with 2 to 1 side slopes. A diversion dike formed from the excavated materials was constructed on the right side of the channel to confine the water in the channel and to act as a protection for the upstream cofferdam. The dimensions of this entrance portion of the channel were so chosen that the diversion flows did not exceed scouring velocities along the silt and sand banks.

The portion of the channel which curves and cuts through the ridge to the left of the dam is excavated in

rock. The grade through this stretch drops about 10 feet below that of the upper channel, and the bottom width narrows to 50 feet. The side slopes are 1 to 1 and the channel has a depth in excess of 175 feet. The downstream portion of the channel, which forms the immediate forebay for the control structures, widens so as to provide a basin about 500 feet long and 350 feet wide in front of the structures. Except for the side slopes and bottom immediately upstream from the structures, the channel is unlined.

The power plant intake structure is a mass concrete gravity dam over 500 feet long, varying from a minimum height of 135 feet to a maximum height of 163 feet, and having a base up to 133 feet wide. The cross section is of conventional shape, having a nearly vertical upstream face and a 0.7 to 1 downstream face. The crest of the structure carries a 30-foot roadway, and a transformer and take-off tower deck is integral with the toe of the structure. This structure contains five 22-foot diameter steel penstocks spaced at 60-foot centers, through which the water passes to drive the turbines in the powerhouse. Each penstock opening is protected by a semicircular trashrack structure, and a 17.5- by 34.66-foot fixed-wheel gate operated by a hydraulic hoist controls each penstock entrance. The penstocks extend from this structure to the power plant, which is placed along the original river edge immediately downstream from the main dam embankment and opposite the intake structure. The intake and power plant structures join at their downstream ends with an angle of 28° between the axes of the two structures. The power plant is a semioutdoor-type concrete structure 114 feet high and 615 feet long, containing five 62,200-horsepower, Francis-type turbines powering five 45,000-kv.-a. generators.

The spillway and outlet structure, placed across the downstream end of the channel, is a concrete gravity structure about 355 feet long, consisting of a center overflow or spillway section, and flanking abutment sections each of which contains a radial gate outlet. The structure rises 221 feet above its foundations, which is 170 feet below the normal water surface. The overflow section is an ogee-shaped dam, surmounted by three 50- by 50-foot fixed-wheel gates, with their sills at the ogee crest elevation 597 and supported between 10-foot-thick piers. The piers rise to support the hoist deck 100 feet above the crest. The outlet openings on each side of the overflow section are 22 feet wide and 19 feet high, the sills of the openings being at elevation 542, 55 feet below the crest of the spillway. The 22- by 19-foot radial gates which control the flow through these openings are housed within chambers formed inside the abutment sections.

The combined spillway and outlet structure terminates in a 246-foot-wide stilling basin which accommo-

Only a thin sheet of water escapes under the spillway gates at Davis Dam. Silhouetted against the sky is the housing of the hoist mechanisms for the gates. The steel towers at the left are part of the power plant take-off structure, supporting the electrical transmission lines as they cross to the switchyard.

dates discharges from both the spillway crest and outlet openings. The spillway and stilling basin are designed for a flow of over 175,000 cubic feet per second at normal water surface elevation 647. The stilling basin is 170 feet long which, together with the ogee dam, results in a 380-foot-long structure, measured in the direction of flow. The floor of the stilling basin is at elevation 460, 137 feet below the spillway crest and about 50 feet below tailwater elevation.

The gravity wall flanking the left side of the forebay is about 400 feet long, is roughly triangular in cross section, and is necessary to dam a low place in the natural rock ridge forming this left side. The wall is placed about 40 feet back from the top of the 1 to 1 excavated channel slope. To minimize seepage through the rock beneath the wall, the adjacent slope of the channel is paved with an anchored and reinforced concrete lining. The outside face of the gravity wall is backed up with a rock-fill embankment to form a 30-foot-wide berm.

The spillway gate openings are bridged by a 30-foot-wide deck so as to complete a continuous roadway across the operating structures.

About 1,100 feet upstream from the spillway structure, the forebay channel is crossed by a 420-foot-long, twin-span deck plate girder bridge, having a 30-foot roadway and two 5-foot walkways. The central pier supporting the spans is placed in the center of the channel, and measures 152 feet from the channel floor to the top. This bridge carries the Arizona-Nevada State highway which crosses the main dam embankment.

CONSTRUCTION

A contract for the construction of Davis Dam and appurtenant structures was awarded to the Utah Construction Company in June 1942. After a considerable amount of preparatory work and a commencement of the forebay channel excavation, the contract was terminated early in 1943 because of the war emergency.

Work was resumed in April 1946 when the same contracting firm was awarded the contract for the dam's completion.

Because of the schematic arrangement of the structures construction was confined to three phases, dictated by the requirement for handling the river flows during the construction period. The first phase included the digging of the diversion channel and the construction of the foundation portions of the forebay structures in preparation for the diversion of the river flow. During the second phase, while the water was being diverted through the newly created channel, the original river channel was unwatered, the earth dam was built, and the upper portions of the forebay structures across the new channel were completed. The final phase was the closure of temporary openings which were left in the spillway structure to handle the river flow while the second phase work was being accomplished.

The first construction phase involved the excavation of over 3,800,000 cubic yards of earth and rock material, a large portion of which was stock-piled for later use in the dam embankment. About one-half of the excavated rock material was sorted into two sizes by being passed through grizzlies and then stock-piled separately. Excavation averaged over 10,000 cubic yards per day, and at times as many as 25,000 cubic yards of material were handled in a 24-hour period. In doing the work the contractor employed 4 power shovels, forty 10- to 13-cubic-yard dump trucks, and 11 scraper-earth-movers.

After completion of the excavation, the foundations for the concrete structures were intensively grouted to tighten the shattered and jointed rock.

Concrete placement in the lower portions of the forebay structures and the forebay lining was commenced in January 1948 and six months later was completed to a height sufficient to permit the diversion of the river through the new channel and through six 13-foot-wide temporary openings in the spillway crest structure.

In the second construction phase, the river was diverted into the new channel by the construction of cofferdams both upstream and downstream from the dam foundation area. The upstream cofferdam, completed first, was constructed by building a timber pile trestle across the river from which earth and rock materials were dumped to barricade the river channel. The downstream cofferdam was constructed by similar methods. After the cofferdams had been built and made relatively watertight, the dam foundation surface was pumped dry and a subsurface drainage system of 3,300 well points was installed to dry up the river-bed materials so that excavation and backfilling of the cut-off trench could begin.

The river closure was completed in July and August of 1948, and the first embankment materials were placed during September. The cut-off trench was excavated and ready to receive the first layer of backfill late in October 1948. In April 1949, 10 months after diversion of the river, the dam embankment, totalling 3,800,000 cubic yards, had been completed to its full height. There were 770,000 cubic yards of embankment placed during the month of February 1949, which at that time established a record for embankment placement in a large earth dam.

During this second phase of construction, while the turning of the river and the construction of the dam embankment held the spotlight, work on the forebay structures continued and construction of the power plant was started. Concrete, forms, reinforcement, and other construction material required to build the spillway and intake structures were handled by two Whirley cranes with 160-foot booms, traveling on an S-shaped construction trestle 90 feet high and 1,084 feet long. The trestle adjacent to the intake structure had a branch to serve the power plant. Two railroad tracks passed along the trestle and terminated at the 8-cubic-yard capacity concrete mixing plant on the Arizona river bank about one-fourth mile south of the structures. Transported on railroad cars, the buckets of concrete were picked up and spotted by the Whirleys. Construction of the forebay structures and power plant involved the placing of about 600,000 cubic yards of concrete and over 25,000,000 pounds of steel penstock, structural steel, and reinforcement steel.

The final phase of construction involved the closure of the six temporary diversion openings and the completion of the spillway crest. During this stage the river flow was passed through the radial gate outlet openings. This work was accomplished from January to June 1950.

At peak periods, up to 1,300 workers were employed by the contractor, and some 150 by the Bureau. The contractor's camp, on the Nevada side of the river a few miles below the dam, was equipped with all the facilities usually established for a small town of about 3,000 population. A part of the Government camp on the Arizona side of the river was built as a permanent installation for housing the operation and maintenance personnel for the dam and power plant.

In addition to the two trestles which crossed the river at the cofferdam locations, two access bridges were provided downstream from the dam site and a haul bridge spanned the upper end of the diversion channel.

CONCRETE

Aggregate for concrete was produced from natural deposits located on the Nevada side of the river some 3 miles downstream from the dam site. As a normal

Modern construction techniques speed today's giant dams. This construction trestle was arranged so that the Whirley cranes, one of which was at work on the right, could reach any portion of the work. The spillway and outlet structure was crossed by the trestle, and the penstock intake structure began to take shape in the foreground.

precaution against alkali aggregate expansion in the completed concrete structures, type II, modified, low alkali cement was used in all concrete mixes. As an extra precaution, and to reduce the initial heat of hydration in the concrete, a calcined Puente shale pozzolith admixture was used to replace 20 percent by weight of the portland cement.

The construction specifications required that the placing temperature of the concrete was not to exceed 80° F. Several measures were adopted to achieve this condition. As an initial step the contractor provided a wooden shed over the aggregate stock piles at the mixing plant to protect them from the intense heat of the desert sun. Sand was not shaded, however, since its insulation qualities prevented heat penetration to any great extent. During the hot summer months a refrigerating plant was used to cool the mixing water to a temperature of 40°. At times, as much as 100 pounds of chopped ice per cubic yard of concrete was also used to bring the temperature of the mix down to the prescribed limit.

All mass concrete was cooled after placing by circulating river water, which seldom exceeded 65° F., through embedded tubing placed on the rock foundations and on top of each 5-foot lift. Such cooling was continued until the average temperature of the surrounding concrete was within 5° of the cooling water temperature.

ADDITIONAL CONSTRUCTION WORK

The contract for construction of the earth dam and appurtenant structures was essentially completed in November 1950, and water storage was started that month. The first turbine and generator unit in the power plant was completed and power generation commenced in January 1951.

As excavation work progressed under the original contract, it became apparent that certain weaknesses, not completely disclosed by preliminary investigation, existed in the rock downstream from the spillway structure to the extent that it (the rock) was not considered capable of resisting the high-velocity flows created at the end of the bucket apron as designed. Ac-cordingly, a longer, hydraulic-jump, stilling basin was substituted for the original spillway bucket. Because the longer basin could not be constructed without interfering with the contractor's downstream cofferdam and construction plant, and in order to avoid postponement of the initial diversion, it was decided to defer the completion of the stilling basin until all other construction was finished and the river flow could be diverted through the power plant turbines. This stilling basin completion work was undertaken by separate contract in 1951 and, in addition to construction of the reinforced concrete basin, involved excavation of considerable river channel material downstream from the spillway and in the river channel below the dam and power plant.

GREEN MOUNTAIN DAM

COLORADO-BIG THOMPSON PROJECT, COLORADO

By Judson P. Elston

The definition of "upper basin," stated in the Colorado River Compact of 1922, is as follows:

The term "upper basin" means those parts of the States of Arizona, Colorado, New Mexico, Utah, and Wyoming within and from which waters naturally drain into the Colorado River system above Lees Ferry, Ariz., and also all parts of said States located without the drainage area of the Colorado River system which are now or shall hereafter be beneficially served by waters diverted from the system above Lees Ferry.

From this it is clear that the possibility of diverting water from the Colorado River watershed to another watershed was recognized.

The Continental Divide forms the dividing line between the eastern half and the western half of the State of Colorado. The Rocky Mountains, highest in the United States, supply the major portion of the total run-off of the State from melting snows during April, May, June, and July. Unfortunately, the water resources of Colorado as provided by nature were unevenly distributed both geographically and seasonally and, in order to preserve and stabilize the agricultural industry of the State, required some means of redistribution.

As the first scheme for diversion of Colorado River water to another watershed, the Colorado-Big Thompson project contemplated the delivery of water to the fertile agricultural area deficient in rainfall lying in the South Platte River Valley in the northern part of Colorado. Green Mountain Dam, as the first construction feature of the Colorado-Big Thompson project, is a monument to the vision of men whose dream was to assist nature to supply irrigation benefits where it had neglected to do so.

The project plan called for diversion from the headwaters of the Colorado River through the Continental Divide and into the Big Thompson River, whence the water would be led via canals, tunnels, conduits, and reservoirs to lands on the eastern slope. The development of hydroelectric power from the energy of the water in its descent of the eastern slope of the mountains was an additional benefit. The law authorizing the project stipulated that certain construction would be necessary in order that all rights and interests of Colorado River water users on the western slope would be protected. The law further stated that:

The Green Mountain Reservoir, or similar facilities, shall be constructed and maintained on the Colorado River above the present site of the diversion dam of the Shoshone power plant above Glenwood Springs, Colo., with a capacity of 152,000 acre-feet of water, with a reasonable expectancy that it will fill annually.

In other words, the water diverted to the eastern slope was to be replaced by the storage of Colorado River waters not otherwise being used or available for diversion.

The State of Colorado interposed no legal obstacles to the construction of the Colorado-Big Thompson project nor to the diversion of Colorado River water to the eastern slope. In arriving at the quantity of water required for replacement, it was assumed that all irrigable lands along the main Colorado River would be irrigated as well as all such lands on the tributaries.

The most suitable location for a replacement reservoir for the benefit of western slope water users was on the Blue River as it was the closest available site to the area where irrigation replacement was necessary. Such a location was also convenient for the development of power to fit into the other power developments of the project and the utility companies of the State.

RESERVOIR

The Green Mountain Reservoir has a capacity of 152,000 acre-feet of water, 52,000 acre-feet of which is replacement for project diversions and 100,000 acre-feet of which is available for power purposes. Irrigation outlet capacity is 1,000 cubic feet per second, and power

The impounded water of Blue River stretches away to the foot of the distant hills, held back by Green Mountain Dam, sharply outlined by the dark water of the lake.

outlet capacity is 1,500 cubic feet per second. The spill-way capacity is 25,000 cubic feet per second.

The reservoir is located on the Blue River, about 16 miles southeast of Kremmling, Colo., in a canyon north of the Gore Range of the Rockies. The reservoir extends up the Blue River for a distance of 7 miles and has an area of 2,125 acres. The hills bordering the northeast side of the reservoir are, in general, more abrupt than those on the southwest side and include a high outcropping of rock called Green Mountain, from which the dam takes its name. The area which was submerged by the reservoir consisted of hay and grazing land, with sagebrush and some scattered timber.

DAM SITE

Four potential dam sites within a distance of 1¼ miles were studied, and foundation investigations were carried out at the two upstream sites. The topography and geology were quite similar at all of the sites. A concrete gravity dam was considered at one of the sites farthest down the river; but at the upstream sites only earth- and rock-fill dams were considered. Total costs as well as the unit cost per acre-foot of storage dictated

selection of one of the upstream sites. While the lowest cost per acre-foot of storage was at the adopted site, the controlling feature in selection was the overburden of gravel at the left abutment of the second-best site. The danger of not obtaining a good cut-off to bedrock on the left abutment excluded this site.

Nearly all the area at the dam site finally selected was covered by overburden. On the right abutment, it consisted of soil and slide rock up to 35 feet thick. On the left abutment, it consisted of glacial till, gravel, and boulders, with a maximum thickness of 65 feet. The rock floor was made up of several different types of rocks, the principal members of which were sandstones and shales and igneous intrusions of the porphyry type. No important faults were found and smaller slip-shears or seams presented no special difficulties. The sand-stones were hard and durable. The softer shales, though not suitable as foundation for a masonry dam, were tight and satisfactory for construction of an earth dam.

THE DAM

The plan and section of the embankment as shown on the accompanying drawing were followed with only

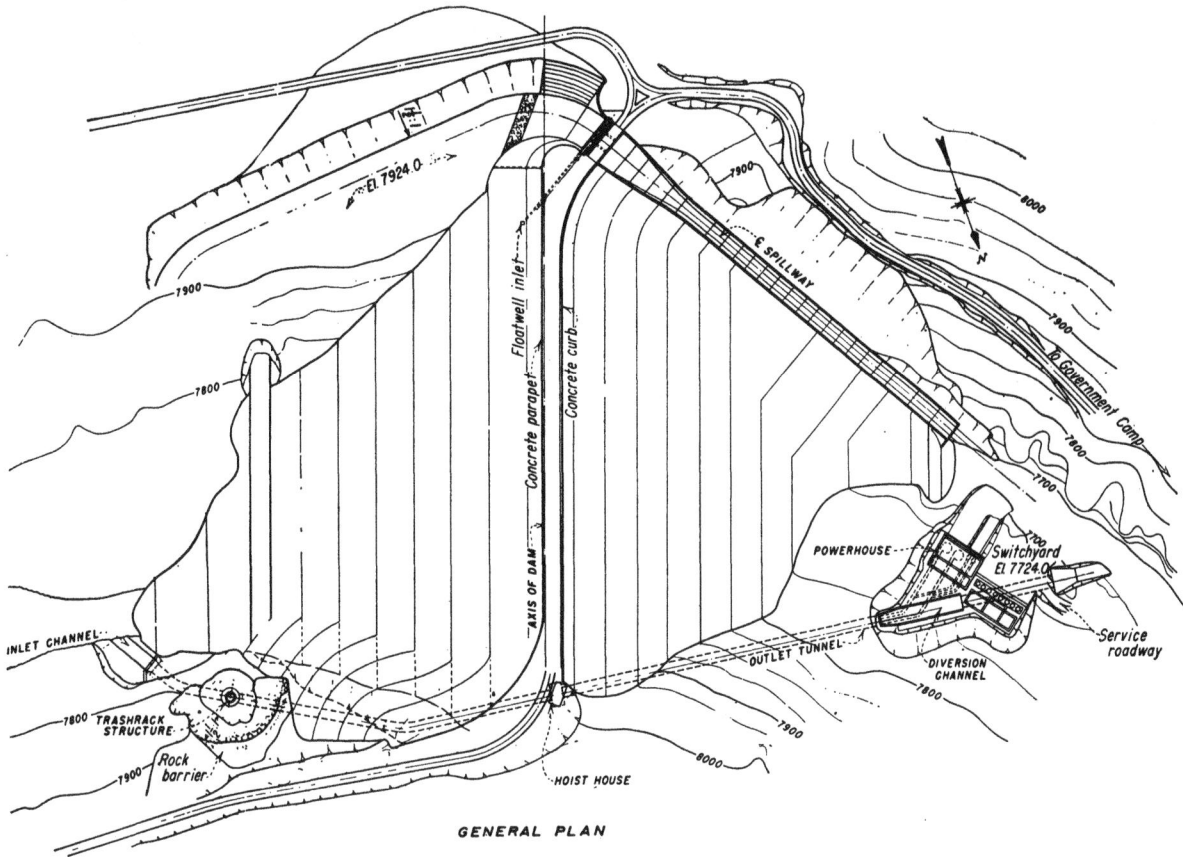

GENERAL PLAN

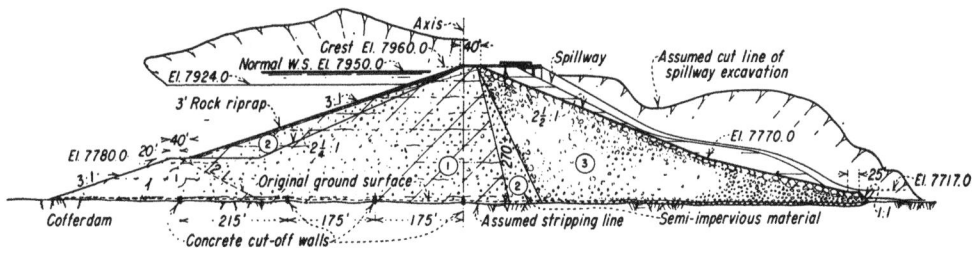

SECTION IN RIVER CHANNEL

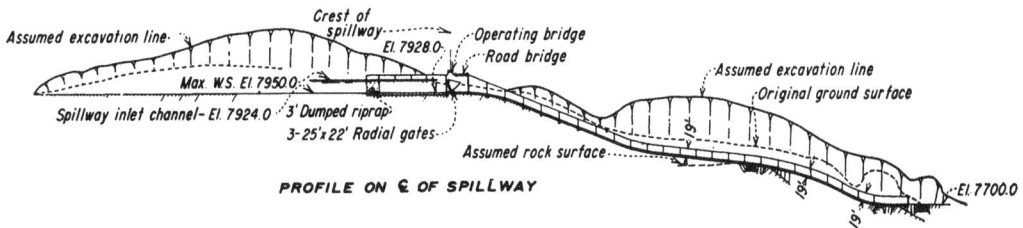

PROFILE ON ℄ OF SPILLWAY

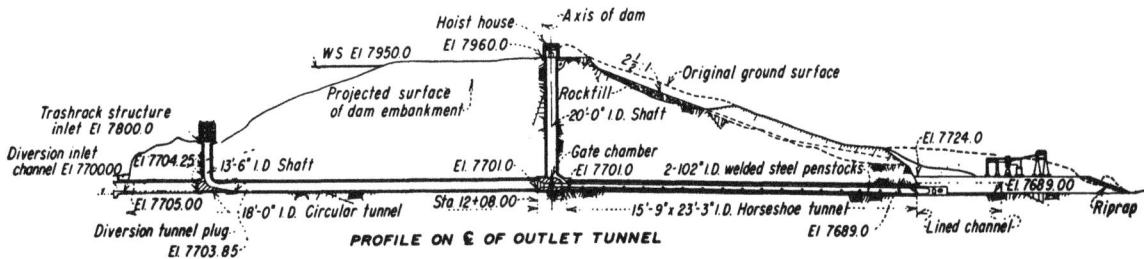

PROFILE ON ℄ OF OUTLET TUNNEL

Down this spillway chute plunge the overflow waters from Green Mountain Reservoir. At the foot of the chute, the power plant and switchyard stand on the right and, across the bridge on the far bank of the gorge, stand the buildings of the government camp.

minor modifications during the actual construction. From the lowest point in the river gorge the dam extends upward 309 feet to the 40-foot-wide roadway crest, with an average height of 274 feet above stream bed. The slopes limiting the zones of the various materials within the body of the embankment were so adjusted as to utilize available materials to the best advantage. The volume of the compacted earth fill and dumped rock fill placed in the dam amounted to a little over 4,316,000 cubic yards. The upstream zone 2 material was about one-ninth the volume of the earth materials in the dam. The impervious zone 1 material, classified as a screened glacial drift and made up of sand, silt, and clay, occupied about five-ninths of the volume. Cobble and rock fill constituted the downstream one-third of the volume of the dam. Construction of a sluiced cobble and rock-fill inverted filter, 15 feet thick, was considered necessary under the downstream rock-fill section of the dam. In case any leakage occurred, this filter was intended to prevent washing of fines, such as clays, silts, and sands, from the foundation into the rock fill. Four concrete cut-off walls

were located in the foundation rock of the dam, one of them in the upstream cofferdam area.

SPILLWAY

The spillway for the dam is located on the left abutment and consists of a concrete-lined open channel designed for a maximum discharge of 25,000 cubic feet per second. Release of water from the reservoir is controlled by three radial-type gates, 25 feet wide and 22 feet high. The outlet channel for the spillway is about 1,000 feet long and for better flow conditions is made up of a series of vertical curves, dropping from elevation 7928 at the spillway crest to elevation 7700 at the discharge end of the channel.

OUTLET WORKS

The entrance to the outlet works is located in the reservoir on the right abutment and consists of a 15-foot 5-inch diameter concrete-lined vertical intake shaft protected by a trashrack tower within which can be installed a 20-foot 5-inch timber cylinder for the emergency unwatering of the intake shaft and tunnel. The intake shaft is connected by a 90° bend with a concrete-lined 18-foot diameter tunnel driven through rock. The tunnel leads to a gate chamber located in rock and connected to the top of the dam through a 20-foot diameter vertical shaft. The gate chamber contains two 102-inch ring-seal slide gates which are for emergency control of the flow of water into the steel penstocks leading to the powerhouse and irrigation release pipes. Downstream from the gate chamber the tunnel is horseshoe in shape with maximum dimensions 23 feet high and 16 feet wide. This horseshoe tunnel contains two 102-inch diameter penstocks, each of which at the powerhouse is reduced to 84 inches in diameter for connection to the turbines. Two 50-inch outlet pipes branch from the 102-inch pipes near the powerhouse and are connected to 42-inch needle valves which control irrigation releases.

POWER PLANT

A reinforced concrete powerhouse is located on the east bank or right side of the Blue River, immediately below the dam. The building is 67 by 97 feet and houses two 15,000-horsepower turbines and the two 42-inch outlet valves. The plant is operated under a variable head of water in the reservoir of from 120 feet to 255 feet. Maximum horsepower capacity is reached under a head of slightly over 200 feet, and best efficiency is obtained while operating under an effective head of between 200 and 220 feet.

DIVERSION OF RIVER

In designing the dam it was necessary to provide for some means of diverting the Blue River, at least temporarily, from the dam site to permit excavation and construction on the lowest parts of the foundation. This problem was handled by using the outlet tunnel to take care of river flows. During construction, the outlet tunnel was connected with a temporary tunnel upstream from the intake structure and with an open channel downstream from the end of the outlet tunnel. Shortly before completion of the dam, the intake of the temporary upstream diversion tunnel was bulkheaded to permit the construction of a concrete plug at its lower end, thus connecting the outlet tunnel to the permanent intake shaft. During the same period, the diversion channel below the outlet tunnel portal was removed, and the ring-seal gates were installed in the gate chamber. With this work completed, the flow of the river was under control.

CONSTRUCTION CAMPS AND SCHOOL

A number of contractors went together to submit bids, with the result that five bidding combinations were involved. Bids were opened on October 12, 1938, and Warner Construction Company of Chicago, Ill., was found to have submitted the lowest proposal of the five bids at a figure of $4,226,206. Notice to proceed with construction was issued to the Warner Company on November 30, 1938. Housing and other facilities to provide comfortable living for construction forces had to be constructed at the dam site. The Bureau of Reclamation camp, located on the left side in the river below the dam, was in an attractive setting in the river canyon. The Government camp and its utilities, and a steel arch bridge for access to the powerhouse, were constructed by Government forces and were designed for permanency so that not only engineers engaged on construction but people to be later employed on operation and maintenance of the powerhouse would be properly housed. The Government camp included residences, dormitory, an office and laboratory building, a concrete garage and shop, a steel frame warehouse, and temporary houses, together with all necessary utilities, streets, and walks.

Because the construction site was remote from centers of population from which an adequate supply of labor might be drawn, the contractor felt it necessary to construct and maintain camps for the accommodation of a large number of employees. Three camps, known as the tunnel camp, the headquarters camp, and the trailer camp, were constructed. The tunnel camp was initially used for the accommodation of the tunnel workers and later for men used on other construction work. During the early stages of the work, trailers and small shacks

and tents of the contractor's employees were distributed over the Government reservation in such a manner as to make it impossible to enforce sanitary regulations. The contractor took steps to improve the situation by graveling the streets, providing a sewer and sewage-disposal system, constructing a water supply system with a hydrant at every street intersection, building two community bathhouses with laundry rooms and toilets for men and women, and installing electric lights for the camp. The United States Government, watchful over the welfare of all people employed on the construction, particularly in regard to the educational advancement of their children, required in the construction specifications that the contractor provide school facilities and instruction up to and including the twelfth grade, or high school, for all employees on the project without charge. The contractor employed three teachers—one for the first four grades, one for the fifth to eighth grades, and one for the high school. The school buildings were constructed from former bunkhouses, with adequate windows and sanitary facilities.

CONSTRUCTION—OUTLET TUNNEL

Tunnel excavation was started on December 9, 1938, from the upper end, and the tunnel was holed through on May 15, 1939. The work was started at the upper or inlet end of the tunnel because of the necessity of first removing about 70,000 cubic yards of common overburden excavation and about the same amount of rock. The rock through which the tunnel was excavated consisted of formations of sandstone, shale, and porphyry intrusions. Blasting powder was used throughout, with 30-percent powder being used in the shale and sandstone and 60-percent powder in the porphyry. After excavation, rock in the tunnel required the support of 6-inch, steel I-beam ribs, weighing 12½ pounds per foot and spaced from 16 to 72 inches apart. Steel liner plates to provide ground support between the ribs totaled some 180,000 pounds. The type of drill round was a modified V cut, which was changed to suit rock conditions. From 44 to 52 holes were drilled per round in the tunnel upstream from the gate chamber, and as many as 76 holes per round were necessary in the penstock tunnel leading to the powerhouse. Drilling was usually done in 8-foot rounds. Muck was loaded into Diesel-powered trucks by an electrically operated slusher scraper mounted on a narrow-gage track. Mining methods were used in sinking the 20-foot diameter gate chamber shaft for a distance of 150 feet. First an 8- by 8-foot-square pilot shaft was sunk to intersect the tunnel at the gate chamber. The pilot shaft was then enlarged from the top down, the muck being drawn off as necessary from the bottom and removed from the tunnel.

Lining of the gate chamber shaft with 18 to 30 inches of concrete, depending on the soundness of the rock

Excavation of river-bed material disclosed this ancient "plunge basin," in effect a prehistoric waterfall. The depths of these rock pockets were carefully backfilled with compacted earth.

encountered, followed the excavation down in 10-foot lifts. The circular section of the tunnel has a minimum thickness of concrete of 18 inches and the horseshoe section has a minimum thickness of 24 inches, with the amount of concrete per linear foot of tunnel ranging between 4½ to 6½ cubic yards.

The rock outside the tunnel and shaft linings was consolidated and strengthened by injecting, under pressure, a mixture composed of water and cement into holes drilled into the rocks through pipes set in the concrete. Relatively little cement grout was forced into the undisturbed shale rock; however, where grout holes penetrated porphyry and sandstone in the gate-chamber and inlet shafts, it was possible to pump large quantities. Pressures used in the tunnel and shafts varied from 50 to 200 pounds per square inch. Several large crevices under the trashrack structure were grouted initially with a sand-cement grout and later regrouted with neat cement grout.

All necessary work in the outlet tunnel was completed on December 2, and, on December 13, 1939, the river was diverted through the tunnel by means of a temporary cofferdam across the river upstream. The cofferdam consisted of steel sheet piles interlocked with one another and driven to foundation rock under the river. The sheet piles were supported on each side by dirt, gravel, and other waste overburden materials removed from the dam site prior to actual construction of the dam.

CONSTRUCTION—DAM

At the time of its commencement, Green Mountain Dam was the highest and largest earth-fill dam undertaken by the Bureau of Reclamation, and the highest earth-fill dam for continuous reservoir operation in the world.

Stripping of unsuitable material from the dam site was started on May 12, 1939, and was, for the most part, completed by the fall of 1941. The total vertical depth of stripping varied considerably and reached a maximum of 78 feet on the left abutment, 92 feet on the right abutment, and 45 feet in the river channel. Average depth of excavation through the river channel was about 10 feet. All usable materials were stripped separately and either hauled upstream from the dam site and stored in stock piles for later application or, where feasible, placed immediately in the dam. The unsuitable materials such as weathered and badly broken shales, sandstones, and decomposed shale were wasted at the upstream end of the dam or along the river bank.

A "plunge basin," as it is geologically known, was discovered while excavating for the dam foundation. This formation was encountered on the foundation 37 feet below the surface of the earth. Some 30 feet wide by 100 feet long, the basin had been carved out of a hard porphyry rock and then filled with about 2,000 cubic yards of river sediments consisting mostly of gravel. In appearance the plunge basin greatly resembled a dry waterfall. This geological oddity being directly beneath the dam, it was necessary to fill it with selected materials and compact them to the same density as required for the rolled-earth dam sections. Filling the plunge basin increased the maximum height of Green Mountain Dam from 272 feet to 309 feet.

Four reinforced concrete cut-off walls discourage seepage beneath the dam embankment at its contact with the rock foundation. Prior to concreting the cut-off wall footings, 2-inch diameter pipes were supported vertically in the excavated trenches at spacings of not more than 10 feet. After the concrete in the footing trenches had set, holes were drilled into the foundation rock through the grout pipes to depths as great as 150 feet. Mixtures of neat cement and water were forced into these holes at pressures ranging from 30 pounds per square inch in the shale to 100 pounds per square inch in the porphyry. Supplemental grouting of the foundation rock benches on the left abutment between the cut-off walls was performed to reduce the seepage water and to localize excess flows so they could be more effectively handled in sumps and drains.

Earth materials available from the abutment stripping and from borrow pits were suitable for the earth-fill section of the dam but contained a large percentage of glacial cobbles and boulders. Getting the material from borrow pits to the dam was not only a problem of transportation down steep canyon slopes, but also one of separation into two main classes of fill—rock for previous fill, and fines for impervious rolled embankment. The solution was to combine transportation and processing with a downhill haul from borrow pit to the separation plant, gravity flow of materials through the plant, and more downhill haul of the separated materials to the fill. After some experimenting, the contractor located the separation plant high on the canyon wall, low enough for the downhill haul from the pit and also high enough to transport processed materials to final position with downhill haul.

At the pit, material was loaded into bottom-dump tractor trucks by shovels. The trucks reached the top of the plant via a circular trestle and slowed down just long enough to straddle and dump into a divided receiving hopper before returning to the pit. Material under 3 inches was diverted to a receiving hopper for the fines, while the rejects and larger rock continued on down into the end of the chute. The hopper for fines was fitted with slide gates for discharge into bottom-dump tractor-trailers traveling on a winding road of fairly easy grades leading to the upstream and center parts of the dam, where the fines for the impervious core of the dam were deposited. The rock chute contained endless slack chains with 3-inch links. As the chains were slowly rotated, the under side of the loops—that is, the side next to the rock in the chute—moved downward, permitting the rock to dribble out of the end of the chute into waiting end-dump trucks. By stopping the movement of the chains, flow of rock was stopped. Good haul roads were maintained down the side of the canyon to the downstream section of the dam for these trucks. The trailer-dumps deposited the impervious material in windrows to be spread with bulldozers, after which the material was rolled and compacted with sheepsfoot rollers. Natural water content of the fines was usually near the maximum specified, thus requiring but little sprinkling. Tests showed the fill to be unusually well compacted, with wet samples running as high as 156 pounds per cubic foot and averaging 150 pounds, and with moisture content of the material averaging slightly less than 9 percent.

Special attention was paid to the contact between the impervious fill material and the foundation rock. Rock faces were cleaned and the material thoroughly tamped against them. One of the interesting construction features was the use of emulsified asphalt for treatment of

Handling of earth and rock material was simplified by this ingenious processing plant. Excavated material from the borrow pits was dumped into the screening and separation plant from the trestle. Finer material, suitable for the impervious fill, was loaded into tractor-drawn dump trailers on the upper level, while rock slid down the chute to be loaded into trucks for transportation to the downstream pervious section of the dam.

the shale foundation which occupied much of the dam site and which disintegrated when exposed to air for a prolonged period. The procedure was to blow off all the loose material down to a clean, tight surface and then spray the solid shale with emulsified asphalt.

The total amount of seepage water that flowed into the dam foundation area, powerhouse foundation, and spillway was comparatively small, but it entered at many places. Most of the water that made its way into the foundation came in the 1940 season during the first of the embankment construction. This water was removed by pumping or through gravity pipelines. A number of small seeps along the dam abutments in the canyon walls were taken care of by bailing until sufficient compacted earth was placed to smother them completely. In a few locations, a pump sump or well was made from a bottomless oil drum with perforated sides, or from a large piece of pipe set over or close to the seep, around which was placed clean gravel. Grout pipes were placed at the bottom of each sump and, after the sumps had served their purpose, the pipes were capped and filled with grout when 30 feet of earth fill had been placed over them.

In order to observe settlement of the embankment during and after construction, four telescoping pipe settlement gages were installed in the embankment.

CONSTRUCTION—SPILLWAY

Stripping of top soil and other unusable materials from the spillway chute section was done in the fall of

1939. Earthy, glacial drift materials were stock-piled and then later used in the rock-fill sections downstream. All excavation was completed by the spring of 1941. To prevent possible damage from hydrostatic uplift pressures under the spillway, tile drains and free-draining gravel blankets were constructed, and steel anchor bars were embedded in the foundation rock prior to placing the spillway concrete. An effort was made to force grout into a line of holes drilled below the bottom of the spillway cut-off trench, but the rock was very tight and little grout was injected. The first concrete was placed in the cut-off trench across the lower end of the spillway channel. Placing of concrete for the channel floor was started in August 1940, and continued except for the cold weather seasons until 1942. The steep grade, severe climatic conditions, and lack of skilled labor made placing of the floor sections a tedious job. The placing of concrete in gate and inlet structures of the spillway was started in August 1942, and all concrete had been placed by the spring of 1943. Installation of the 25- by 22-foot spillway radial gates was deferred because of the war. Delivery of the gates was not made until March of 1944, and installation was essentially completed during that year and the gates put in operation.

CONSTRUCTION—POWER PLANT

Removal of the earth material down to bedrock over the area to be occupied by the powerhouse was carried on simultaneously with similar excavation for the outlet channel. The necessary rock excavation for the switchyard was done at the same time and by the same methods as used on the powerhouse foundation. Excavation for the powerhouse proper consisted of removal of the porphyry rock to a depth of 68 feet.

Excavation of rock was begun in February of 1940 and the first concrete was placed in September. By the end of the working season the draft tube forms had been encased with concrete and the floor over the sump was finished. Progress on the superstructure was good

during 1941, and this feature was completed in October of that year. The 50-ton crane was erected during the winter in anticipation of machinery installation the next spring. The concrete encasing the outlet pipes, those parts of the main penstocks within the powerhouse itself, the scrollcases, and the generator bases was placed in the fall and winter of 1942.

The fabrication, testing, and installation of the 102-inch penstock pipes which carry water to the powerhouse and to the outlets for irrigation releases were slowed by material shortages and priority difficulties due to the war. Some minor changes in design were made to avoid prolonged delay, and, fortunately, it was possible to put the power plant into operation in the summer of 1943 and ease the demands for electricity by war industry in the Denver area.

In 1942 it was determined that, by lowering the river bed over a distance of approximately 1,600 feet downstream from the powerhouse, the annual power output of the plant would be increased by about 3,500,000 kilowatt hours. Accordingly 27,000 cubic yards of material were excavated, the work being completed in April of 1943.

CONCLUSION

Green Mountain Dam is carrying out the purpose for which it was intended. From Granby Reservoir on the Colorado River, huge pumps lift water to the west portal of the tunnel under the Continental Divide through which the water flows by gravity to storage reservoirs on the eastern slope. Green Mountain Reservoir stores water for release to water users downstream on the Colorado River. Such releases, at a rate to offset the effect of diversion of water from Granby Reservoir to the eastern slope, require a draw-down of Green Mountain Reservoir of approximately 135 feet each year in advance of the flood season. More than 80,000,000 kilowatt hours of electrical energy are generated yearly at the dam and transmitted to utility companies throughout the State.

BOYSEN DAM

BOYSEN UNIT, BOYSEN DIVISION, MISSOURI RIVER BASIN PROJECT, WYOMING

By Richard T. Larsen

Boysen Dam was one of the first dams whose construction was initiated as a part of the far-flung Missouri River Basin project. It is situated at the upstream end of the spectacular and imposing Wind River Canyon in central Wyoming. In 1908, a Wyoming business man named Asmus Boysen constructed a small concrete slab-and-buttress dam about 1 mile downstream from the site of the new dam, which provided enough head and storage capacity to serve a small hydroelectric plant. After silting of the small reservoir had depleted storage to the point where power generation was no longer feasible, either practically or economically, the power plant was abandoned and the old dam was breached to decrease the encroachment of flood water on the property upstream. The present dam takes its name from its predecessor.

Consideration of the physical requirements of the present dam and reservoir, which provides storage capacity for power as well as space for silt deposition and temporary storage of floods, led to selection of the new site upstream from the old dam. Estimated construction costs and the availability of construction materials pointed to the economy of an earth- and rock-fill dam with the spillway, outlet works, and power plant of concrete construction.

Two main arteries of transportation through central Wyoming—United States Highway No. 20 and the Chicago, Burlington and Quincy Railroad—traversed a portion of the reservoir and occupied strategic positions near river level through the dam site. Relocation of about 4 miles of the highway was necessary, and was carried out under a separate contract prior to construction of the 13-mile railroad relocation and construction of the dam and appurtenant works. Construction of the dam and relocation of the railroad were included in a single contract. An interesting feature of the railroad relocation is the 7,130-foot tunnel, which begins above the reservoir water surface and descends to

emerge adjacent to the right abutment of the dam as shown in the drawing, thereby serving in effect as a tube through the reservoir. This expedient shortened the required length of the railroad relocation some 5 miles.

Construction of the dam and appurtenant works was initiated in September 1947, concurrently with construction of the railroad relocation. Excavation for and construction of the spillway structure was possible without removal of the existing railroad, and the major portion of the spillway was completed before railroad traffic was transferred to the relocated line and the old line abandoned.

As may be observed from the drawing, the spillway is controlled by two radial gates, each of which is 30 feet wide by 25 feet high, and is operated by an individual electrically driven gate hoist. When in the down or closed position, the top of the gates is at elevation 4725 (normal water surface), at which elevation the volume of the reservoir is approximately 820,000 acre-feet. Of this quantity, 260,000 acre-feet are for inactive storage, 410,000 acre-feet are for conservation storage, and 150,000 acre-feet constitute a joint-use space normally used for conservation but subject to evacuation for flood storage when necessary.

With the onset of a flood, it is planned, after evacuation of the joint-use storage space, to operate the spillway gates in such a manner as to restrict the total discharge (powerhouse, outlets, and spillway) to a uniform rate of 20,000 cubic feet per second until the rising reservoir water surface reaches the maximum water surface elevation of 4752 feet, at which point the gates will be opened as much as necessary to hold the reservoir water surface at elevation 4752. Consequently, most of the water entering the reservoir during large floods (those for which the inflow exceeds 20,000 cubic feet per second) will be stored temporarily and released after the flood is over. The volume of the reservoir between

127

Panoramic view of Boysen Dam, power plant, and reservoir showing the spillway discharging.

the normal water surface elevation of 4725 feet and the maximum water surface elevation of 4752 feet is approximately 673,000 acre-feet. With the method of spillway operation already described, this volume is sufficient to contain a flood with a peak inflow rate of 160,000 cubic feet per second and a 15-day volume of 1,300,000 acre-feet.

A concrete-lined circular tunnel, 28 feet in diameter, constructed at river bed elevation directly beneath the spillway and discharging into the spillway stilling basin, provided for diversion of the river during the construction of the dam embankment. After the tunnel had served its purpose as a diversion outlet, it was plugged with concrete for 90 feet immediately beneath the spillway gates, and the downstream end filled with concrete to complete the spillway discharge channel sloping down to the stilling basin.

With the river flowing through the diversion tunnel, the river bed between the upstream and downstream cofferdams was prepared for the dam embankment foun-

Spillway construction was near completion at Boysen Dam at the time of this photograph. The diversion channel was later closed and work began on the power plant at the left of the spillway channel.

Selected earth material was brought from the borrow pits, spread on the dam embankment, and compacted in layers by special tamping rollers.

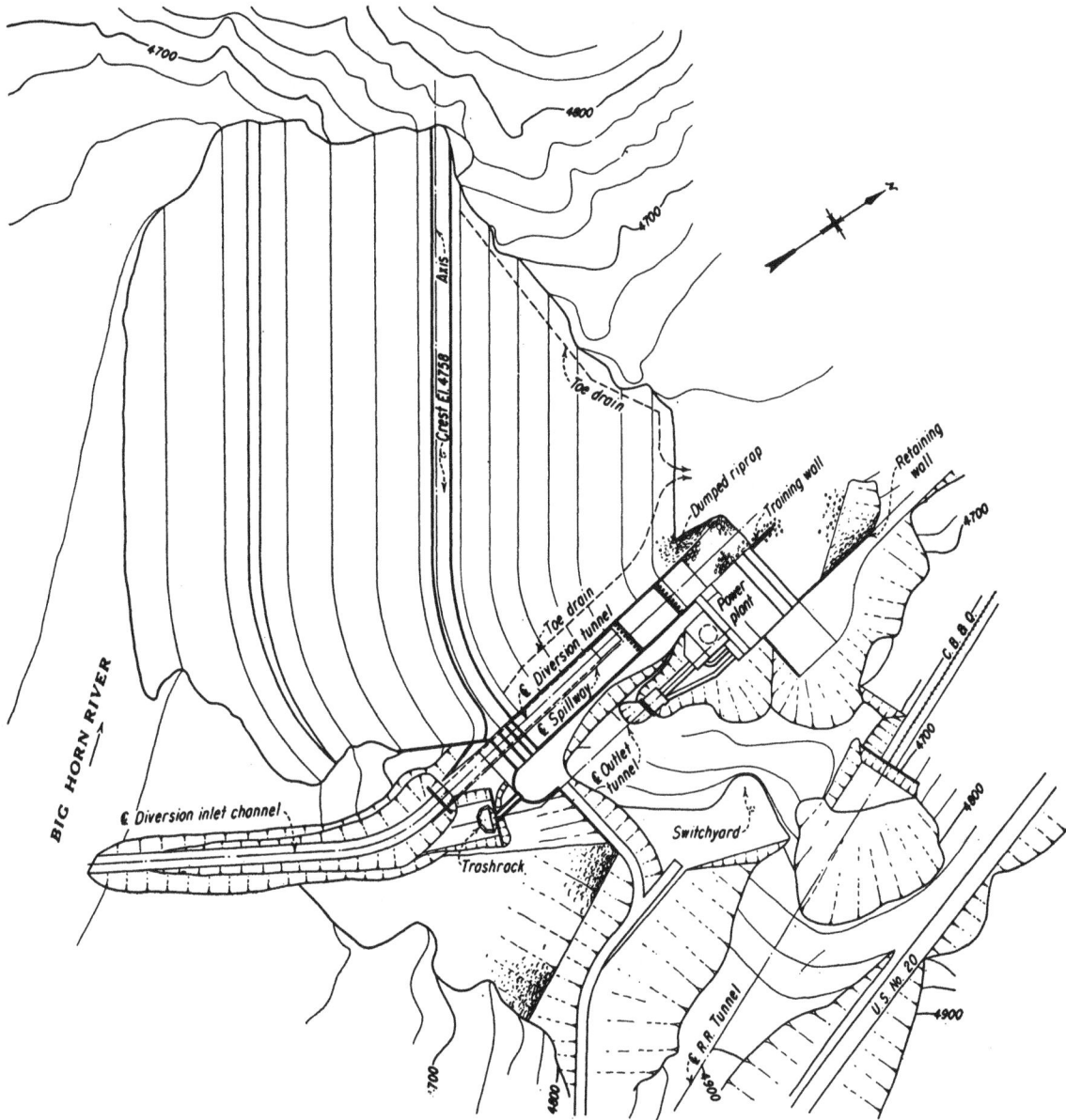

GENERAL PLAN

SCALE OF FEET

PROFILE ON ℄ OF SPILLWAY

PROFILE ON ℄ OF OUTLET TUNNEL

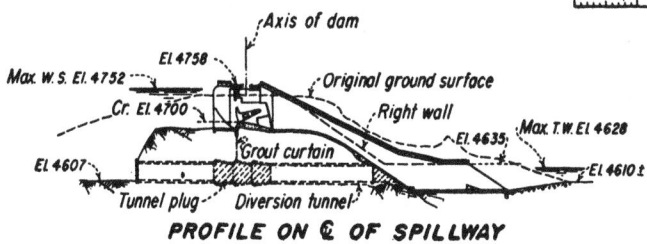

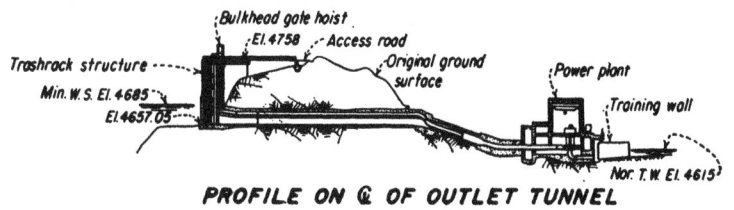

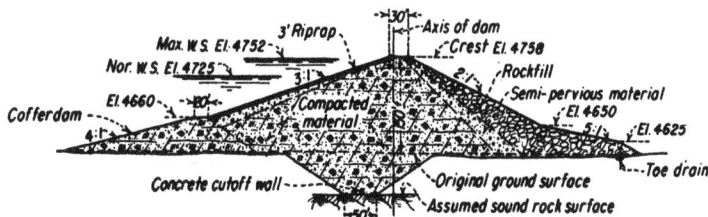

SECTION IN RIVER CHANNEL

Boysen Dam, power plant, and reservoir from high on the right abutment. The dam is storing 638,000 acre-feet of water for downstream irrigation, flood control, and power production.

dation. A cut-off trench was excavated to bedrock, which was reached at a depth of about 80 feet below the river bed. Five stages of well points on both sides of the cut-off trench stabilized the steep excavation slopes and maintained the water table well below the excavation and backfill levels as those operations proceeded. Curtain and blanket grouting below the elevation of the cut-off trench and a concrete cut-off wall keyed into the rock prevent seepage and loss of water through the foundation.

Beginning with the backfilling of the cut-off trench, selected earth and rock materials were hauled from borrow pits in the reservoir area. The materials for the impervious zone 1, or water barrier, comprising the major portion of the dam, were compacted to 6-inch layers by tamping rollers weighing at least 20,000 pounds (10 tons) per drum. The materials for the zone 2 transitions from the impervious portion of the dam to the stabilizing downstream rock-fill portion, and from the impervious zone to the upstream riprap blanket, were compacted in 1-foot layers by the treading action of tractors weighing a minimum of 40,000 pounds (20 tons). The zone 2 transition between the impervious zone and the riprap protective blanket prevents the reservoir wave action from washing the fine materials of the main portion of the dam out through the coarser

riprap. The actual performance of the embankment will be deduced from readings of 36 piezometer tips and two cross-arm settlement installations embedded in the embankment, which will give information on seepage and settlement, respectively.

With transfer of traffic to the relocated railroad in October 1950, work was initiated on the outlet works and power plant. A portion of the short tunnel on the old railroad immediately adjacent to the spillway was utilized to contain the 15-foot diameter steel power penstock and the 66-inch diameter steel outlet pipe. The penstock and pipe are embedded in concrete through the old tunnel, and the portions of the pipe and penstock between the tunnel and the powerhouse are anchored with massive concrete blocks and covered with earth materials.

The power penstock serves two 7,500-kilowatt generators, and also serves a 57-inch diameter outlet pipe which bypasses the generators. An interesting feature of the power penstock is provision for future installation of a surge tank if operation indicates that the protection provided by such a tank is needed.

Both the 66-inch diameter outlet pipe and the 57-inch diameter generator bypass discharge through 48-inch diameter hollow-jet valves, protected by emergency ring follower gates. The combined discharge capacity of the two outlet valves when the reservoir is at elevation 4725 is approximately 1,325 cubic feet per second.

Finished, Boysen Dam and power plant required the excavation of 1,070,000 cubic yards of material to make way for the structures, and the placement of more than 1,300,000 cubic yards of earth fill, nearly 235,000 cubic yards of rock fill and riprap, and about 60,000 cubic yards of concrete. More than 2,800,000 pounds of re-inforcement steel, and approximately the same quantity of other metalwork, went into the structures. Total cost of the dam and power plant was $7,200,000.

Boysen Dam's height of 230 feet places it tenth in rank of the Bureau of Reclamation's earth- or rock-fill dams. The capacity of Boysen Reservoir, 1,493,000 acre-feet, places it tenth in rank in the list of Bureau reservoirs.

MEDICINE CREEK DAM

CAMBRIDGE UNIT, MISSOURI RIVER BASIN PROJECT, NEBRASKA

By George W. Mattson

Medicine Creek Dam is a part of the multiple-purpose Missouri River basin project which was conceived for the purpose of regulating the flow of the Missouri River and its tributaries. The Missouri River basin project is comprehensive in scope, involving developments in an area comprising approximately 500,000 square miles. The floods caused by spring run-off and summer cloudbursts affect not only the Missouri River basin area, but inflict damage along the Mississippi River to its mouth in Louisiana.

The construction of dams in the Missouri River basin plains area presented problems different from those encountered in building dams in the mountainous regions where canyon walls provide substantial rock abutments and the foundation rock is easily accessible. The weakly cemented sedimentary formations necessitated extensive study and skillful application of the latest developments in soil mechanics. Deposits of loess, a fine-grained, wind-deposited material occurring to considerable depth in the abutments of some dam sites, required special studies and tests to determine the necessary treatment. Undisturbed samples of loess indicated a porosity of 50 percent and the probability of considerable settlement upon being saturated by the impounded waters of the reservoirs. Layers of permeable material in the foundations had to be excavated in some instances to depths of about 70 feet to contact bedrock and provide adequate cut-offs. These problems at Medicine Creek Dam were solved successfully by using large-capacity earth-moving equipment and by developing improved methods of construction controls which permitted the accelerated program of handling and compacting large quantities of embankment materials.

Medicine Creek has its origin in the Nebraska Sand Hills, about 100 miles from its confluence with the Republican River at Cambridge, Nebr. It has a drainage area of about 680 square miles. Although the drainage area is not exceedingly large, rainfalls of considerable

intensity, produced by unusual storm conditions, have in the past wrought havoc along the creek and river valleys below the dam. The most disastrous flood occurred in 1947 when 13 lives were lost at Cambridge, Nebr., and a peak flow of 120,000 cubic feet per second was recorded. In 1948, a less severe flood was reported with a peak flow of 40,000 cubic feet per second. It was the flood of 1947 that accelerated the design and construction of Medicine Creek Dam.

RESERVOIR

Medicine Creek Dam is located on Medicine Creek about 9 miles northwest of Cambridge, Nebr. The reservoir surface area at normal water surface, which is the top of the irrigation pool, is about 1,900 acres and at that elevation the reservoir extends upstream a distance of about 8½ miles from the dam.

The reservoir provides 102,700 acre-feet of temporary storage above the crest of the portion of the spillway

From the air, Medicine Creek appears as a striking example of the manner in which an earth dam is made to fit the ground surface at the site.

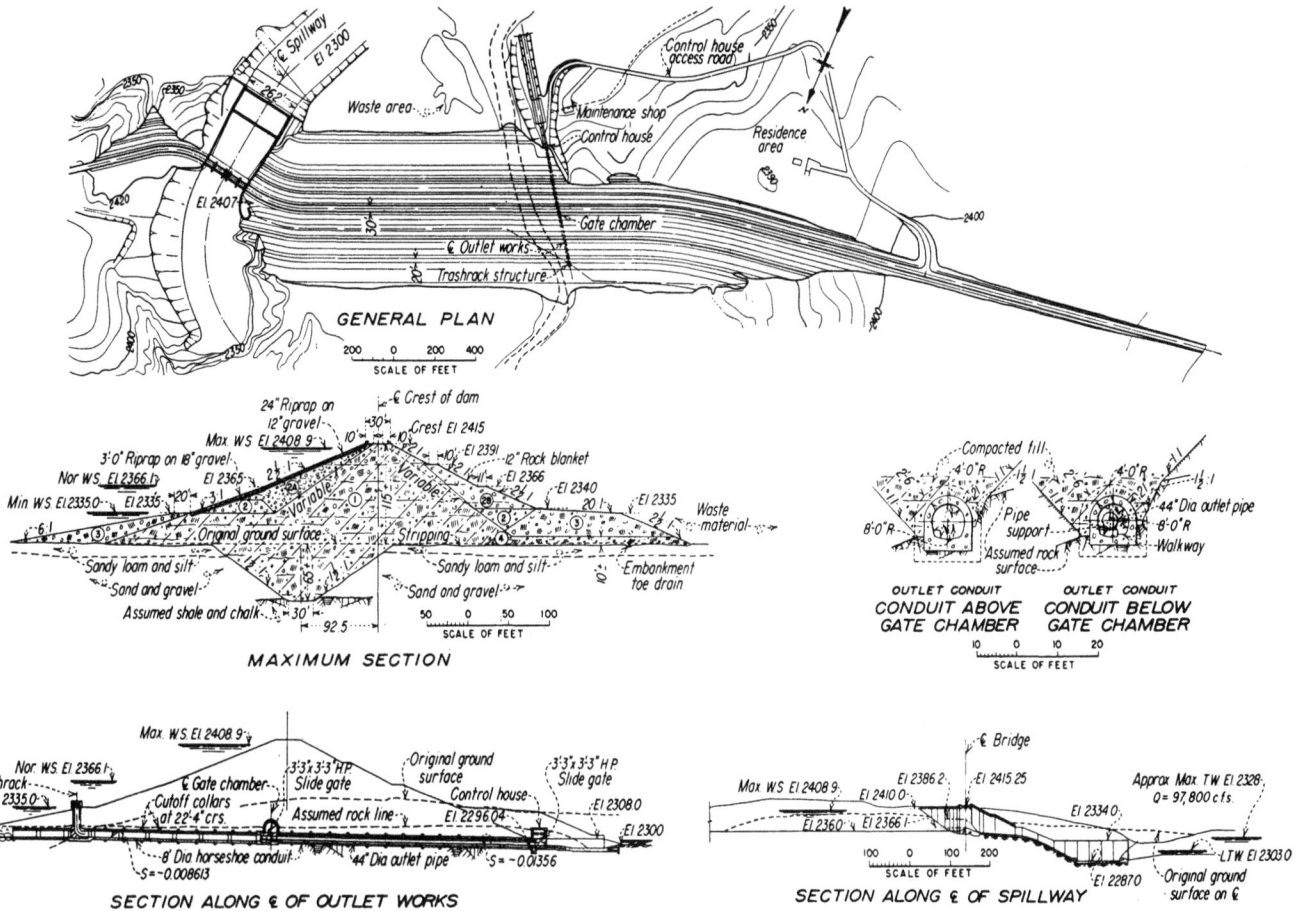

GENERAL PLAN

MAXIMUM SECTION

OUTLET CONDUIT
CONDUIT ABOVE
GATE CHAMBER

OUTLET CONDUIT
CONDUIT BELOW
GATE CHAMBER

SECTION ALONG ℄ OF OUTLET WORKS

SECTION ALONG ℄ OF SPILLWAY

ment. The desired consolidation was accomplished by using a system of ponding within dikes on level areas and by spraying the steeper slopes of terraces with rotating sprinklers.

In order to increase the narrow width and to protect the easily eroded loess slopes of the right abutment, the upstream slope is covered with an impervious blanket, a minimum of 10 feet in thickness. This impervious material extends by means of the right abutment cut-off trench to bedrock so as to seal off the previous stratum overlying the Niobrara formation. The blanket is covered with riprap to protect it against aggressive wave action when the winds blow from the north or northwest.

The spillway is located in the left abutment on the Ogallala and Niobrara formations. These formations are strong enough to support the structure but, being susceptible to rapid erosion, are protected by a concrete lining throughout the length of the spillway. The spillway is a concrete-lined chute type consisting of the crest structure, chute, and stilling basin.

Considerable model study was required to develop a hydraulically satisfactory right wing approach wall and involved progressive refinements in seven additional designs after the initial one.

The uncontrolled spillway has two crest elevations. The lower or moderate flood control portion of the crest is 13 feet wide, at elevation 2366.1, and is situated

The loess of Nebraska required consolidation before it could bear the weight of the dam without subsidence. At Medicine Creek, the foundation was consolidated by flooding flat ground with pools and by spraying the slopes which had been scraped into terraces.

Medicine Creek Dam's spillway has a combination of semi-controlled and uncontrolled crest. As shown, the narrow center portion is lower than the broad sections on either side. Until the reservoir level reaches the broad portion of the spillway crest, flow will be at a rate no greater than the normal capacity of the river channel. Should the level rise higher, the discharge will be at a greater rate until the reservoir surface falls to the crest level again, when the small channel will carry the discharge.

in the center of the spillway between the two central piers; the remainder of the spillway crest is 200 feet wide, at elevation 2386.2, and passes the less frequent but greater flood waters. The discharge of the moderate flood control crest is 3,750 cubic feet per second, with the reservoir water surface at elevation 2386.2. A flow of 4,000 cubic feet per second was assumed as the creek channel capacity. Thus, the floods which discharge only over the lower 13-foot crest will not exceed the creek channel capacity. Even the disastrous flood of June 21–22, 1947, would pass over the lower crest without encroaching on the higher crest and would be confined to a flow within the capacity of the creek channel.

The spillway was designed to accommodate a maximum flow of 97,800 cubic feet per second with the maximum reservoir water surface elevation 2408.9. Theoretically, this elevation would be reached by a flood which has a maximum peak of 200,000 cubic feet per second and a 3-day volume of 300,000 acre-feet.

The spillway crest is of mass concrete. The chute downstream from the crest structure has reinforced concrete cantilever and counterforted walls and the stilling basin has reinforced concrete counterforted walls. Riprap is placed immediately downstream from the stilling basin to prevent erosion of the channel at that point.

OUTLET WORKS

The outlet works rest on firm shale at the base of the dam a short distance to the right of the natural channel of Medicine Creek, and about 1,600 feet to the right of the spillway. The outlet works consist of a trashrack intake, a 6½-foot diameter drop inlet, an 8-foot diameter horseshoe-shaped conduit, a gate chamber, a 44-inch diameter outlet pipe installed in the conduit downstream from the gate chamber, a control house, and stilling basin.

An opening was left in the base of the drop inlet from which a temporary conduit extended upstream for stream diversion during construction. The temporary conduit was closed by a concrete plug after diversion was no longer necessary.

Discharges from the outlet works are controlled by a high-pressure slide gate, 3 feet 3 inches square, in the control house. An emergency high-pressure slide gate of the same size is located in the gate chamber, and can be used to shut off flow through the 44-inch diameter pipe during inspection or for repair of the pipe or the control gate at its downstream end.

The stilling basin, which dissipates the energy of the discharged water before it enters the downstream channel, is designed to handle flows in excess of the flow of 382 cubic feet per second which occurs when the reservoir water surface is at normal elevation 2366.1. Since the bottom of the channel is 6.4 feet higher than the basin floor, and the channel is 12 feet wider than the basin, a transition section conveys the flow from the stilling basin to the outlet channel. Six different stilling basin models were tested to determine the most desirable arrangement.

CONSTRUCTION

Invitations for bids for construction of the dam were issued November 3, 1947. Award of the contract was made on January 15, 1948, to the C. F. Lytle Company of Sioux City, Iowa, and the Amis Construction Company of Oklahoma. Actual work was started on March 23, 1948, on the access road, and stripping of the dam foundation was commenced on April 8, 1948. The dam was completed and accepted on December 9, 1949.

Diversion of the creek was accomplished by shifting the channel to accommodate construction progress and passing the stream through a gap in the dam. The cut-off trench on the east side of the valley was excavated and backfilled while the creek was diverted through an old river channel in the right abutment. Upon completion of this part of the trench to an elevation above assumed flood waters, the creek was rerouted along the east side of the valley, and the west half of the cut-off trench completed. When the dam embankment had been constructed to a predetermined eleva-

tion, and the construction of the spillway had advanced to such a stage that all flows could be passed without damage to any part of the structure, operations were initiated to effect closure of the diversion gap in the dam. A cofferdam was quickly built by moving stockpiled earth into the channel above the dam. The impounding of water had begun. The river channel in the gap was cleaned and embankment placing commenced. During this period of impounding sufficient water to raise the reservoir water surface to the outlet works sill, it was necessary to pump 3.1 cubic feet per second into the creek below the dam to satisfy the legal requirements for stock water and for the maintenance of aquatic life. Personnel of the State Fish Hatchery were on duty to preserve the fish stranded by the reduced flow of water below the dam.

In anticipation of future consolidation of the dam foundation and embankment, a camber with a maximum of 1.5 feet was provided over the crest length. The crest was surfaced with gravel to provide a roadway and concrete posts with guardrails were installed along the crest for protection of traffic over the dam.

Cement (low alkali, type II) and concrete aggregates were unloaded from railroad cars on a siding near Cambridge, Nebr., and hauled 8 miles to the dam site.

QUANTITIES AND COST

The principal quantities involved were as follows:

Earth embankment:
    Zone 1—Impervious material_____cubic yards__ 1, 982, 298
    Zones 2A and 2B—Semipervious
        material _____do____ 438, 905
    Zone 3—Mixture of fine and coarse
        material, not suitable for 2A
        and 2B_____do____ 186, 768
    Zone 4—Sluiced sand and gravel_____do____ 68, 412
                                                         _____
        Total_____do____ 2, 676, 383
Riprap: Upstream slope of embankment____do____ 62, 900
Gravel blanket:
    Upstream slope of embankment_____do____ 30, 636
    Downstream slope of embankment_____do____ 19, 542
                                                         _____
        Total_____do____ 50, 178
Water for irrigation of right abutment
    of dam_____gallons__ 33, 000, 000
Concrete:
    Spillway_____cubic yards__ 26, 991
    Outlet works_____do____ 2, 578
                                                         _____
        Total_____do____ 29, 569
Reinforcement steel_____pounds__ 2, 033, 142

The total cost of the dam and access road was approximately $7,000,000.

ARCHEOLOGICAL INVESTIGATIONS

Archeological investigations during the stripping and excavation operations were conducted by three organizations; namely, the Smithsonian Institution, the University of Nebraska, and the Nebraska Historical Society. A large part of an entire village was uncovered in the right abutment and more than 11,296 specimens were obtained. These specimens included utensils made of pottery; tools of bone, stone, and shell, and remains of foodstuffs, such as animal bones and mussel shells. Additional village sites were uncovered in the vicinity of the dam, in the borrow area, and along Medicine Creek within one-fourth mile of the dam axis.

Investigators found vegetable materials, such as corn, sunflower, and squash seeds. These discoveries were valuable in establishing the approximate dates of Indian occupation hundreds of years ago at the site.

Studies were made of trees in the area to obtain a tree-ring sequence that would help establish the exact age of the archeological discoveries. The collected specimens, photographs, and information obtained by the three groups provided additional data on the prehistoric inhabitants of western Nebraska.

SUPERIOR-COURTLAND DIVERSION DAM

BOSTWICK DIVISION, MISSOURI RIVER BASIN PROJECT, NEBRASKA-KANSAS

By Maurice E. Day

Severe droughts, combined with the greatest flood on record, severely crippled the agricultural economy of the lower Republican River Basin and surrounding territory during the decade from 1930 to 1940. Although widespread public interest in the storage and diversion of water from the Republican River had prompted studies by various agencies over a long period, these years of hardship intensified interest in the area and made evident the necessity of stabilizing agriculture in the region. The Bostwick division was developed, in view of this need, to provide irrigation water for about 96,000 acres of land in Nebraska and Kansas, most of which was formerly devoted to dry-farming. The division is a part of the flood control and irrigation development of the Missouri River basin.

The Superior-Courtland Diversion Dam is located on the Republican River in southern Nebraska about 3 miles west of Guide Rock. The Courtland Canal on the south side of the river is designed to carry 751 cubic feet per second and the Superior Canal, on the north side of the river, 139 cubic feet per second. These canals serve approximately 53,600 acres of land in Nebraska and Kansas.

DAM SITE

The Republican River at the dam site winds through a broad, shallow valley cut into the surrounding rolling tableland. With the exception of the two canal headworks, the dam is founded on shale which was overlaid by coarse to fine sand in the river channel and coarse to silty sand on the banks. The depth of overburden varied from 10 feet at the lowest point in the river channel to approximately 20 feet on the flood plains. The most economically suitable design for a dam at the site, promising the maximum safety and reliability, proved to be a concrete overflow weir combined with

earth dikes at each end which confine the anticipated flood discharge to the overflow weir and sluiceways. The uncontrolled overflow crest combined with the two 20-foot sluiceways was chosen to give the minimum obstruction to the great quantities of floating debris carried by Republican River floods.

The Harlan County Reservoir upstream from the site provides storage for flood control and for irrigation waters to be diverted by the dam. The intervening 1,400 square miles of drainage area contribute the design flood of 42,000 cubic feet per second which has been estimated to have a frequency of about once in 50 years. While the storage capacity of the diversion dam reservoir is negligible, the backwater caused by a flood of this magnitude required protection of the Chicago, Burlington and Quincy Railroad on the north bank of the river and relocation of a small amount of county road on the south bank. Willow Creek, a small tributary of the Republican River, formerly emptied into the river upstream from the site. Economic studies led to the permanent diversion of Willow Creek around the north end of the dam. Many of the local roads and bridges near the dam site are inundated during normal flood periods, thus separating the two headworks by many miles of road. To overcome this situation, a permanent cableway connecting the north and south ends of the dam gives the operator, whose quarters are located north of the dam, emergency access to the Courtland Canal headworks on the south bank of the river. The cableway is also used for silt sampling and other measurements in the river channel at the dam.

THE DAM

The structural height of the dam is about 41 feet measured vertically between the base of the stilling pool

The overflow weir of the Superior-Courtland Diversion Dam stretches between the headworks of the Superior Canal in the foreground and the headworks of the Courtland Canal across the river.   A cableway, visible against the sky at the upper left, crosses the river to provide access to both headworks during periods of high water.

and the top of the nonoverflow section. The overflow weir is 420 feet long and has a vertical height from the foundation to the crest of about 20 feet. The weir has a vertical upstream face and a rounded crest. Spillway water energy is dissipated at the downstream toe by a slotted bucket developed by model tests. A short upstream apron is required for stability, and upstream and downstream cut-off walls are provided to prevent undermining and to increase the length of the water percolation path beneath the dam. A continuous gravel pocket is provided at the upstream face of the downstream cut-off wall to collect the seepage water. The overflow weir is divided by vertical contraction joints into 12 monoliths, each 35 feet in length.

In the Courtland Canal headworks at the south end of

the weir, five 10- by 5-foot top-seal radial gates control discharge into the canal. The top-seal gates were used because of the extreme fluctuation of the upstream water level. The south upstream wing wall provides a cut-off to firm shale in front of the headworks to prevent undermining and to reduce the percolation of water under the structure and around the end of the dam. The wing wall extends about 44 feet upstream from the headworks as a protection against erosion and connects with the south dike. The south dike is slightly over 3,400 feet long and prevents the flow of water over the south flood plain. The upstream slope of the dike is relatively flat except near the dam where the slope is steepened and protected by riprap. Where feasible, the flat slope was used in preference to the steeper riprapped slope because of the high cost of riprap. All dike slopes not riprapped are seeded to native grasses.

The south sluiceway is located between the south end of the overflow weir and the Courtland Canal headworks. The flow through the sluiceway is controlled by a 20- by 12-foot radial gate. A steel sheet-pile training wall whose top is level with the crest of the overflow weir extends upstream from the left wall of the sluiceway. The location of the training wall was determined by model tests so that the normal flow is forced to pass in front of the headworks and then into the sluiceway. With both the sluiceway and the headworks operating, the effect of this type of flow, in this particular case, is to reduce the amount of sediment taken into the canal and thus to reduce the annual maintenance. The sill of the sluiceway and the channel bottom immediately up-

stream is at elevation 1632. The floor of the headworks structure is 3.55 feet higher, which also reduces the sediment intake. The excess energy of flow through the sluiceway is dissipated by the south stilling pool, which consists of a flat slab at the proper elevation to promote the formation of a hydraulic jump, and side walls to confine the flow. The south downstream wing wall connects the stilling pool with the river bank. The river bed and bank downstream from the stilling pool are protected by heavy riprap.

The Superior Canal headworks at the north end of the weir has one 10- by 5-foot top-seal radial gate. The headworks, sluiceway, stilling pool, wing walls, and training wall at the north end of the weir are similar in arrangement to those at the south end. All gates are controlled by electrically operated hoists, but the head-

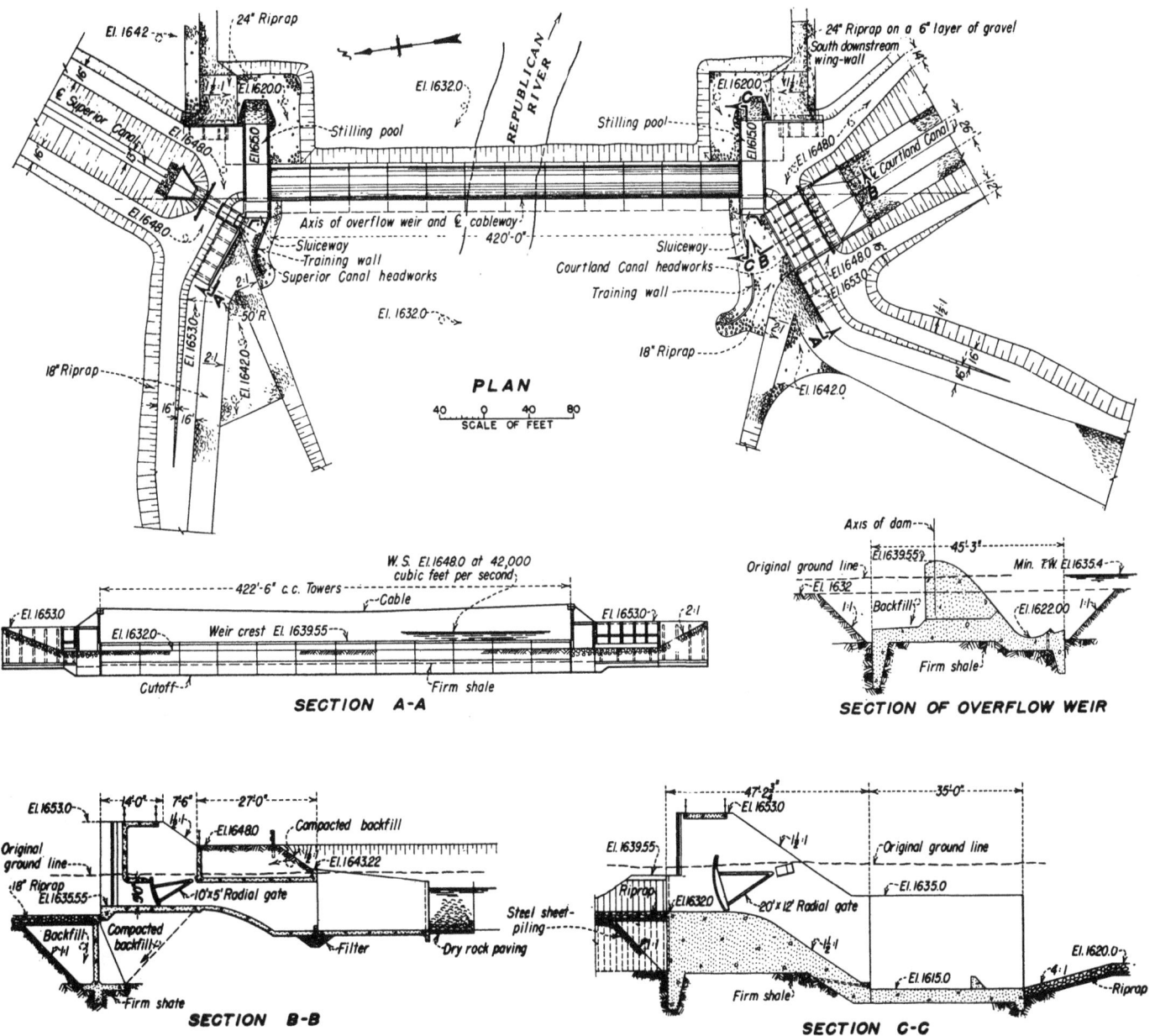

PLAN

SCALE OF FEET

SECTION A-A

SECTION OF OVERFLOW WEIR

SECTION B-B

SECTION C-C

The overflow weir of Superior-Courtland Diversion Dam crosses the Republican River between the headworks of the Courtland Canal, in the upper part of the picture, and the headworks of the Superior Canal, which crosses under the Willow Creek channel change by a siphon at the bottom.

works gates can also be operated manually. The north dike, which protects both the railroad and the canal, is approximately 4,200 feet long and 2,500 feet of its length is protected by riprap.

The Superior Canal is carried under the Willow Creek channel change by a 66-inch precast concrete pipe siphon. The caretaker's house is located near this siphon. The access road to the Superior Canal headworks from Guide Rock follows the left bank of the canal and crosses over the Willow Creek channel change by means of a timber bridge.

CONSTRUCTION

The construction contract was awarded to Knisely-Moore Company of Douglas, Wyo., on March 2, 1949. Preparatory work began in April and actual construction was started July 15. The river was diverted over the south flood plain. Excavation for some of the structures was facilitated by the use of well points to control the ground water level. A large part of the shale excavation for the overflow weir was accomplished by loosening the top layers with a rooter and blading off the loosened shale to the foundation elevation. The deeper shale excavations were made with the aid of light blasting. A shale saw was used in excavating the cut-off walls to obtain minimum over-excavation. Two saw cuts were made and, after light

blasting, the material was removed with a back hoe shovel. The surface of the shale excavation had to be coated soon after exposure to prevent the shale from drying out and weathering prior to the placement of the concrete.

The first concrete was placed August 27, 1949. Concrete operations continued through the winter, using heated water and aggregates and a mixer heater. The concrete was maintained at the proper temperature for curing by canvas covers and oil-burning salamanders. Concrete sand was obtained from Cowles, Nebr., and aggregate from Lincoln, Kans. Compaction of dikes and backfill was obtained by using sheepsfoot rollers and mechanical tampers. The riprap was hauled by truck from quarries near Franklin, Nebr., and placed by dragline.

On July 8, 1950, a storm of high local intensity produced a flood of approximately 42,000 cubic feet per second at the dam. This once-in-a-50-year-period flood, occurring almost at the completion of construction, tested the dam rather severely. At the time, some riprap remained to be placed and construction of the outlet transition of the Courtland Canal headworks was nearing completion. Some scour occurred on the river banks below the dam and in the new Willow Creek channel and the Courtland Canal headworks transition was inundated, but the dam stood the test very well. The contractor was granted an extension of time because of the flood.

The dam was dedicated on August 17, 1950, and was accepted by the Government on September 15, 1950. The total cost of the dam was approximately $1,550,000.

Construction of the monolithic overflow weir of Superior-Courtland Dam proceeded in 35-foot-long blocks built alternately. The heavy black line outlining the weir block and also appearing on the face of the training wall in the background is a rubber waterstop that prevents seepage through construction joints.

DUNLAP DIVERSION DAM

MIRAGE FLATS PROJECT, NEBRASKA

By Joseph A. Hufferd

The Dunlap Diversion Dam is located on the Niobrara River, about 8 miles downstream from Box Butte Dam, and about 25 miles south of Chadron, Nebr. Its function is to divert 220 cubic feet per second of water from the Niobrara River for the irrigation of the Mirage Flats project. A constant supply of water for diversion is maintained by the controlled outflow from the storage reservoir formed by Box Butte Dam.

PROJECT HISTORY

The Mirage Flats project comprises 12,000 acres of irrigable land in Sheridan County, Nebr., about 8 miles south of Hay Springs. The project, authorized under the Water Conservation and Utilization Program, was approved by the President April 26, 1940. This approval, together with the subsequent construction, concluded a long struggle for survival waged by residents of the area against ravages of drought.

The Sioux Indians were in possession of the area until 1877. Then, when the gold rush to the Black Hills began, the Government placed the Indians on the Pine Ridge Reservation in South Dakota. In 1878, with the Indian menace removed, cattlemen began to establish large ranches on the open range. However, settlers seeking homesteads in 1883 began to confine the operations of the cattlemen to smaller areas.

When good crops were obtained by the first homesteaders, migration to the area was greatly stimulated. By 1885 there was a homesteader on nearly every quarter section. This period of good crops and rapid settlement was followed by a severe drought, culminating in the extremely dry years of 1893 and 1894 when a total failure of all crops was experienced, resulting in general abandonment of homesteads and an exodus of the population.

The remaining settlers, recognizing their need for supplemental irrigation water, organized a mutual company in 1895 and constructed a system to irrigate approximately the area of the present Mirage Flats project, but abandoned it when fire destroyed two large flumes. The loss of interest that resulted was due to the fact that the natural fertility of the soil was sufficient to produce good crops during favorable years, but not good enough to allow the accumulation of a backlog fund with which to repair and operate the irrigation system. Under these unprofitable conditions, the settlers sought "greener pastures" in other places, leaving much of the land vacant.

Until 1920, the Mirage Flats area was almost entirely under one ownership, the Peters-Williams Ranch. In 1920 Peters and Williams began a campaign to bring settlers into the area by dividing their ranch into small farm units. This promotion attracted a Danish colony from central Nebraska. Consequently, the present population is predominantly Danish.

When the "dry 30's" struck, it again became evident that irrigation was essential if the existing farm economy was to survive. Several efforts were made to irrigate the area, all of which failed. Then, in 1939, a project investigation was made by the Bureau of Reclamation, resulting in the authorization and approval of the Mirage Flats project.

DESIGN

Subsequent to preliminary investigations, two sites for the diversion dam were considered. The first was downstream from Cottonwood Creek, requiring that the dam be designed to pass the flood discharge of Cottonwood Creek in addition to the outlet discharge from Box Butte Reservoir. The second site was upstream from Cottonwood Creek, requiring a siphon to carry the diverted water under the creek. In a comparative study, the latter plan proved more economical and was adopted.

Dunlap Diversion Dam is at the head of the Mirage Flats Canal. The headgate structure for the canal is at the right, the overflow weir at the left, and between the two, partially hidden, is the sluiceway.

Foundation conditions in the vicinity of the site selected for the diversion dam were practically uniform over an extensive area. The foundation comprised an outwash of silty sand and fine gravel, 20 to 50 feet thick, as overburden on the Arikaree Formation, which is a soft sandstone. Test holes were made at two points along the proposed axis of the dam to determine whether the concrete spillway structure could be placed on the bed formation, but the test hole logs showed that the overburden was too deep; consequently, a spillway design was adopted which was suitable for the sand and gravel material encountered at shallow depth.

The spillway is a semireinforced concrete weir of the ogee type, constructed in two monoliths. The right (looking downstream) monolith is integral with the right abutment. The left monolith abuts the sluiceway, but is separated from it by a painted construction joint. The spillway was designed to pass 2,000 cubic feet per second with the upstream water surface about 4 feet above the weir crest.

At the downstream toe of the ogee spillway, a 24-inch-thick concrete slab, with dissipator blocks, provides a basin for dissipating the energy of the spillway water by the formation of a hydraulic jump. Ultimate protection from scour is obtained by the concrete sheet piling cut-off at the downstream end of the jump apron; however, primary protection for the river bed downstream from the apron is provided by a 3-foot layer of riprap composed of durable rocks, varying in size

from 1 to 10 cubic feet. The "leaching" of the fine materials from the foundation is prevented by a layer of gravel beneath the riprap.

A reverse filter, under the downstream apron at the toe of the weir, provided with ejector pipes, serves as an outlet for seepage water, thereby partially relieving the hydrostatic uplift pressure under the jump apron. The gravel in the filter was selected and arranged in layers in a manner that would prevent the "leaching" of the fine fraction of the foundation material and the consequent formation of a "pipe" by backward erosion.

An upstream apron with a concrete sheet piling cut-off provides the additional length of percolation path required under the spillway to prevent high exit pressure at the reverse filter and excessive uplift pressure beneath the downstream apron. The 2,160 linear feet of 6- by 24-inch concrete sheet piling used for the upstream and downstream cut-offs was necessary because of the shortage of steel and timber piling during World War II. Also, the aggregate required for the concrete sheet piling was readily available near the dam site.

The sluiceway is a reinforced concrete structure equipped with a 10- by 10-foot radial gate, which is regulated by a hand-operated hoist installed on the operating bridge over the sluiceway. All of the flow in the river in excess of that required for irrigation, but not exceeding 500 cubic feet per second, is passed through the sluiceway so that sediment accumulated in the vicinity of the headworks will be removed by the

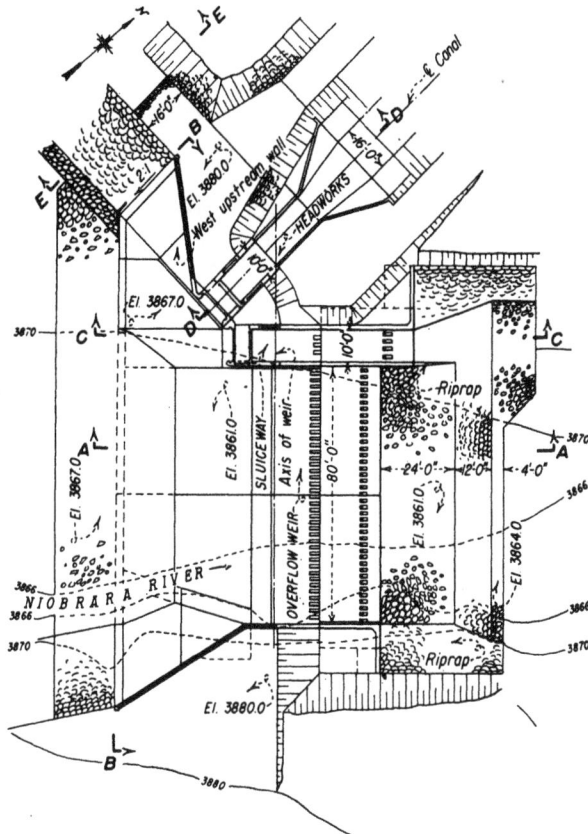

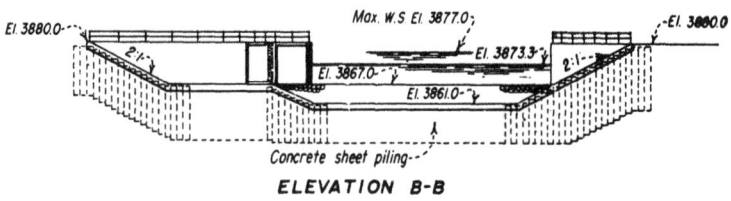

ELEVATION B-B

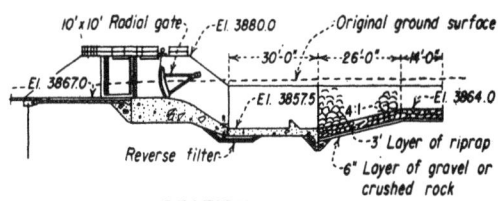

SECTION C-C

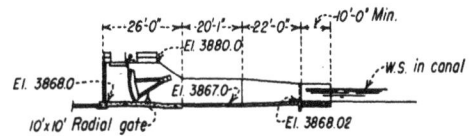

SECTION D-D

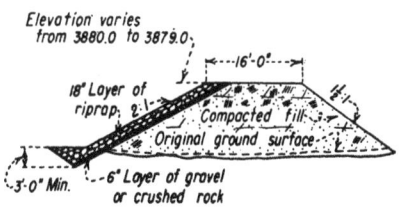

SECTION E-E

PLAN

SCALE OF FEET

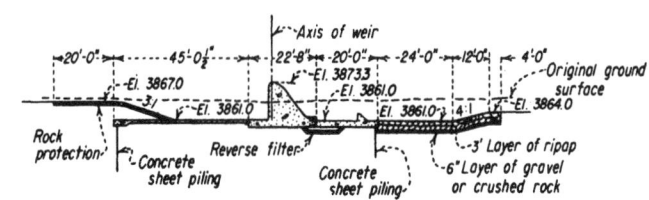

SECTION A-A

sluicing action. The sluice gate sill, at the elevation of the river bed, was placed 1 foot below the headworks gate sill to allow some accumulation of silt between periods of sluicing.

The high concentration of discharge through the sluiceway during the sluicing periods produces a higher hydraulic jump than the one formed below the spillway, thereby requiring a deeper stilling pool downstream from the sluiceway gate.

The headworks was proportioned to produce velocities of about 5 feet per second with the 10- by 10-foot radial gate entirely open and the water surface in the diversion reservoir at the normal operating level. During periods of high water in the diversion reservoir, the gate will be partially closed, producing high-velocity flow under the gate and a hydraulic jump down-

stream from the gate. The slab below the headgate was depressed to form a shallow stilling basin to confine the jump when such conditions are encountered. The downstream transition is designed partly as a retaining wall and partly as canal lining (paving). A similar design was employed for the upstream wing walls adjacent to the left side of the headworks and the right abutment of the spillway. Some economy was realized in the use of the thinner concrete pavement sections because of the ease of placing pavement as compared with the forming of retaining walls.

The closure dike was constructed of earth materials available in the immediate vicinity which were suitable for compacting to form a dense impervious embankment. The freeboard of the dike varies from 3 feet near the headworks to 2 feet at the closure with the left

The overflow section of Dunlap Diversion Dam, with the sluiceway at right center, and the embankment portion of the dam stretching away to the right.

river bank, thereby creating a low point where the water can overflow in the event a flood occurs which is appreciably greater than the design flood. This precaution will localize any failure, due to such cause, at a point sufficiently remote that no damage will be done to the concrete structures and repairs can be readily made when the water surface in the river has returned to normal.

CONSTRUCTION

Construction was commenced in November 1944 by the contractor, Malcolm G. Long. Diversion and unwatering operations included construction of a bypass channel for the river on the left bank and installation of a well-point system completely around the dam site at about the elevation of the natural river bank. One hundred and sixty well points were installed at 5-foot intervals with bottom elevation 12 feet below the base of the spillway weir.

Excavation for the headworks was started when the site was unwatered, followed by excavation for the sluiceway, spillway, and upstream and downstream aprons. Rough excavation for all structures was made with a ¾-cubic-yard, Diesel-powered dragline which loaded the materials into 2-cubic-yard dump trucks for disposal, and a tractor equipped with a 2-cubic-yard rotary-type scraper. All excavated materials were stock-piled in convenient locations to be used later in backfilling and embankment or dumped in lower areas and used for landscaping purposes.

Concrete aggregates were shipped to Hemingford from the Strasbough pit at Scottsbluff, Nebr., by rail and hauled to the diversion dam by truck, where they were stock-piled on level ground adjacent to the mixing plant on the right bank of the river near the abutment of the dam. The south (right) bank of the river is about 10 feet higher than the north bank at the dam,

which afforded a convenient location for the mixing plant as the concrete could be discharged from the mixer and conveyed to the forms without the use of a hoist. All concrete was mixed in a 10-cubic-foot, horizontal type mixer and transported in two-wheel buggies. For the concrete in the floor of the headworks, sluiceway, cut-off wall, and footing for the northwest wing wall of the sluiceway, two buggies in a truck were loaded directly from the mixer and hauled to the area of the placement where they were pushed along runways to the point of discharge. All other concrete was transported in buggies directly from the mixer. Because of war shortages, plywood for the lining of forms was not readily available; however, enough plywood was obtained for the forms for the energy dissipator blocks downstream from the overflow weir and sluiceway. For all other forms, shiplap only was used with a consequent requirement for special concrete finishing methods. A steam heating plant was used to heat the aggregate and for the warming of concrete under canvas and frame enclosures.

Concrete piling was manufactured by the contractor at Scottsbluff, Nebr. One face, both ends, and the tongue and groove were cast in wood forms, the other face being left open for concrete placement. Immediately after placement of concrete, the piling was subjected to steam curing, using saturated steam at temperatures of 100° to 130° F. for a period of 72 hours. The piling was then hauled to the diversion dam by truck.

The reverse filter was constructed during the months of January and February 1945, as required by the concrete placement schedule. The sand and gravel were secured in the proper size gradations from the Strasbough pit. Excavation for the filter and placement of the material were performed by hand methods.

Riprap placing operations were started in May by placement of 36-inch riprap below the downstream apron, followed by 12- and 18-inch upstream riprap. The gravel for the 6-inch blanket beneath the riprap was screened from a natural deposit of sand and gravel located about 2½ miles northwest of the dam and transported to the site by dump trucks. Rock suitable for riprap was found about 7 miles from the dam in a ledge of rock varying in thickness from 12 to 18 inches under about 2 feet of overburden on top of a small knoll. Most of the rock was loosened by blasting, then loaded by hand-pushing over a ramp into trucks for transportation to the site. During the final stages of excavation, a ¾-cubic-yard dragline was utilized in stripping and loading operations.

Materials for backfill were moved into place by a bulldozer and, because of their moist condition, considerable compaction was secured by routing construction traffic over the fill where feasible. Materials for compacted backfill, previously selected and stock-piled from excavation for structures, were deposited in layers 6 inches in thickness, moistened, and then compacted by hand with the aid of power hammers.

Materials for compacted fills (embankments) were selected and stock-piled during excavation for structures or selected from borrow pits. The materials were either moved directly into place by a bulldozer or excavated and transported by means of a tractor and a 12-cubic-yard scraper, then leveled with a bulldozer. The materials were then watered and each lift was rolled with a tractor and compaction roller with a minimum of six passes before the succeeding lift was started.

SUMMARY

The Dunlap Diversion Dam is typical of the many small diversion dams constructed by the Bureau of Reclamation to bring life-giving water to fertile soils in semiarid regions. Following the delivery of the first irrigation water to the project in 1946, many veterans of World War II found new homes in the area. Long-term loans have eased the financial burden, while the increased productivity of the soil provides the potential income that will enable each settler to pay his portion of the project cost as he builds up his homestead.

Though the mirage of high, flat-topped buttes, which the early settlers of the area saw above the horizon and from which the project takes its name, is still there, it is obscured by the new vision of prosperity shared by the modern pioneers who tend the soil.

IMPERIAL DAM AND DESILTING WORKS

ALL-AMERICAN CANAL SYSTEM, BOULDER CANYON PROJECT, ARIZONA-CALIFORNIA-NEVADA

By Arthur W. Kidder

Imperial Dam and desilting works, a feature of the Boulder Canyon project, is constructed on the Colorado River 17 miles above Yuma, Ariz. Construction was completed in 1938 at a cost of $9,956,800. The purpose of the feature is to provide diversion and desilting facilities for 15,155 cubic feet per second of water required in the All-American Canal on the California side of the river and 2,000 cubic feet per second in the Gila Gravity Main Canal on the Arizona side.

From the All-American Canal, water is bypassed into the Yuma Canal which previously had been supplied by water diverted at Laguna Dam, about 5 miles below the Imperial site. Laguna Dam has, therefore, become practically obsolete except as a tailwater control for the new dam. Aside from supplying the Yuma project via the Yuma Canal, the All-American Canal provides irrigation water for the Imperial Valley which formerly received water from the Imperial Canal that runs for a considerable portion of its length through Mexican territory. Provision is also made for supplying San Diego with 155 cubic feet per second of water from the All-American Canal.

The diversion capacity of 2,000 cubic feet per second to the Gila Canal is for irrigation of the Gila project in Arizona. Headworks for an ultimate diversion of 6,000 cubic feet per second were constructed, but it was later decided to divert most of this water further upstream for use in central Arizona.

Although Imperial Dam created a reservoir of 85,000 acre-feet capacity and a surface area of 7,500 acres, this storage was not required by the project, nor considered a project feature, as it was anticipated that the reservoir would soon become filled with silt. The total drainage area above the dam is 187,000 square miles,

but the run-off from 167,000 square miles is controlled by Hoover Dam and an additional 10,000 square miles is controlled by Davis and Parker Dams. The uncontrolled drainage area contributing to the flow at Imperial Dam is, therefore, only 10,000 square miles. With this limited drainage area and the equalizing effect of the immense reservoirs in the river above the Imperial site, it is estimated that the maximum flood at this location will probably not exceed 150,000 cubic feet per second.

DAM SITE

Several possible dam sites were under consideration for many years but the Imperial site, formerly known as the Cocopah site No. 3, was finally selected. The alluvial flood plain at this location has an elevation of 160 feet above sea level and a width of 2,300 feet between rocky abutments. The alluvium consists chiefly of very fine sandy silt with some layers of clay and of coarse sand or gravel. It is probable that the depth of this waterborne material is about 200 feet. The rocky abutments, rising at the sides of the flood plain, consist of old formations of gneiss and schist which have been much contorted and fractured so as to show little continuity. Overlying the uneven gneiss and schist is a layer of rhyolite in evidence on both sides of the river.

The dam site was investigated by sinking a number of test holes and pits to make sure that no strata of plastic mud lay under the location of the dam. From some of these test holes, undisturbed samples of the underlying material were taken and tested to determine the load carrying ability and percolation factor. These test data, in conjunction with theoretical analysis and model studies, pointed to the probability of seepage

The desilting basin of the Gila Canal headworks, with the canal leading away at right angles, is at the left, the overflow section next, in the center, and at the right, the sluiceway, headworks, and the desilting basins for the All-American Canal.

beneath the dam. Test piles were also driven and tested both for load carrying capacity and uplift resistance.

THE DAM

The dam is of the slab-and-buttress type, with a total length of 3,430 feet, inclusive of nonoverflow sections, headworks, gate structures, sluiceway, and overflow spillway. The structural height of the dam is 46 feet and the normal water surface of the river is raised 23 feet. All gate structures are founded on pile foundations to prevent any chance of unequal settlement. The piles are battered (inclined), some upstream and some downstream, so as to carry the horizontal as well as the vertical loads in pure compression and tension. At both abutments, several units of the dam are

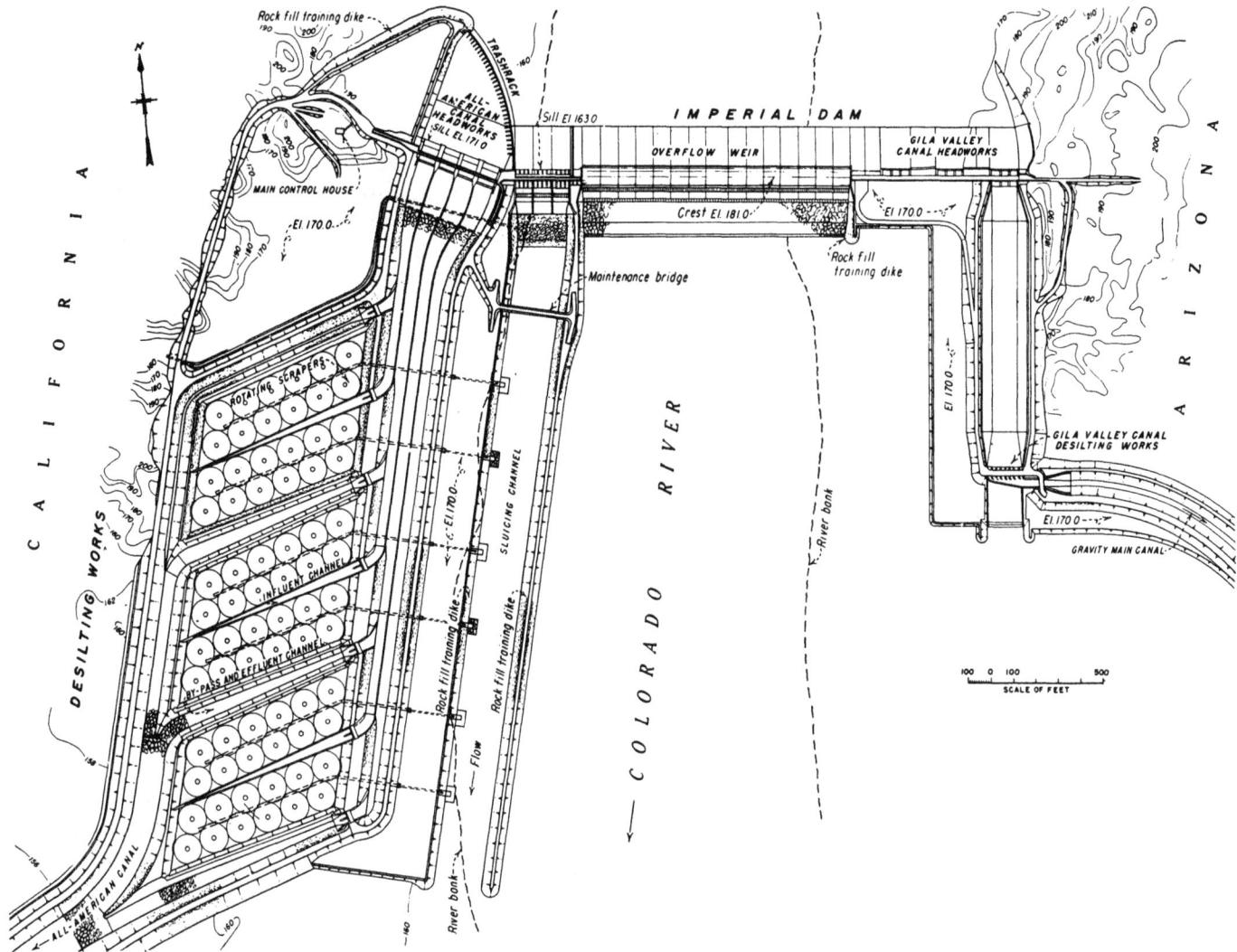

founded on rock; otherwise the dam is of the floating type, resting on the sandy bed of the river and divided into units to give some flexibility. Rubber joint seals are used between the units. A certain amount of settlement is also allowed by a mastic filler seal at the top of the sheet piling used as cut-offs under the structure.

The dam is provided with an upstream apron under which are three rows of sheet piling of sufficient depth to inhibit percolation of water beneath the dam. The apron is tied to the floating dam by steel reinforcement.

A prominent feature of the design is the provision for drainage of any water that does percolate under the dam. This underdrainage was designed to reduce the uplift pressure under the dam and the downstream apron, resulting in a greater safety factor against sliding of the dam and a considerable saving in the required thickness of the apron. The drainage system consists of a pipe surrounded by a filter to prevent the fine river-bed sand from being washed out from under the structure. This filter is made up of four layers of sand and gravel graded in size so that the material of

each layer will hold the material of the next outer layer from being washed into the drain. The pipe drain discharges through ejectors in the downstream slab of the overflow weir. These ejectors, which operate by means of the aspirator effect of the overflowing sheet of water, maintain a water pressure under the dam considerably below that of the flood water level downstream. It was computed, by means of the tests and studies previously mentioned, that the maximum seepage under the dam might be as much as 4 gallons per minute per foot of dam.

The overflow or spillway section of the dam is 1,198 feet in length with an ogee downstream slab. With sluice gates open, the estimated maximum flood of 150,-000 cubic feet per second may be discharged with a depth of 8.5 feet of water over the crest. For design, the maximum water surface was taken at 10 feet above the crest. The nonoverflow structures were given an additional freeboard of 6 feet. The overflow weir has a heavy downstream apron with dentated sill and rip-

The great roller gates in the headworks of the All-American Canal dwarf the man standing before them.

rap to protect the stream bed below. The design of the apron was based upon extensive model studies.

The standard units of the overflow weir are 78 feet 6 inches long with four buttresses at 20-foot centers, leaving cantilever slab overhangs at both ends. The slabs are monolithic with the buttresses and are designed as continuous beams, the fixed-end moments approximately balancing the moments from the overhanging cantilevers, thus relieving the buttresses of any appreciable bending moment.

The sluiceway is between the overflow section of the dam and the All-American Canal headworks and has twelve 16-foot-wide by 7-foot-high radial gates with sills set 16.5 feet below normal water surface. The gates therefore require top seals. The total design discharge of the 12 gates under normal water level conditions is 31,000 cubic feet per second and, at maximum flood stage, 42,500 cubic feet per second. These sluice gates are automatically controlled by a float switch so as to maintain the normal water surface in the reservoir. The sluiceway discharges into a channel into which the sludge from the All-American desilting works flows. The sluice water thus helps to carry the sludge downstream.

The All-American Canal headworks has four 75-by 22-foot roller gates controlled from two gate houses or from the main control house near the west end of the dam. A trashrack curved on a 799-foot radius and 575 feet long is constructed upstream from the headgates. The location of the trashrack and the concrete wall upon which it was founded was determined from model tests giving the best hydraulic conditions and excluding the maximum amount of silt from the headgates. The trashrack is cleaned with a mechanical rake which

consists of a motor-driven traveling gantry equipped with a motor-operated hoist and a rake unit. The trash is dumped into trash cars which travel along the top of the trashrack structure.

Three gate structures are provided on the Arizona end of the dam for the Gila Canal diversion. Each gate structure has three 35-foot 8-inch by 14-foot 6-inch radial gates provided with top seals. It is now expected that only one set of gates will ever be used. From the gate structure, the water passes into a desilting basin through which it flows at a velocity of 0.7 foot per second (for maximum diversion) and where it drops a large percentage of the silt load. Periodically, the diversion gates at the lower end of the basin are closed and the adjacent sluice gates opened to increase the velocity through the basin to about 15 feet per second and thus sluice the deposited silt out into the river channel below the dam.

ALL-AMERICAN CANAL DESILTING WORKS

Although the water issuing from Parker Dam is practically free from silt, it quickly picks up its full load from the bed of the stream. This silt, if allowed to remain in the water diverted at Imperial Dam, would soon obstruct the flow in the canals and have to be removed. It was originally estimated that the desilting works, costing $1,500,000, would save a yearly expenditure of $1,000,000 for excavation of silt from the canals.

The desilting works consists of three basins, each of which is approximately 500 by 800 feet in plan and 12.5 feet deep with an influent channel through the center. Provision is made for the addition of a fourth basin if found necessary when the ultimate diversion of 15,000 cubic feet per second is reached. Effluent channels leading to the All-American Canal are built between basins and at both ends of the row of three basins.

The headgate structure of the All-American sluiceway anchors the California end of the overflow section of Imperial Dam.

From the headworks, the diverted water flows through the inlet canal, which is divided by concrete sheet piles into four channels. This channelization was deemed necessary in order to maintain nonsilting velocities at times of partial diversion when one or two of the basins are closed down. The inlet canal delivers water through twin 21- by 17-foot radial gates to the tapering influent channels extending down the centers of the basins. Similar gate installations make it possible to bypass the flow directly to the effluent channels and hence into the All-American Canal. Normally, however, the water enters the influent channel from which it is distributed uniformly to the half-basins on either side through unique vertical slots along the walls. These slots serve the dual purpose of: (1) reducing the velocity and therefore the turbulence of the water entering the basins; and (2) distributing the inflow into the basins uniformly both as to depth and width of basins.

Each half-basin is provided with 12 center-drive rotating scrapers 125 feet in diameter. Each scraper, with a peripheral speed of 30 feet per minute, continually feeds the settled silt into a central collecting trench from which it flows through sludge collecting pipes into the river. Provision is made for sampling and measuring the sludge discharge. An interesting feature of the rotating scrapers is the overload limiting arrangement. If the silt load becomes too great the hinged arms of the scraper automatically rise so as to scrape less deeply into the silt. At the same time there is a visual warning of the overloaded condition. The 72 motor-driven rotating scraper mechanisms may be controlled individually from the respective basin control house. They may also be controlled in groups of 24 from the central or main control house near the west end of the dam. Scraper overload indicators, water level indicators, and gate controls, all at the main control house, centralize the complete operation of the diversion and desilting works.

CONSTRUCTION

The contract for construction of the dam and desilting works was awarded in November 1935 to the Morrison-Knudsen Company, of Boise, Idaho, the Utah Construction Company, of Ogden, Utah, and the Winston Brothers Company, of Minneapolis, Minn., at their joint bid of $4,374,240. Work was started in January 1936.

The California abutment, All-American Canal headworks, and sluiceway on the west side of the river, and the Gila Canal headworks, Arizona abutment, and dike on the east side of the river were all constructed back of cofferdams. Then, by opening the cofferdam at the Gila headworks and building another cofferdam across the river, the flow was diverted through two gate structures of the Gila headworks during the construction of the overflow weir.

Excavation for and construction of the All-American desilting works was carried on simultaneously with the construction of the dam. The accompanying drawing shows the general plans of the dam and desilting works.

EASTON DIVERSION DAM

KITTITAS DIVISION, YAKIMA PROJECT, WASHINGTON

By Lewis K. Maires

❯❯

Easton Dam is located below the junction of the Yakima and Kachess Rivers, three-fourths of a mile northwest of Easton, Wash. It is situated on the eastern slope of the Cascade Range in timbered country subjected to heavy snowfalls.

The diversion structure is unique in that it is capable of maintaining practically a constant lake level for diverting water into the Kittitas Main Canal while passing floodwaters or releasing varying amounts of irrigation water over the crest of the dam. The irrigable area of the Kittitas division, for which the diversion dam primarily was constructed, comprises 72,000 acres.

Two transcontinental railroads, the Northern Pacific and the Chicago, Milwaukee, St. Paul and Pacific, traverse the southern shore of the lake. The Sunset Highway crosses a neck formed in the upper end of the reservoir.

Water enters the reservoir created by Easton Dam from three sources: Keechelus Storage Reservoir, located on the Yakima River about 11 miles above Easton Dam; Kachess Storage Reservoir, located on the Kachess River about a mile above Easton Reservoir; and the natural flow of the Yakima River between Keechelus and Easton Dams.

The total drainage area is 185 square miles and the average annual run-off is about 560,000 acre-feet. Floods in the Yakima watershed occur in the months of November to June, inclusive, and are caused by excessive rainfall or rapid melting of snow, or a combination of these conditions. No discharge records are available at this point prior to the construction of the Kachess and Keechelus Reservoirs, which very largely control the flood run-off.

The reservoir created by Easton Dam is 1½ miles long and one-half mile wide, covering an area of 240 acres and containing 4,000 acre-feet of water. To prevent damage to the two transcontinental railroads, the maximum water surface is limited to elevation 2181.

During 1924, the dam site was tested by drilling and was examined by Mr. Kirk Bryan, a geologist of the United States Geological Survey, who reported as follows:

The diversion dam is to be built in a narrow gorge carved in a massive, but impure quartzite that is a part of the Easton schist. The rock is suitable for the foundations of any structure and its joints can be successfully grouted. * * * There is an old deep channel beneath the gravel plain north of the dam site and moderate leakage through this gravel may be anticipated. No danger to proposed or existing structures is involved. The Easton schist, as previously mentioned, is one of the older rocks of the Cascade Mountains. Normally, it is a silvery-gray or green rock with thin layers of quartzose material separated by micaceous minerals. In this form the rock is usually in thin crumpled plates gashed with quartz veins. At the dam site, however, the rock is a massive greenish-gray quartzite composed mostly of quartz but contains also various accessory minerals to which the color is due. Immediately east of the dam site the normal schistose rock is found.

THE DAM

The dam is a straight gravity, concrete, overflow structure, having an overall length of 248 feet and a maximum height of 66 feet. It raises the water surface 43 feet above the normal low-river stage. To maintain a lake level sufficient to divert 1,320 cubic feet per second into the Kittitas Main Canal and also to pass the varying amounts of water required for the lower divisions of the Yakima project, an automatically controlled, structural-steel drum gate is mounted over a float chamber in the spillway crest. The gate, which is 64 feet long and 14.5 feet high, maintains the desired water level within a three-quarter-inch variation. For flood conditions, the spillway has a discharge capacity of 13,000 cubic feet per second when the gate is in its lowered position.

Two 12-foot-wide by 11-foot-high radial gates in an intake structure located at the south abutment of the dam regulate the flow of water into the Kittitas Canal.

Easton Diversion Dam diverts water from the Yakima River to the Kittitas Main Canal. Water cascading over the spillway against a tree-covered background makes Easton one of the more handsome of small Bureau dams.

The intake structure is also a transition to a 12-foot 3-inch diameter, concrete-lined, horseshoe tunnel, 303 feet long, passing beneath the Northern Pacific Railroad. The tunnel is terminated by an appropriate transition connecting with a concrete-lined section of the canal.

For the purpose of sluicing silt and passing water through the dam at low reservoir stages, sluiceways are provided at each end of the dam near the bed of the river. The sluiceways are controlled by electrically operated 4.8- by 6-foot, air-vented, cast-iron gates, protected by trashracks. Each trashrack is 7 feet 3½ inches wide by 45 feet 6 inches high.

A reinforced-concrete fish ladder, designed and constructed to conform to the requirements of the Washington State Department of Fisheries, is located on the north shore of the river. The fishway has 20 bays 10 feet long by 6 feet wide and risers of 2 feet 1 inch. The bays are separated by 6-inch concrete baffle walls 3 feet 6 inches high. To prevent the attempts of the fish to turn back after entering the fishway below the dam, an ingenious device called a "fish guard" is placed on top of the baffle walls of each bay. The fish guard is made up of a grid of short bars having a quarter of a circle bend at one end. Control of water through the fishway is obtained by the use of stop planks set in grooves at each end of the ladder. In conjunction with the stop planks on the lower end of the fishway, a movable fish guard, operated by a hand hoist, is placed in a separate groove in the concrete. This device may be raised or lowered to any desired river level.

In a memorandum dated December 16, 1927, L. E.

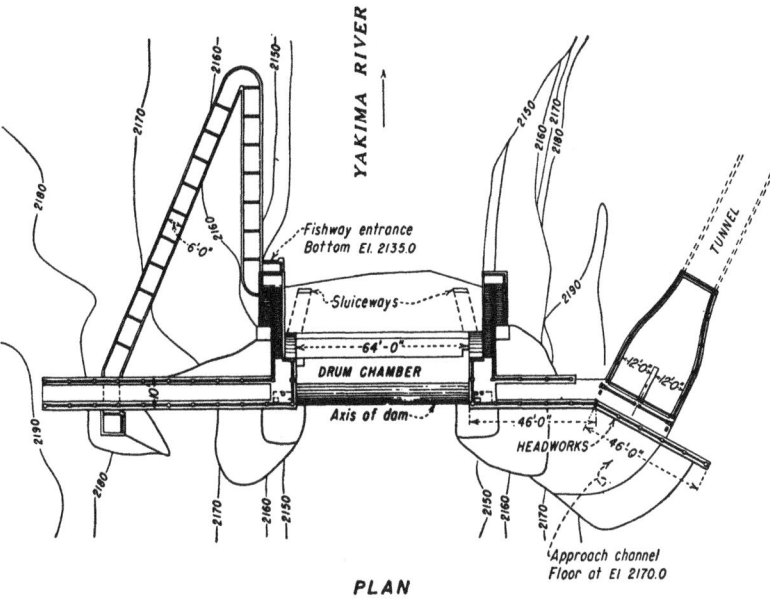
PLAN

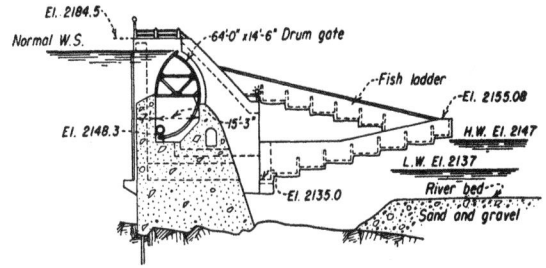
SECTION THRU SPILLWAY

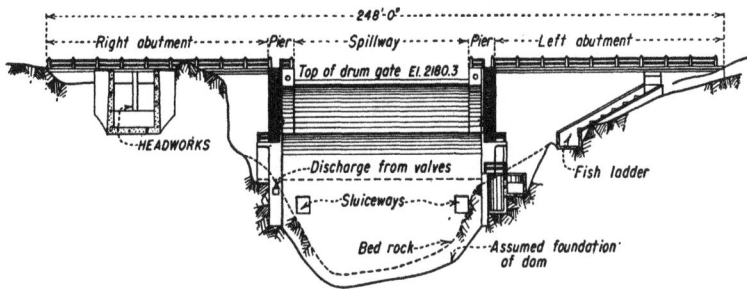
DOWNSTREAM ELEVATION

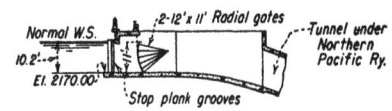
SECTION THRU HEADWORKS

Mayhall, General Superintendent of Hatcheries, Department of Fisheries and Game, Seattle, Wash., makes the following statement:

Salmon ascend the rivers as a result of their instinct to reach their place of birth, the spawning beds in the upper reaches of the river, and when they meet an obstruction or dam they will make desperate efforts to get over, and on their way. By taking advantage of their desperation, they can be tricked into jumping small waterfalls into the entrance of the fishway, but when confined in a small place between the concrete walls of a fishway, they are quick to realize it is not their natural element, the big open river, so they become panicky and rush out, unless held in a fishway by the trapping device.

The fish ladder has been found to operate successfully without the use of cover screens. A gallery inside the dam provides a means of passing from one side of the river to the other and also for access to the automatic, float-controlled mechanism actuating a 16-inch balanced valve which controls the water pressure in the drum-gate float chamber. A hand-operated emergency valve, controlling the drum gate, is also located in a chamber adjacent to the gallery.

CONSTRUCTION

The dam was built under a contract dated March 16, 1928, and awarded to C. F. Graff, of Seattle, Wash. The contractor commenced construction but later, due to financial difficulties, relinquished the job to one of the sureties, Hans Pederson, who completed the work on October 11, 1929.

On September 20, 1928, the gates of Kachess and Keechelus Reservoirs were closed and work was started on the building of two cofferdams. The upper cofferdam was first built as a rock-filled crib with timber sheeting extending a few feet into a 10-foot stratum of gravel and boulders. The lower cofferdam consisted of a rock-filled section. Notwithstanding the use of additional reinforcements consisting of sack dams, both cofferdams leaked badly through the gravel stratum. To unwater the foundations it became necessary to drive interlocking steel-sheet piling and to cover the cofferdams with clay blankets. The water in the river was first diverted by means of a 12-foot-wide by 9-foot-high timber flume placed on the north side of the river. Later the river was diverted through a temporary open-

Construction of Easton Diversion Dam in 1929 followed the established sequence of concrete placement in alternate blocks, carefully keyed together. Curing of concrete previously placed is carried on under the cover of tarpaulins.

ing left in the dam. As construction progressed further, the opening was closed and the river allowed to pass through the sluice gates.

The foundation was excavated into solid rock to a depth of 3 to 4 feet, and the cut-off trench for the dam was excavated an additional 5 feet. The excavated material was removed by stiff-leg derricks.

Grout holes were drilled on the center line of the cut-off trench at 5-foot centers. Drill holes were 25 feet deep in the river channel and tapered off to 10 feet on each river bank. Where the exposed foundation showed signs of seams, additional grout holes were drilled. In the south half of the spillway section, downstream from the cut-off trench, 20-foot grout holes were drilled on 10-foot centers, both ways. The maximum amount of grout needed in any one hole was 38.5 cubic feet. The average requirement of grout per hole was 4.75 cubic feet. Before actual grouting was started, the foundation was covered with a 5-foot layer of concrete. When the concrete had partly set, the grout

pipes extending through the concrete were loosened and slightly raised above the top of the rock foundation. Grout under a maximum air pressure of 100 pounds per square inch was then forced into the seamy rock and into voids that existed between the rock and the concrete blanket.

A concrete testing laboratory, established at Ellensburg, Wash., made numerous tests on strength, permeability, and economy of the mixes of concrete to be used in the dam. The sand and gravel used in the concrete was shipped from Steilacoom, Wash. The height of concrete lifts varied from 4 to 10 feet. Metal-lined forms were used on concrete surfaces exposed to view.

The center portion, or spillway section, of the dam was constructed first and the abutments later. The drum gates were assembled in place. Alignment of the steel castings making up the gate hinges was obtained by placing the anchor-bolt nuts on each side of the casting. After the erection of the structural-steel sections of the gate, it was possible to set and grout the cast-iron wall plates in correct position with respect to the drum-gate side seals, and to concrete around the gate hinges and seal castings.

Main items of the contractor's equipment consisted of one ¾-cubic-yard steam shovel; forty ½-cubic-yard side dump cars; 2 steam dinkey locomotives; 1 guy and 2 stiff-leg derricks; a stationary air compressor, 1,190 cubic feet per minute; one 200-horsepower motor; and one 1-cubic yard concrete mixer.

For convenience in receiving materials, a spur track was constructed from the Northern Pacific Railroad yards at Easton to the dam site. From 40 to 80 men were employed on the construction.

The following are the principal quantities pertaining to construction of the dam:

Excavation	cubic yards	7,000
Concrete	do	5,800
Reinforcement steel	pounds	60,700
Structural steel	do	320,200
Drilling grout holes	linear feet	2,200
Grout	cubic feet	509

GRAND VALLEY DIVERSION DAM

GRAND VALLEY PROJECT, COLORADO

By Charles R. Burky

The Grand Valley Diversion Dam is the diverting structure for the main canal of the Grand Valley project in western Colorado. The dam is situated about 8 miles northeast of Palisade, Colo. The diversion is on the right, or west, bank. For some distance above this point the river flows in a canyon; but at Palisade the canyon walls recede and the river enters Grand Valley, the scene of extensive irrigation development.

The first canal on the project was completed by private interests in 1884. High water in the year 1909 damaged the crib dams used by the irrigation districts, and in 1910 the districts contemplated building a new dam. Negotiations were already under way, however, between the Water Users Association and the Federal Government, and on August 19, 1913, the Secretary of the Interior gave authority to proceed with the construction of the dam.

The drainage area above the dam is 8,055 square miles. Based on values obtained from the recorded and estimated data of the gaging station a short distance above the dam on the Colorado River near Cameo, Colo. (1914–45), the maximum annual run-off (1914) was 5,475,000 acre-feet; the minimum annual run-off (1934) was 1,859,000 acre-feet; and the average annual run-off was 3,505,000 acre-feet. The maximum discharge was 52,400 cubic feet per second on June 16, 1921, and the minimum flow was 700 cubic feet per second in 1939. From gage heights at Fruita, the flood of July 1884 is estimated to have been more than 50 percent greater than that of 1921.

On account of the nearby tracks of the Denver and Rio Grande Western Railroad, on the right bank of the river, it was necessary to pass flood waters at no greater elevation than was used for supplying the headgates during low discharge. In some seasons it was necessary to divert all the water possible; consequently excessive leakage had to be avoided. To meet those conditions, the roller-type gate was favored over radial or fixed-wheel gates on account of the much wider waterways provided and the simplicity of operation. At the time of construction there were but two roller-crest dams in the United States.

Diamond drilling had been employed on several sites as early as 1908 and 1909, and this method was therefore used in the investigations. Holes a short distance above and below the adopted site indicated that rock would not be encountered at a depth of less than 30 feet; so the dam was designed as resting on a gravel base.

DESCRIPTION OF STRUCTURE

The dam consists of an ogee weir with concrete aprons upstream and downstream from the weir, a cut-off wall under the weir, and curtain walls at the ends of the aprons. The weir is surmounted by roller gates to regulate the water height. At the right end is a

The headworks of the canal occupy the foreground, and the dam extends across the river, gate hoist houses occupying alternate piers. The main line of the D. & R. G. W. Railroad is situated on the near side of the canyon, and U. S. Highway Routes 6 and 24 on the far side.

SECTION A-A

sluiceway between the weir and the outlet gates to carry off silt.

The dam is 542.5 feet long between abutments. The cut-off wall under the weir extends to elevation 4758, and the top of the weir is at elevation 4782. The surface of the downstream apron is at elevation 4774 and a heavy riprap is provided at the end of the apron. The river bed at the weir had an original elevation of about 4774.

There are six roller gates, each 70 feet long by 7 feet 1¾ inches in diameter and provided with an extension arc which rests on the weir and which places the top of the roller 10 feet 3 inches above the weir when in the lowered position. Across the sluiceway at the right end of the dam is a roller gate, 60 feet long, provided with an extension arc which gives a height of 15 feet 4 inches above the weir. The sluiceway weir is 5 feet lower than the main weir.

Rollers extend about 2½ feet into recesses in the piers and operate on tracks embedded in the side of the recesses at an angle of 20° with the vertical. Around each end of the roller is a toothed rim engaging a toothed rack fixed parallel with the track. The rollers are operated by means of a chain, one end of which is fastened to and partly encircles the roller, the other

being wound around a drum in the hoist house on top of the pier. The ends of the rollers are sealed by means of oak strips attached to flexible diaphragms near the ends of the cylinders. A wooden strip, fastened to the extension arc, provides a bottom seal when the roller is down.

Each roller is hoisted from one end, and each alternate pier is surmounted by a hoist house containing the hoists for two rollers. Hoists for the 70-foot rollers are driven by 10-horsepower motors and the hoist for the 60-foot roller is driven by a 20-horsepower motor. Each 70-foot roller weighs about 40 tons and requires a maximum chain pull of 25 tons. The 60-foot roller weighs approximately 53 tons and requires a maximum chain pull of 45 tons. The 70-foot rollers can be raised from the lowest to the highest position, a distance of 16 feet, in 15 minutes. Chains are of special design and are made of heavy pins and links.

The sluiceway is 60 feet wide with the sill at elevation 4777, 5 feet below the main weir and 8 feet 4 inches below the sill of the headgates. For design, it was assumed that a velocity of about 2¼ feet per second at the entrance to the sluiceway would carry silt into the canal. The sluiceway will take a deposit of about 2½ feet at its entrance before this velocity is reached.

When the deposit of silt reaches this depth, the roller is raised and the channel cleared by flushing the mud down the river. In operation the rollers are raised about once a week.

The outlet structure at the west end of the dam contains nine gates, each 7 feet square. These gates are operated by a line shaft, installed on the head-wall platform, driven by a 3-horsepower motor. The gate sill is at elevation 4785.25 and the normal high-water surface in the canal for 1,425 cubic feet per second is 4791.50. Eight of the gates can take the full discharge with a velocity of about 4 feet per second.

With an assumed flood surface at elevation 4792.5, it was computed that a river discharge of 50,000 cubic feet per second could be taken by the six weir bays and the 60-foot sluiceway, and that an additional 1,600 cubic feet per second could be carried by the canal intake. A flood of approximately 50,000 cubic feet per second occurred the summer following completion of the dam and was carried by the six weir bays, with the water at the elevation of the top of the sluiceway roller gate (4792.33) which was down at the time.

CONSTRUCTION OF DAM

The river was handled by first completing the sluiceway and head wall at the west end, during low water, and excavating a channel above and below the sluiceway to facilitate the flow. Cofferdams were then built above and below the dam, east of the sluiceways, and auxiliary cofferdams were built parallel to the river flow, creating basins to reduce seepage. Between these cofferdams the different portions of the weir and aprons were completed. When the 70-foot rollers were in place and raised, the main cofferdams were demolished and a cofferdam built to enclose the sluiceway so that this portion of the dam could be completed.

Foundation excavation was begun at the west end on August 27, 1913, using a small team outfit. By the first of January 1914, the excavation necessary for the head wall and sluiceway was completed and by May 1914, the concrete was placed. The cofferdams were then completed and the rest of the excavation continued.

The Orchard Mesa Canal, occupying the east abutment of the site, was diverted by a wooden flume. Ex-

The training wall in the foreground separates the flow through the sluiceway at the left from the main current of the Colorado River.

cavation at the east end was started by a dragline with a 30-foot mast and a 60-foot boom. After 3,000 cubic yards were excavated, a flood occurred October 3, 1914, which took out the east end of the upstream cofferdam and put the dragline out of commission. On the main weir, excavation was carried down to elevation 4768 with teams. Below this elevation, which is the bottom of piers and main weir, the excavation of the cut-off trench was carried on by hand with the aid of skips handled by a cableway. Water was held to approximately elevation 4768 by drainage to a main sump which was located downstream from the first bay east of the sluiceway. Auxiliary pumps were used to keep the water out of the cut-off trench. Solid rock was encountered in the trench at the east and west ends. Excavation for the upstream curtain wall was made without trouble, but the excavation for the downstream curtain wall was hampered by an excessive flow of water.

The first concrete was poured at the west end on January 9, 1914. The mixing water was heated and, during the coldest weather, rough sheds were erected over the concrete and heated by open fires. For the first part of the job, the concrete aggregates were obtained from a gravel pit three-quarters of a mile below the work on the west side of the river, where a temporary screening and mixing plant was erected. In May 1914 the temporary plant was dismantled, and erection of the permanent mixing and screening plant on the east side of the river was started. The permanent plant was completed in August of that year. It consisted of machinery for crushing rock, rolling sand, and elevating aggregates into bins from which they could be dumped into the mixer hopper by gravity. The crusher, of the gyratory type, which with the sand rolls was shipped from the Strawberry Valley project, was served with broken sandstone from the nearby quarry by means of a stiff-leg derrick with 40-foot mast and 60-foot boom, handling a 5-cubic-yard skip. The aggregate storage bins had a capacity of 130 cubic yards of sand and 270 cubic yards of broken stone and emptied into measuring boxes provided with gates discharging directly into the mixer hopper.

The mixer hopper was set flush with the cement-storage floor. On this floor was a 36-inch gage track running through the side of the building and about 50 feet downstream, where it ended directly under the cableway. Cement was received at the railroad siding on the west side of the river and unloaded onto skips. The skips were carried across the river by the cableway, deposited on flat cars on the 36-inch gage track, and taken to the storage room. About two cars of cement could be held in the storage room and about five or six cars could be stored in the cement warehouse at the siding on the west side of the river.

The gasoline engine mixer used at the temporary plant was moved to the permanent plant. Just beneath the discharge spout of the mixer was a 36-inch gage track leading to two tracks 6 feet apart, under the cableway. The mixer discharged into a 2-cubic-yard, tilting-dump concrete bucket standing on a flat car. Three batches filled the bucket and it then was rolled under the cableway where it was picked up and transported where desired. Water was supplied to the mixer from an 800-gallon wooden tank on the hillside, 50 feet above the mixer. Water was pumped to the tank from the river.

A pile bridge was constructed upstream from the dam, running from the outer wall of the sluiceway to the east side of the river. A truss bridge was built across the sluiceway. This bridge was used for access to the construction plant on the east side of the river and also for constructing the upstream cofferdam.

A power plant was constructed at Cameo, some 6 miles above the dam, and a transmission line built to the dam where a substation reduced the voltage from 16,500 to 2,200. The 2,200-volt line operated a 150-horsepower motor, direct-connected to a 100-horsepower, 250-volt, direct-current generator which supplied the power used during construction.

The general method of construction was to unwater certain areas of the dam site, dig trenches for the cut-off wall and curtain walls, and fill them with concrete. Pier foundations were then prepared, and concrete placed up to the elevation where nose and side forms for the piers were required. Then the forms were built up to elevation 4794, and recesses for the roller gates, racks, and guards formed at the proper locations. Concrete in the piers was placed in lifts of 4 or 5 feet per shift. Forms were removed in 2 to 4 days, and the concrete was pointed up and washed with thin cement grout. The aprons were usually placed in about three sections between piers.

All the work was done by Government forces.

INSTALLATION OF ROLLER GATES

Since the piers were constructed before the racks were received, recesses were left for installing the racks and anchor bolts. The racks were positioned by adjusting bolts attached to iron bars held by a wooden frame and, when in proper position, a thin grout was poured around the steel support and anchor bolts and allowed to harden.

Rollers were shipped in five sections. Rollers for the second, third, and fourth bays east of the sluiceway were installed by the cableway. The two end sections were swung into place and the proper teeth of racks and rims were engaged. Intermediate sections were then brought into place, lined up, and the ends drawn together with pulling jacks. Each roller was tested with a wye level before riveting started. When these three rollers were completed and raised, the cofferdam protecting them was demolished and the cofferdam needed for protection of the two rollers at the east end was repaired.

For the installation of the two east-end rollers a trussed beam, 8 by 18 inches in cross section and 74 feet long, was placed on the piers 9 feet downstream from the roller gate sill. Six cables were dropped from the beam to hold sills for false work which supported the gate sections during erection. The beam was an improvement over the method used on the first three rollers.

The most westerly of the 70-foot rollers was next erected. This bay was protected by the sluiceway wall and by cofferdams built upstream and downstream from the first pier east of the sluiceway. The roller was assembled and raised and, after temporary sealing strips were attached, the roller was lowered to cut the water off from that bay.

The last roller to be installed was the 60-foot sluiceway roller. On this work the false work for supporting part of the roller sections was built on flat cars running on a 36-inch gage track. This was possible because of the position of the sluiceway apron and greatly facilitated the erection of the roller.

Erection of the service bridge spans was started in May 1915, before the installation of the rollers was completed. Anchor bolts for the cast-iron shoes at the arch ends had been set in the concrete piers. The arches were assembled on the west side of the river and swung into place by the cableway. Latticed handrailing was put up and a temporary floor laid. The temporary floor was later taken up, after which the permanent 3-inch plank flooring was laid.

POWER FOR OPERATION

A power plant was constructed at the west end of the dam to generate electric power for operating the rollers and head gates. This plant consisted of a 4-cylinder, automobile-type gasoline engine, direct-connected to a 25-kilowatt, 250-volt, direct-current generator. The generator charged a storage battery of 108 ampere-hours capacity and was so connected that it could furnish power either direct or through the battery to the direct-current hoist motors. In 1938, a 3-kilowatt hydroelectric generator was installed as a charger for the battery and the use of the gasoline engine was reduced materially. The battery was capable of raising one roller, and it also furnished lights for the dam. By 1949, however, the hydroelectric generator, battery, and most of the other equipment was in poor condition and in urgent need of extensive repairs or replacement. The direct-current generator and the direct-current hoist motors for the rollers were still in serviceable condition. Negotiations with the Public Service Company of Colorado resulted in bringing an 11,000-volt transmission line to the dam to furnish alternating-current power. A 30-horsepower, alternating-current motor was installed to drive the direct-current generator, and a 5-horsepower, gear-type motor was installed to operate the canal head gates. The hoist motors for the rollers were retained in service, being powered direct from the generator. The hydroelectric generator was left in place but not in use. The new equipment was installed during the winter of 1949–50 and at this time the power plant was rewired, using conduits, and new inside and outside lights were installed.

COSTS

At the time this dam was built, costs were very much lower than the present-day cost for similar work. The actual cost of the dam and appurtenant items was a little over $500,000. The quantities of main items of work are given below:

Excavation, dry	cubic yards	25, 051
Excavation, wet	do	31, 010
Embankment	do	4, 863
Concrete, reinforced	do	17, 990
Backfill, dry	do	11, 405
Structural steel, bridges	pounds	97, 331
Structural steel, roller crests	do	743, 119
Riprap	cubic yards	7, 870
Gates	pounds	64, 105
Machinery, electric	do	112, 947

WIND RIVER DIVERSION DAM

RIVERTON PROJECT, WYOMING

By Irwin B. Hosig

⌄⌄

The Riverton project, comprising 100,000 irrigable acres, is located in Fremont County, westcentral Wyoming. The Wind River is a part of the Bighorn and Yellowstone River systems and drains the south slope of the Shoshone Mountains and the northerly portion of the east slope of the Wind River Range, the latter a part of the Continental Divide. The drainage basin has an area of 1,860 square miles above the diversion dam (Wind River Dam). Of this area, a considerable portion has Alpine topography with peaks as high as 13,800 feet above sea level.

The average annual run-off is about 785,000 acre-feet, 80 percent of which comes from mid-May to early August, principally from melting snow. The maximum observed discharge is 12,300 cubic feet per second, occurring in June 1906.

While the irrigation demand synchronizes reasonably well with the flood cycle, the full use of the water supply requires some seasonal storage as well as carry-over storage. The seasonal storage is required in late August and early September to finish off the potato, bean, and sugar beet crops. Storage capacity of 155,000 acre-feet is provided in Bull Lake Reservoir; and Pilot Butte Reservoir, completed in 1926, provides 31,550 acre-feet of storage capacity.

The main canal has an initial capacity of 2,200 cubic feet per second. A drop of 105 feet in the canal from the diversion dam to Pilot Butte Reservoir is utilized to produce electric power which requires year-round operation of the canal; however, the reservoir conserves the winter flow for irrigation purposes in the following year.

DAM SITE

In the vicinity of the diversion dam, the Wind River Valley has an average fall of 25 feet per mile. The valley is 25,000 feet wide and is bordered by sandstone bluffs 40 to 100 feet high. The low-water channel is 120 feet wide and 2.5 feet deep and 8 to 10 feet lower than adjacent ground. Low water surface is at elevation 5552. Bedrock of rather soft sandstones and shales is generally at elevation 5546 or higher, except in an old channel near the south abutment, where it is about elevation 5536. Overlying the bedrock is a 7- to 15-foot thickness of granitic and andesitic cobblestones, which was deposited by the river during flood stage. Except in the present stream bed, these cobblestones are covered with 1 to 3 feet of sands carried higher in the river cross section or brought in by winds and intermittent side streams.

DESIGN

For the dam to function properly, it was required that the low water surface of the river be raised 17.75 feet to permit diverting 2,200 cubic feet per second into the canal with none passing down the river, that coarse materials carried by the river in flood be kept out of the canal, and that the diversion structure be able to pass a flood of 40,000 cubic feet per second safely. It was also required that means be provided for passing the saw logs and railroad tie timbers annually floated down the river by a lumber company operating in the headwater forests, that a fish ladder be provided so that the mountain trout can continue their annual upstream run to spawning grounds, and that a roadway be provided crossing the valley for the benefit of future work on storage developments. The Wyoming State Highway Department cooperated in the last-mentioned matter, building a steel bridge over the spillway section and, more important, incorporating the highway from the rail head at Riverton to the diversion dam in the State Highway No. 287, one of the principal routes to Yellowstone National Park.

The entire width of valley not being necessary for a flood spillway section, the dam is divided into two parts, a concrete spillway section and an earth dike.

165

Trusses of the highway bridge span Wind River at the diversion dam.   In the foreground, the training walls of the logway separate the river channel from the sluiceway.

The lengths of the two sections are 650 feet and 1,650 feet, respectively.   The spillway section is 28 feet wide with a fore (downstream) apron 24.75 feet wide and rests on rock.   A gravity ogee section was adopted as against the Ambursen cellular type in spite of slightly less favorable cost estimates, principally because of the greater suitability of the solid section to the rigors of the northern mountain climate.   The fore apron is intended to contain the hydraulic jump for dissipating the energy of the spilling water and to prevent undermining of the main section.   The dike section has a maximum height of 25 feet with a 5-foot freeboard at maximum flood stage.

Examination of the cobblestone bed under the dike section made it appear certain that an impervious sheet-piling cut-off would be very difficult to construct because of the cobblestones, which ranged up to 18 inches in diameter.   Excavation for a trench to be backfilled with impervious material was also considered very expensive because of trouble with ground water.   It was finally decided that the dike section with a moderate cut-off trench near the upstream one-third point of the base would be satisfactory.   Absolute imperviousness was not necessary and piping was not feared in view of the coarseness of the cobblestone bed.   Events have proven the correctness of this conclusion as no piping has occurred.

During normal high water, leakage of about 5 cubic feet per second through low-lying gravel appears in the borrow pit.   In the early years of operation, the downstream face of the dike also showed dampness.   Silt and wind-blown sand have now accumulated on the upstream slope, tightening it so that the dampness no longer shows on the lower slope.   Prevailing winds confine wave action to the spillway section, and no breaching of the upstream face of the dike has occurred.

Flow into the canal is regulated by six structural-steel radial gates, 10 feet wide and 10 feet high, set in a headworks 70 feet long and 22 feet 6 inches high.   The gates have a top seal against curtain walls 12 feet 6 inches high.   They are lifted by a cable at each side of each gate, winding on drum hoists.   Counterweights to reduce externally applied hoisting forces are housed in slots in the 24-inch partition walls which separate the gates.   In front of the headgates and 4 feet lower than the gate sills is a concrete silt basin, about 70 by 70 feet in plan, in which the major portion of the silt load of the water to be diverted is dropped to be later sluiced out through the sluice gates.

At the side of the silt basin and in line with the axis of the spillway is a battery of four 10- by 12-foot radial gates of the same general design as the headgates. The sills of these gates are at the level of the bottom of the silt basin, 2 feet above the normal low water surface below the dam and 9 feet above the adjacent fore apron. The full canal discharge will flow across the silt basin at a velocity of 8 feet per second when the canal headgates are closed and the sluicegates opened wide. This

is calculated to sweep the basin clean. Both sets of gates are operated by electric motors. Since commencing operation, the pond above the dam has practically silted full. The silt basin is scoured clean annually, usually in October after the close of the irrigation season. Even these scouring periods must be short as canal storage runs the power plant but a few hours. The scoured-out silt lodges below the dam to be removed by next season's flood. While grade retrogression is general to the river, none has occurred at the dam to date, possibly because no severe floods have occurred since its construction. As a result tailwater remains high,

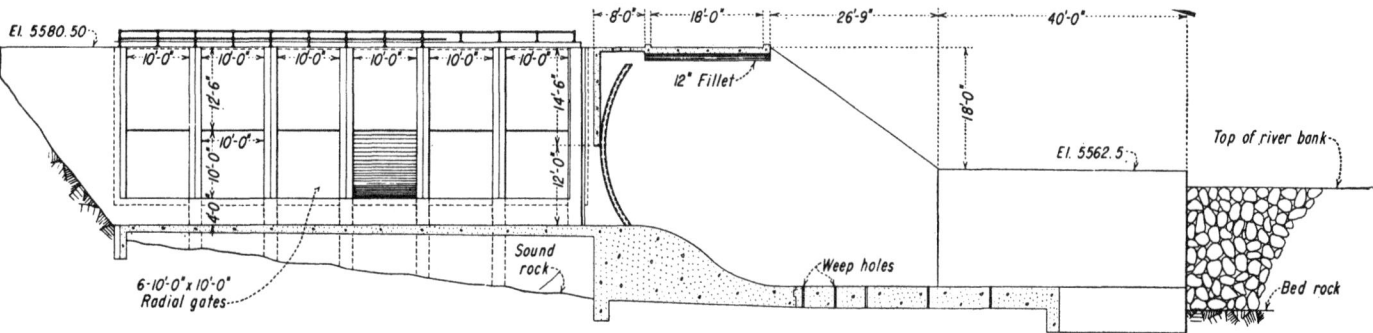

SECTION THROUGH SLUICEWAY

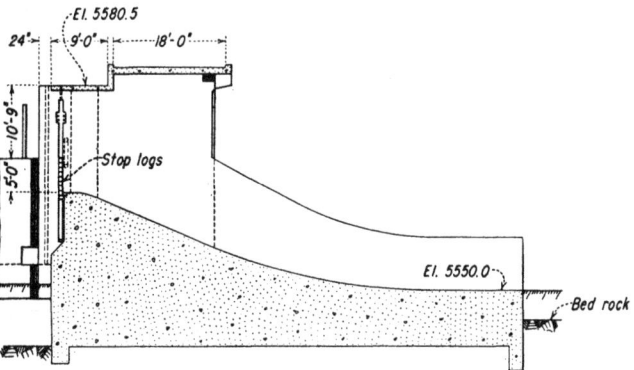

SECTION THROUGH LOGWAY

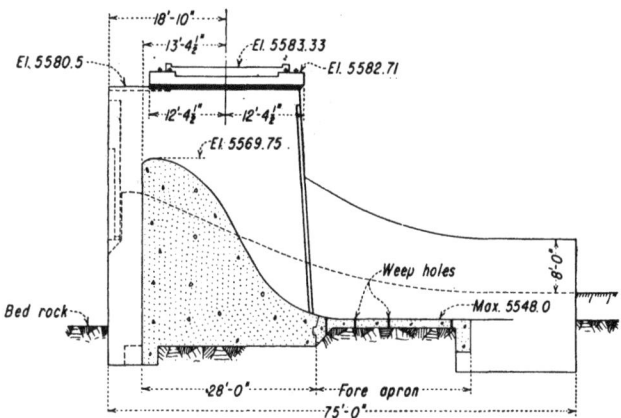

SECTION THROUGH SPILLWAY

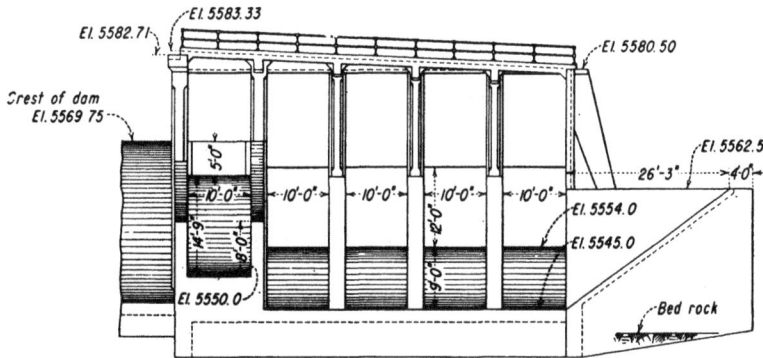

DOWNSTREAM ELEVATION OF SLUICEWAY

Headworks of the Wyoming Canal and the sluiceway of the Wind River Diversion Dam are protected from floating rubbish by the trash boom just visible at the lower right.

the jump at the foot of the ogee section is drowned, and scour below the fore apron, if any, is filled up as high water recedes. The apron is always found covered with a tight layer of gravel.

The logway is a section of flat ogee spillway 10 feet wide, the crest of which is 10.75 feet above the sluiceway gate sills and 5 feet below the normal spillway crest. The section is located adjacent to the spillway. The forebay is separated from the silt basin by a concrete training wall which sets on the edge of the silt basin floor and has the same top elevation as the spillway crest. Timbers and iron pipe set in the wall afford additional guidance to floating timber. When not in opera-

tion the logway is closed by a wooden gate which brings the crest to the level of the spillway crest.

Project development has not yet required diversion at full canal capacity. A principal operating difficulty is caused by slush ice. This is combated most successfully by taking water out of the silt basin above canal grade, through hand-operated wooden gates placed in the stop plank grooves, and sluicing the ice as much as possible out of the pond through the logway opening. In recent years the ice has been moved toward the opening by using electric motors equipped with propellers to create a current in that direction. Compressed air released at a depth of several feet assists in bringing the slush ice to the surface and into the current set up by the propellers.

CONSTRUCTION

Work was started by Government forces in July 1921 and completed May 1923. The working period included one flood season and two winter seasons. The first winter season was unusually severe and caused work interruption of 10 weeks. Before the arrival of the flood season, the headworks, silt basin, sluiceway, logway, and one 27-foot section of the spillway were completed or were above high water so that work could be continued. Also, that part of the dike south of the old river channel was completed, so that the channel could be used as a bypass during construction. After the flood, work was prosecuted continuously to completion except for minor delays occasioned by railway strikes and impassable roads, both of which stopped delivery of materials. The second winter was so mild that concrete work in favorable locations was possible with ordinary winter placing procedure.

The main construction machines were gasoline-driven draglines which were used on the project principally for canal excavation. Two draglines with 1½-cubic-yard buckets and 45-foot booms, and one with 2½-cubic-yard bucket and 75-foot boom were interchanged between canal excavation and dam construction. These machines cut diversion channels, stripped foundations, acted as traveling cranes for moving forms and light equipment, loaded the dike materials into wagons and concrete aggregate onto screening equipment, all as required. The concrete mixer was a ½-cubic-yard machine, spotted at various points as work progressed. Concrete distribution was mainly by ½-cubic-yard cars, operated by hand on a narrow-gage track built on a trestle.

The cobblestone admixture was not run through the mixer, but placed as "plum" rock brought to the trestle from the screening plant or dike building job in ½-cubic-yard cars pulled by a gasoline hoist. Generally, excavations were opened wide enough to avoid the use of sheeting. Unwatering was by means of gasoline-driven centrifugal pumps and hand-power diaphragm pumps.

The principal item of rock excavation was the cut-off trenches, which were 2½ feet wide and 3 and 5 feet deep below the spillway and fore apron bottoms, respectively. To avoid shattering the adjacent rock, line drilling on both sides of each trench was necessary. Holes were spaced about 6 to 8 inches on centers and a channeling effect was produced with a special tool fitted to a pneumatic hammer. Light blasting in center holes broke out the channeled material. Excavation in two lifts was found best for the 5-foot trench, which had one side 7 feet deep.

The dike was built with a mixture of soil and cobblestones found in the river bottom. The material was loaded by dragline, from borrow pits on the downstream side of the dike or from spillway excavation, into 3-cubic-yard dump wagons hauled by 10-ton caterpillar tractors, two or three wagons constituting a train according to lift. The borrow pits were wet, and as a result of the method of excavating and loading, the top sand and lower cobbly gravel arrived on the job wet and thoroughly mixed. The mixture was spread in layers by a standard road grader pulled by horses.

Large cobblestones were worked to the outside faces by horse-drawn stone boats. The travel of placing and leveling equipment produced sufficient compaction and a dike was produced which was more rodentproof than if clay material had been used. The deficiency of riprap on the two faces was made up with cobblestones, hauled from the river bed and the concrete aggregate screening plant. The latter was operated without crushing apparatus as the river-bed material ran sufficiently high in moderate sizes to make crushing unnecessary. Washing of sand and gravel was, however, necessary.

The principal construction items were:

Excavation	cubic yards	42,278
Stripping	do	19,961
Embankment	do	99,153
Backfill	do	8,563
Riprap	do	7,495
Concrete, plain	do	12,175
Concrete, reinforced	do	5,037
Gates	pounds	208,975

The total cost of the dam was approximately $483,800.

DESIGN OF DAMS

By Max Ford, concrete dam design; Jack W. Hilf, earth dam design; Maurice E. Day, design of diversion dams; Everett H. Larson

❯❯

From the passage of the Reclamation Act in 1902 until the end of 1952, the Bureau of Reclamation designed and constructed 122 storage dams, including 31 of concrete or masonry and 91 of earth, and 73 diversion dams. While some of these dams are comparatively small by present-day standards, many are massive structures which rank among the world's largest. Numbered among the concrete structures are Hoover, the world's highest dam; Grand Coulee, the largest concrete structure in the world; and Shasta, which is higher than Grand Coulee and contains more concrete than Hoover. Foremost among the earth dams is Anderson Ranch, the world's highest earth-fill dam.

The construction of major Bureau dams, beginning with Hoover Dam, has been done by what might be called assembly line methods. Construction equipment has been enlarged and improved to the point where progress on major jobs is at phenomenal rates. At Grand Coulee Dam, for example, 20,000 cubic yards of concrete were placed in the dam in 1 day, and at Bonny Dam over 1,000,000 cubic yards of rolled earth fill were placed in 1 month. This evolution in construction methods has been accompanied by a similar evolution in design methods and procedures. Bureau personnel responsible for the design of major dams are divided into specialized groups, each with responsibility for a particular element or elements of design, and the overall progress is closely integrated so that the designs are completed, materials procured, and construction operations accomplished at the proper time.

This article presents a discussion of present-day methods and procedures used by the Bureau of Reclamation in the design of major dams. Also included as being intimately related to design are brief discussions of investigations for selection of site and materials of construction, requirements for diverting the river during construction, and methods of foundation treatment. Construction of concrete dams is not discussed; but for earth dams, methods of construction and construc-

tion control are discussed as well as procedures of design since they are so closely related as to be virtually interdependent.

TYPES OF DAMS

Major Bureau dams are usually constructed entirely of concrete or of earth. In certain instances, however, it has been found economical to construct a dam partly of concrete and partly of earth embankment. Concrete dams have grace and beauty, are inherently watertight, and are readily adaptable to openings in the dam such as those required for outlet pipes. They require smaller quantities of materials for their construction than do earth dams, and the preparation of smaller areas of foundation. On the other hand, the construction of most types of concrete dams requires sound rock foundations, which are available at comparatively few sites. Earth dams, although they are not as spectacular as the graceful concrete structures, are as safe when constructed by modern, approved procedures, and may be constructed with materials which are available at or near the site. They have a further great advantage in that they can be constructed on poor rock and earth foundations that would be unsuitable for most types of concrete dams. These advantages have led to greatly increased use of earth construction for high dams.

The two principal types of concrete dams constructed by the Bureau are gravity dams and arch dams. The term gravity dam applies to a solid masonry or concrete dam which depends primarily on its weight for resisting the water load, while the term arch dam is used to designate a solid masonry or concrete dam which is curved upstream in plan, and transmits the major portion of the water load by arch action to the canyon walls. In addition to these two types of concrete dams, the Bureau has designed a few buttress-type dams. These dams have a sloping upstream face resting on buttresses or walls that transmit the weight of the impounded water on the face to the foundation. Because the large amounts of formwork and labor required for its construction ordinarily make a buttress-type dam uneconomical, comparatively few of these dams have been built. Grand Coulee and Shasta Dams are examples of the gravity type of dam; Hoover, Seminoe, and Hungry Horse Dams are arch-type designs; while Bartlett and Stony Gorge Dams are of the buttress type.

Earth dams may be classed according to method of construction as hydraulic or rolled earth fill. Earth dams have been constructed for centuries by hand methods, but the introduction of hydraulic methods of construction about 80 years ago revolutionized the art of earth dam building. Many hydraulic-fill dams have been constructed since 1868, using a stream of water to perform all the major operations. In recent years hydraulic-fill construction has been used infrequently, having been largely replaced by the rolled-fill type of construction wherein the excavation, transportation, and placement of materials are performed by mechanical equipment. In the United States, costs of rolled-fill dams now compare favorably with those of hydraulic fills because of the availability and efficiency of large earthmoving equipment developed primarily for construction of highways and airports, and because of the development of better methods of control of the embankment construction. Generally, a wider variety of soil can be used and closer construction control is possible with rolled-fill dams than with hydraulic-fill dams. The Bureau of Reclamation has used rolled-fill construction for more than 90 percent of its earth dams and has not had occasion to build hydraulic-fill dams since 1930.

SELECTION OF LOCATION AND TYPE OF DAM

The selection of the most feasible site for a dam and reservoir will depend on their intended purpose, whether irrigation storage, power production, flood control, silt retention, navigation, diversion for irrigation, or some combination of these and other uses. The choice of site will also be affected by foundation conditions, the width of the river, the width of the canyon or the flood plain, and the location of highways, railroads, buildings, cultivated lands, and other property within the area that would be covered by backwater. There are often two or more alternative sites for a storage dam where the topography will provide both a crossing narrow enough for an economical dam and sufficient reservoir capacity upstream to satisfy storage requirements. The most satisfactory of these alternative sites must be selected from the standpoints of safety, economy, and convenience of construction. For a diversion dam the

principal factor affecting selection of the site is the elevation and location of the lands to be irrigated.

The selection of the type of dam to be constructed is closely related to the selection of the site. From the standpoint of topography, narrow canyons with good rock bottoms and abutments are generally preferable for construction of concrete dams. Very narrow canyons are especially adaptable to arch dams since arch action is economically effective for ratios of crest length to height which are less than about 5 to 1. Sites in broad valleys with gradually rising abutments or sites with excessive depth to bedrock usually require some type of earth dam.

Perhaps the most important factor involved in the final location of a site for a dam or in selecting the type of dam for a particular site, is the foundation condition. In general, a hard rock foundation is satisfactory for any type of dam of any height, provided there is no unfavorable jointing or possibility of movement in any existing faults, and provided the foundation can be adequately sealed and consolidated. Rock foundations of high quality are essential for arch dams since they depend on the abutments to absorb the thrusts of the water load. Rock foundations are considered necessary for all but the lowest gravity dams, the desired quality of the rock depending somewhat on the size of dam to be built. Buttress and related types of dams, because of their comparatively light weight, can be constructed on foundations that do not have sufficient strength to support gravity dams. Certain sections of Imperial Dam, a slab-and-buttress structure, are supported on concrete piles driven into the alluvial fill, and the hollow, ballasted, floating-type overflow section is supported directly on a compacted fill of selected materials. Earth dams are often built on solid rock, but foundations of fractured rock, gravel, sand, silt, and clay, or almost any combination of earth materials can be used to support these structures if the materials are properly consolidated and cut-off trenches or other suitable barriers are constructed to prevent excessive percolation. Conditions which are likely to result in excessive percolation through the foundations of a concrete dam would favor the alternative construction of an earth dam. The wide base thicknesses of earth dams create long percolation paths through the foundation, which reduce percolation quantities.

The feasibility of a dam site depends on the proximity of sources of suitable materials for construction of the dam and its appurtenant works. Selection of the type of dam to be constructed depends to some extent on the relative availability of concrete aggregates and materials for earth embankments, and the relative cost of these materials delivered to the site. In case materials for both concrete and earth designs are located at considerable distances from the dam site, a reinforced-concrete buttress type of dam may be the most economical since it requires smaller quantities of material for construction than do other types.

Climatic factors sometimes influence the selection of the type of dam to be constructed. It is well known that concrete often spalls at its surface when exposed to freezing and thawing action. Concrete gravity dams and thick arch dams are little affected in their strength and stability by such spalling, provided their surfaces are repaired from time to time as needed. However, spalling of concrete in thin arch dams and buttress dams might affect the safety of such structures. This is because even small decreases in the effective thicknesses of thin arch dams or buttress dams would increase the stresses considerably. Also, the reinforcement of a buttress dam might become exposed, with consequent danger of corrosion. For these reasons, earth dams or concrete dams of relatively thick sections have been favored for construction in severely cold climates.

When an existing highway location will become inundated due to the construction of a dam and the creation of its reservoir, it is necessary that the highway be rerouted. It is usually possible to reroute the highway over the dam, thereby avoiding a more expensive crossing elsewhere. Gravity dams, thick arch dams, and earth dams are easily adapted to the construction of a highway over the top of the dam, but thin arch dams and buttress dams require considerable additional construction to include this feature.

Sometimes topographic conditions and requirements for water release at a site favor the construction of specific types of appurtenant features such as spillways and outlet works, and these, in turn, may affect the selection of the type of dam to be built. These factors are discussed in the portion of this article devoted to spillways and outlet works. Also, the method of river diversion most suited for a particular site may restrict selection of the type of dam to one adaptable to this method, as discussed in the portion of this article on river diversion.

INVESTIGATIONS

The construction of a large dam involves a serious public responsibility for the engineers charged with designing and executing the work. Safety must be the prime consideration since failure of the structure might release a catastrophic flood, entailing loss of life and property. The Bureau of Reclamation takes great precautions to avert failure of its dams, and important among these safeguards are the comprehensive investigations leading to selection of adequate sites, use of suitable methods of foundation treatment, and the proper choice and treatment of construction materials.

Investigations for design and construction of a dam begin in the earliest stages of planning of a project when studies are made to determine whether the project as a whole, including the dam and its appurtenant works, is feasible from engineering and economic standpoints. As a part of these planning studies, the dam site is tentatively selected and mapped and its geological characteristics explored. Sources of materials for construction of the dam and its appurtenant works are also tentatively located. Records of precipitation and run-off are studied to determine the quantity of water available for the project, as well as the magnitude and frequency of floods for use in determining spillway requirements. Based on the information obtained in the planning studies, selection is made of the type of dam to be constructed, and preliminary designs and estimates are prepared and submitted as a part of the project report to the Congress of the United States. After the Congress has approved construction of the project and has appropriated funds for beginning the work, more intensive and detailed investigations of the foundation and construction materials are made to obtain all the information necessary for final design and preparation of construction drawings.

In the early stages of investigations, preliminary information on foundation conditions is often obtained from a study of the known geology of the region, supplemented by surface mapping of rock outcrops and the drilling of a few test holes. When the project has advanced to the final design stage the preliminary information serves as the basis for more extensive and detailed explorations of particular features of the foundation, including the continuity, soundness, and physical properties of the rock; the nature and distribution of the foundation soils; and the elevation of the ground-water surface. Samples of rock are obtained by means of core borings (see figs. 1 and 2), and the presence of fractures, seams, or fissures in foundation rock is investigated by testing each drill hole with water under pressure and measuring the water loss. Overburden materials and earth foundations are explored by hand augers, test pits, and various types of boring machines, depending on the kind of information required. All materials encountered in these explorations are identified and classified, and representative samples are obtained for laboratory tests of permeability compressibility, and strength. Profiles such as that shown in figure 3 are drawn depicting geologic conditions along the axis of the dam and at other locations as desired.

The search for suitable materials begins in the earliest stages of the investigation, with examinations of existing road cuts and stream beds and the drilling of a few test holes in areas near the dam site. Aerial photographs and geologic maps are of considerable aid in

Figure 1.—Core drilling to determine foundation conditions at Hungry Horse Dam site.

Figure 2.—Cores from 4½-inch drill hole in foundation area of Friant Dam.

selecting areas to be explored. Explorations are extended until a complete inventory has been made of all likely sources of concrete aggregates and embankment materials, including rock for slope protection and for rock fill, located within reasonable distances of the site. These explorations provide sufficient data for preparation of preliminary designs and estimates.

Prior to the preparation of final designs, detailed field and laboratory studies are made of the more promising areas or deposits to determine, within fairly narrow limits, the quantity and quality of construction materials available in each deposit. The studies are continued well into the construction period to determine the depth of excavation and the requirements for processing materials. For earth dams, investigations are made of materials from the excavations from spillways, outlet works, and other structures to determine their suitability for use in embankment, because it is often economical to use such materials from these excavations in lieu of borrow.

Auger holes, test pits, and test trenches are used for subsurface exploration of materials deposits. (See figs. 4 and 5.) The location of the ground-water level is recorded and the percentage of water in the deposit obtained. The materials are carefully logged and classified and representative samples are tested in the laboratory. For concrete aggregates, mechanical and petrographic analyses are made and the following prop-

Figure 3.—Geologic profile along axis of Green Mountain Dam.

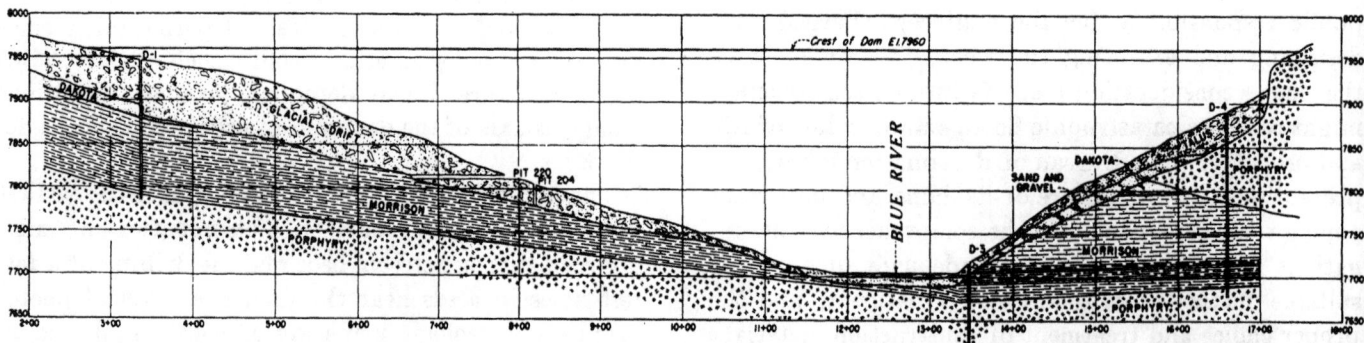

Figure 4.—Test pit in gravel deposit investigated for use in structures of the Columbia Basin project. Material has been screened and segregated into various sizes.

erties are determined: specific gravity, absorption, alkali reaction, soundness, abrasive resistance, structural strength, durability, and the presence of organic solids and soluble salts. Of particular importance is the test for alkali reaction, since aggregates that are otherwise wholly satisfactory may react chemically with the alkali compounds in cement, resulting in serious deterioration of the concrete.

Tests on embankment materials include determination of specific gravity, gradation, compaction characteristics, permeability, shearing strength, and compressibility. Measurements are made of percentages of large rock to determine whether it will be necessary to screen the materials before use in embankments since appreciable quantities of large rock interfere with compaction. Tests are made of the natural unit weight of the material in a borrow pit and compared with the compacted unit weight to determine the amount of shrinkage that will occur from borrow pit to fill.

For the higher earth dams, special studies are made in the laboratory to determine the limits of allowable moisture that should be used in compacting impervious soils so that their strength will not be impaired when subjected to the loads imposed by the fill. Although the desirability of adding moisture to a soil to obtain good compaction was known many years ago, the Bureau of Reclamation was the pioneer in establishing the practice of limiting the moisture content so that an impervious material would consolidate under load without inducing high stress in the fluid contained in the voids between the soil grains. Bureau engineers have shown that these fluid pressures, called pore pressures, act to decrease the shearing strength of soils. The development of the pore-pressure theory has increased the understanding of soil behavior and resulted in improved methods of design and construction, thereby making earth dams of great height economically feasible.

DIVERSION AND CARE OF THE RIVER

Construction of a dam requires that the river be diverted around the site so that the foundation may be unwatered. Several methods are used depending on the type of dam, type of spillway, and space limitations of the river channel or valley floor at the construction site. These methods include diversion of the flow through tunnels driven through the abutments, diversion through openings in the dam itself, diversion over the tops of low sections in the partially completed dam, and diversion around the area in open flumes or in pipes. (See figs. 6 and 7.) Combinations of these methods are sometimes used. Considerable latitude is often given the contractor in selecting the method of diversion of a stream, since the choice will depend to

Figure 5.—Power auger being used to obtain earth samples for testing in the Friant laboratory.

a large extent on his sequence of operations and the rate of progress. Regardless of what method is used, provision must be made to pass the anticipated maximum flow without damage to the structure.

When the river channel or valley is sufficiently broad and large flows must be handled, diversion for concrete dams is usually accomplished over low construction blocks. The flow is first restricted to a part of the channel while the portion of the dam in the remainder of the channel is constructed to an elevation favorable to diversion. The river is then diverted over the top of a completed low block while the portion of the dam is placed in that part of the channel first used for diversion. Following this, the dam is carried to its ultimate height while diversion continues over alternate blocks until use can be made of the outlet works designed for the structure. Under similar circumstances, the economical method of diversion for buttress dams is between buttresses and through a temporary opening in the deck.

Figure 7.—General view of upstream cofferdam and diversion flume at Canyon Ferry Dam site. Excavation equipment is operating in partially unwatered area.

Economy can often be effected in river diversion by the use of the same tunnels for diversion as will later be used for spillways or outlets. The diversion tunnel is so constructed that the upstream portion may be plugged and the downstream portion form a part of the spillway or outlet tunnel. In addition to being economical for both concrete and earth dams under conditions favorable to its use, tunnel diversion normally leaves the entire dam site available for construction operations.

A common method of diversion for earth dams is through conduits under the dam, the outlet works later being built into the conduits as a permanent installation. In cases where the outlet works are located at a relatively high elevation in the dam or its abutment and the cost of a diversion tunnel is prohibitive, provisions are made to divert the river flow through a portion of the channel while a part of the structure and the outlet works are being built. After the outlet works are completed during a period of low flow, the river is diverted through them and the remainder of the structure is completed.

All diversion methods require the construction of temporary dams, called cofferdams, upstream and downstream from the site, to keep the working area dry. (See fig. 7.) The cofferdams are constructed of materials available at the site, such as waste from excavations, or of concrete or timber cribs filled with earth materials. Some cofferdams have been built of steel or wooden sheet piling formed into cells filled with impervious materials for tightness and stability. When an earth dam is to be constructed, it is customary to make the upstream cofferdam a permanent part of the earth dam. The downstream cofferdam, and sometimes the upstream one, is removed after serving its purpose.

Figure 6.—Tunnel intake just after diversion was begun around Anderson Ranch Dam site.

FOUNDATION TREATMENT

The essential requirements for the foundation of a dam are stable support for the structure under all conditions of loading, and adequate resistance to excessive loss of stored water. To secure stable support for concrete gravity and arch dams requires excavation of all materials above the sound rock foundation, followed by treatment of that rock as necessary to consolidate it and make it sufficiently watertight. Foundations of earth materials, as previously stated, require the construction of cut-off trenches, cut-off walls, or other suitable barriers to prevent excessive percolation; they sometimes also require special consolidation procedures to stabilize the material and make it competent to support the superimposed load. Where the overburden is comparatively shallow and is composed of unstable clays, quicksands, saturated silts, or extremely pervious sands and gravels, the entire foundation area of an earth dam may be stripped to bedrock.

After diversion of the river and unwatering of the site, the first step in the preparation of rock foundations, whether for a concrete dam or those portions of an earth dam to be founded on rock, consists of the complete removal of all overburden materials above the rock foundation, including all loose or fractured rock. (See fig. 8.) Blasting is held to a minimum and care is taken to prevent damage to the underlying solid rock. Concrete dams are keyed into the solid rock, and excavation contours for arch dams are made radial or semiradial so that the arch thrust will be approximately at right angles to the rock surface. Immediately before placing concrete on a rock foundation, the surface of the rock is thoroughly cleaned to remove all loose rock fragments, dirt, grout, oil, grease, and other objectionable material. To insure complete bonding and watertightness, a thin layer of mortar is placed on the surface and worked into all irregularities. Essentially the same final treatment is given rock surfaces on which an earth embankment is to be constructed, except that a layer of moist clayey soil is used in lieu of the layer of mortar, and rock surfaces which disintegrate on exposure are either covered immediately with embankment materials or are given a protective coating of bituminous emulsion. Rock foundations for earth embankments require special precautions for control of seepage along the plane of contact between the foundation and the superimposed embankment. These precautions, in addition to removal of all extraneous material, include the construction of a concrete cut-off wall anchored into the rock and extending 5 to 10 feet into the impervious embankment, or provision of an adequate length of contact, or percolation path, between the impervious fill and the rock foundation.

All rock foundations contain seams, fissures, crevices, joints, permeable strata, or faults through which seepage must be prevented or controlled if harmful erosion, excessive uplift pressures, and loss of water are to be avoided. The usual method of treatment is to inject into the foundation under pressure a mixture of cement and water, called grout, through holes drilled into the rock. (See fig. 9.) Grouting operations require supervision by experienced engineers since the injection of fluid under pressure has been known to lift large masses or rock by the principle of the hydraulic jack. For concrete dams, the grouting is usually performed in two stages. The preliminary low-pressure grouting is done before any of the concrete in the dam is placed, to provide a general consolidation of the surface rock and to fill and seal all major surface seams and crevices. Usually, the low-pressure grouting program includes the drilling and grouting of holes throughout an area covered by about the upstream one-third of the dam, and other holes as required to fill all major seams in the general area of the base of the dam. The holes in the upstream area are usually on 20-foot centers and drilled to depths of from 20 to 50 feet. Subsequent to completion of the low-pressure grouting and after sufficient concrete has been placed in the structure, permitting the application of higher pressures, deep holes on a line extending across the channel near the upstream face of the dam are drilled and grouted at high pressure to form the principal grout curtain, or barrier, against seepage underneath the structure. Designs for a major structure usually specify a line of holes on 5-foot centers, but the actual spacing and also the depths will depend on the characteristics of the foundation and the extent to which it accepts the grout. The depth of the holes required for average foundation conditions is usually about 40 percent of the hydrostatic head, but may vary between 20 and 70 percent of this head. Usually a foundation gallery is provided in the dam near the upstream face, from which the drilling and grouting for the high-pressure curtain are done. However, for some low gravity and thin arch dams no gallery is provided, and the drilling and grouting are done from the heel of the dam before water is stored in the reservoir. For most major structures it is customary to extend the foundation gallery into the abutment foundation rock. This facilitates the initial foundation treatment and provides access to the abutments for future foundation treatment should unsatisfactory conditions develop after the reservoir is filled.

The final treatment of the foundation for a concrete dam consists of drilling a line of drainage holes downstream from the high-pressure grout curtain to intercept any water that may percolate under high head through the grout curtain. The drainage holes are drilled after all foundation grouting in the area has been completed,

Figure 8.—Foundation excavation and block construction at Hungry Horse Dam.

Figure 9.—Portable grout machine used at Hungry Horse Dam. Machine forces a mixture of cement and water (called grout) into holes drilled in the foundation.

and the spacing and depth of the holes are governed by the height of the structure and the extent of the foundation grouting. The holes are 3 inches in diameter, spaced on 5- to 10-foot centers, and are drilled to a depth of from 20 to 40 percent of the hydrostatic head.

For earth embankments located on rock, the practice is to drill and grout at least one line of deep holes across the channel to form a cut-off curtain to a depth in the foundation equal approximately to one-half the height of the dam. (See fig. 10.) Holes on 10-foot centers are commonly used. If the surface rock is badly fractured, a trench about 3 feet wide and 3 feet deep is carefully excavated into the rock along the line of the grout curtain and filled with concrete to form a "grout cap." This grout cap acts as an anchor for embedded pipes through which the grouting is done, and also provides a barrier to seepage in the broken surface rock. Shal-

Figure 10.—Curtain grouting in right abutment of Granby Dam.

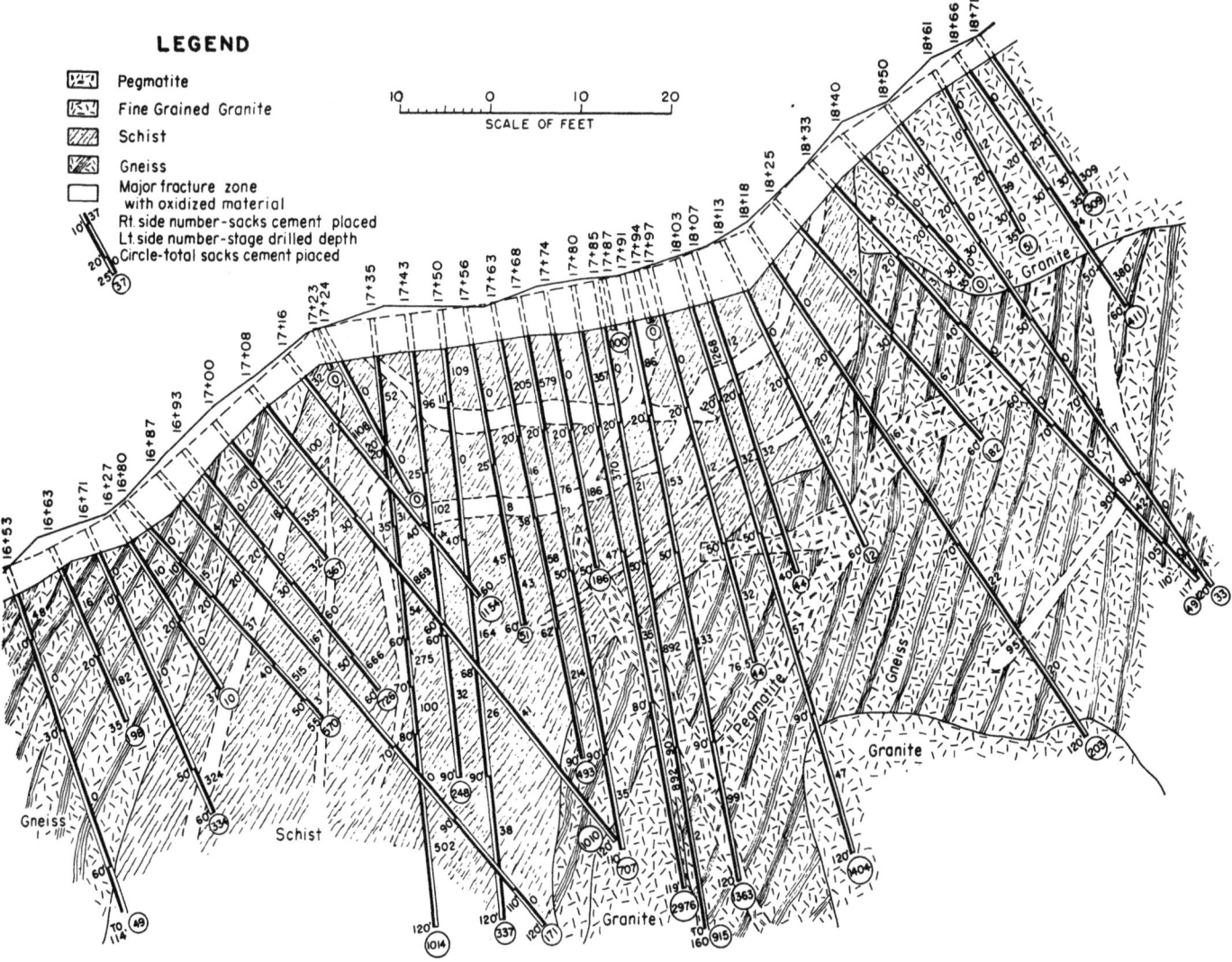

low grouting is sometimes done over large areas of the rock foundation when the nature of the rock requires it.

Earth dams are commonly built on foundations of gravel, sand, silt, and clay. These materials are often unconsolidated and exist in many variations of structural arrangement and combination. For control of seepage in earth foundations the most positive method is to excavate a cut-off trench through the pervious foundation to bedrock or an impervious soil stratum. The cut-off trench is located slightly upstream from the axis of the dam; its side slopes are normally 1½ to 1; and its bottom width is at least 30 feet to permit the use of mechanical equipment in placing and compacting the impervious backfill. Where the bedrock or impervious soil stratum is very far below the ground surface, a positive cut-off by means of a trench becomes impracticable and underseepage cannot be entirely avoided. However, seepage can be reduced and the pressure of the seeping water can be effectively diminished to safe values by lengthening the path of percolation beneath the dam. This is accomplished by constructing a par-

tial cut-off trench or by extending a thick layer of impervious fill on the ground a considerable distance upstream from, and connected with, the impervious zone of the dam. Studies of the effectiveness of such devices are made by the use of graphical methods and laboratory experiments in which the paths of flow of water through the foundations and the loss in the pressure caused by the resistance to flow are determined. Figure 11 shows the foundation flow net obtained by electric analogy experiments for the design of Davis Dam.

Even with a positive cut-off it is virtually impossible to prevent some seepage through the cut-off trench and along the line of contact between an earth fill and its foundation. Although the quantity of this flow is usually small, it is important to provide easy escape of the water to prevent high uplift pressures or springs in the vicinity of the downstream toe. A number of earth dams in this country have failed as a result of excessive pressures from seepage water in the downstream portion of the foundation, through a process called piping. The pressures cause soil particles to be

Figure 11.—Foundation flow net and uplift pressures obtained by electric analogy experiments for Davis Dam

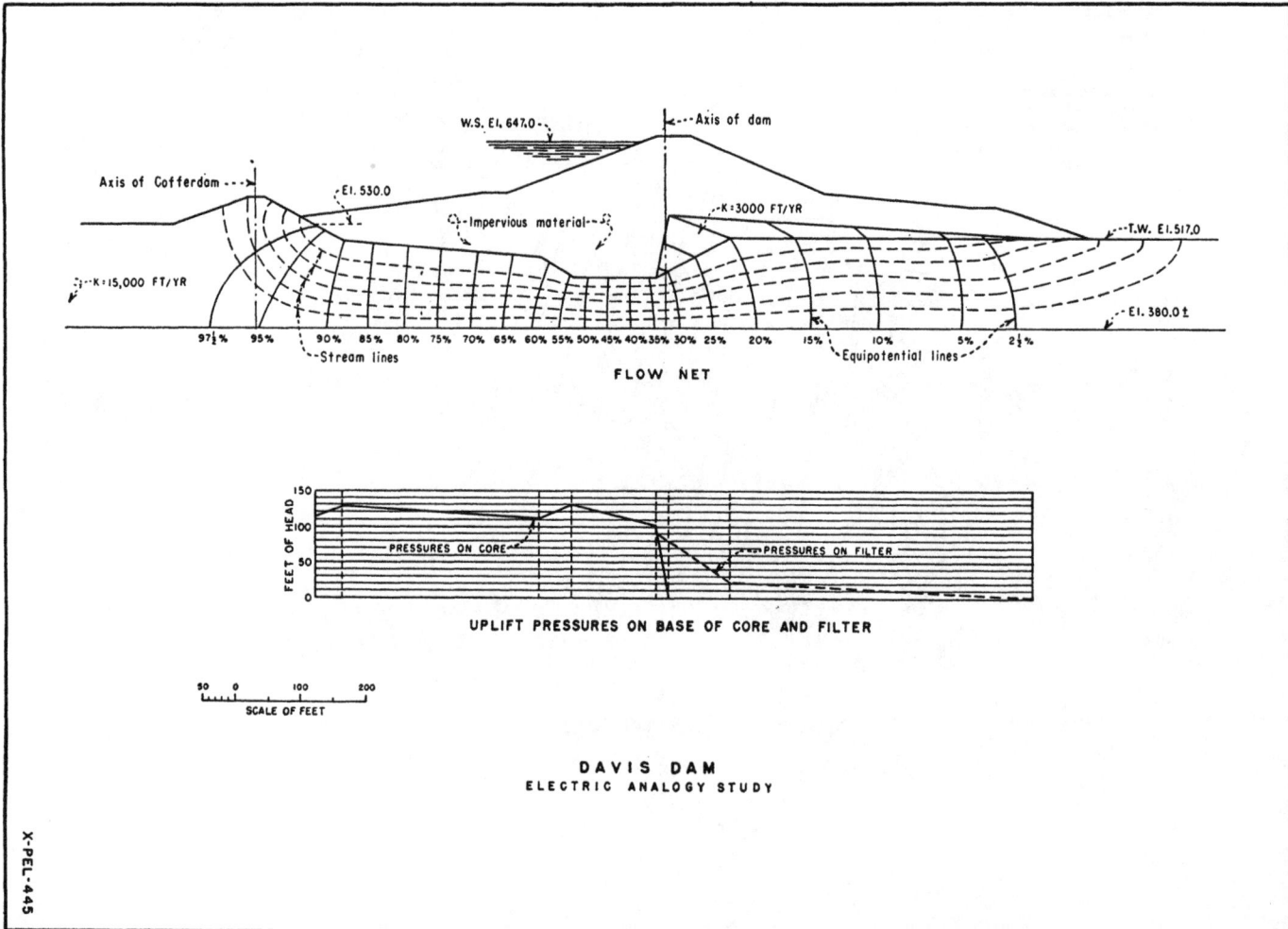

DAVIS DAM
ELECTRIC ANALOGY STUDY

removed, gradually forming an underground channel, or "pipe," which works back to the reservoir, permitting a sudden great flow of water to wash out the foundation. The Bureau minimizes the possibility of piping in its earth dams by providing an "inverted" or reverse filter in the downstream portion of the foundation, which allows dissipation of pressures without removal of soil particles. The filter consists of several layers of graded gravelly material gradually increasing in coarseness and permeability in the direction of flow so that each layer confines the material in the preceding one.

The problem of stability is one that requires special consideration in the treatment of earth foundations. Coarse sands and gravels are usually competent to support the spread load of an earth fill, but fine sands, soft clays, and saturated silts require special treatment to consolidate them. During stripping operations, the stability of earth foundations is improved by removing any pockets of unstable materials such as soft clay, fine sand, or silt, as well as vegetation and organic soil.

Unwatering foundations of saturated fine sand will often serve to stabilize them, and lowering of the groundwater level during excavation of a cut-off trench will often consolidate a considerable area of the foundation. Unwatering operations are generally accomplished by pumping from well points or sumps. (See fig. 12.) Saturated foundations of loose, fine sand sometimes have a tendency to become "quick" when loaded and subjected to vibrations such as earthquakes or large blasts of explosives. These require special study, and extensive treatment is often required to stabilize them.

Foundations of wind-blown silt, called loess, on which many of the dams in the Missouri River basin

Figure 12.—View of cut-off trench at Davis Dam. The collecting pipe along each wall of the trench received water from the rows of well points used to drain that side of the foundation.

rest, require a different treatment from that required for sand. This material, which is usually found in a loose, dry state, resists consolidation until it has become wetted. In order to consolidate the material and preclude excessive deformation and possible failure after the dam is built, loess foundations are irrigated prior to placing fill upon them.

When the foundation for an earth dam consists of soft clay, it is usually necessary to provide special means of preventing lateral movement in the foundation, since soft clays are comparatively low in shear strength. One common method of providing the required resistance to lateral movement is the construction of fills with very flat slopes at the upstream and downstream toes of the dam.

Poor foundation soils which cannot be stabilized are excavated and replaced with stable materials. For example, the foundation material under the buttress overflow section of Imperial Dam was excavated to coarse sand and gravel and replaced with a 12-foot compacted fill of selected materials.

ANALYSIS OF CONCRETE DAMS

Before the final plans and specifications are issued for a dam the structure is subjected to a detailed stability and structural analysis. Following are the principal forces that may act on a concrete dam:

1. Vertical loads due to concrete weight, water pressure, and earthquake accelerations acting on the concrete mass.

2. Horizontal pressures due to reservoir water load and tailwater, and horizontal forces due to earthquake accelerations acting on the concrete mass and the reservoir water and tailwater.

3. Temperature stresses.

4. Pressures from silt loads and earth fills, where present.

5. Ice pressures.

6. Uplift pressures.

Earthquakes are usually assumed to create forces on the dam equal to 0.1 times the force of gravity, acting in horizontal or vertical planes. For major structures located in earthquake zones the forces may be increased to 0.15 gravity. Earthquakes will also cause an increase in the horizontal pressure of the water on the structure, and this increased pressure may be plotted as a parabolic curve instead of the straight-line variation of water pressure alone.

The effect of temperature change is usually disregarded in the stress distribution for gravity dams. However, in the analysis of an arch dam, consideration is given to the probable decrease or increase in temperature from the temperature existing in the concrete of the dam at the time of grouting the transverse contraction joints. For most Bureau dams the temperature existing in the concrete at the time of grouting the joints is lowered artificially to a predetermined point below the minimum stable temperature. When the temperature of the concrete rises after the joints are grouted, the tendency is for the structure to deflect upstream, thus reducing stresses at critical points in the dam.

The amount of pressure exerted by ice is very indefinite and controversial. The pressure is probably due to the thermal expansion of the ice; it may vary with the rate of temperature rise, the thickness and lateral restraint of the ice sheet, and the friction effect of the wind. Pending compilation of additional data and completion of field investigations, a minimum ice pressure equivalent to a horizontal thrust of 10,000 pounds per linear foot is being used by Bureau designers. Appurtenant structures or equipment that might be subjected to ice pressures are often protected by means of a compressed-air ice-prevention system or heaters that prevent the formation of an ice sheet which might cause damage.

Uplift forces are used only in the stability analysis of gravity dams, since for arch dams they are relatively unimportant. Even for arch dams, however, an effort is made to reduce the probable uplift by drains extending vertically through the height of the dam and into the foundation. For preliminary design of gravity dams the uplift pressure is assumed to have an intensity at the line of drains that exceeds the tailwater pressure by one-third of the difference between headwater and tailwater levels, the gradient then extending to headwater and tailwater, respectively, in straight lines. For final design of gravity dams, the pressure distribution is assumed similar to that used in preliminary design, but the pressure intensity at the line of drains is based on results of laboratory experiments using the electric analogy or other comparable method, assuming that the drains are operative and that the grout curtain does not affect the pressure distribution significantly. For both preliminary and final designs the pressures are considered to act over 100 percent of the area. Since earthquake shocks are of short duration, it is assumed that the uplift forces are unaffected by earthquake.

Allowable working stresses used in the design of Bureau dams are conservative. It is felt that public structures of this type should be designed for a maximum degree of safety and long life. Stresses are not usually the controlling factor in the design of gravity dams, but for arch dams the stresses usually do control the design of the structure. Within the past 20 years concrete control and placement methods for Bureau dams have progressed to the point that test cores taken from the structures show the ability to withstand very high compressive stresses. The maximum allowable compressive working stress for use in mass concrete

under extreme loading conditions has recently been increased to 1,000 pounds per square inch, but the working stress in no case is allowed to exceed one-fourth of the maximum compressive strength as determined by tests on concrete cylinders at the age of one year. The maximum allowable shear stress is usually assumed to be about 50 percent of the maximum allowable compressive stress. Concrete is not ordinarily assumed to have strength in tension, but it is recognized that tensile stresses due to twist action and temperature effect probably occur in all concrete dams. Tensile stresses of small intensity are allowed, dependent upon their location and the effect of any consequent cracking on the behavior of the dam.

Stability of the structure against sliding is usually not a problem in the design of arch dams, since the arch is keyed into the abutment walls of the foundation. For gravity and buttress dams, however, the stability of the structure against sliding by shearing on the base is of major concern. In the past, one of the common criteria for estimating the stability of a dam has been the "sliding factor," which assumed that friction is the only horizontal force resisting the downstream force of the water against the dam. However, it has been generally recognized for years that the shear strengths of the concrete and foundation rock contribute a great deal toward the stability of the structure against sliding, and this contribution is presently taken into account by Bureau designers through the use of the shear-friction factor in lieu of the sliding factor. The shear-friction factor may be defined as the algebraic summation of all vertical forces (including uplift) times the coefficient of internal friction for concrete or foundation rock, plus the area times the unit cohesion, all divided by the summation of horizontal forces. The value of the coefficient of internal friction for previous designs has been assumed to be 0.65, but recent laboratory studies have indicated that a value of from 0.8 to 1.0 is probably more nearly correct. The value of the unit cohesion equals the no-load shearing strength of the foundation rock or the concrete, and whichever is the lower should govern the design. The shearing strength of the concrete may be predetermined with some degree of accuracy from known values for previous structures, but the shearing strength of rock may be highly variable, being dependent upon its type and characteristics. Values of shearing strength for Bureau designs have varied from 200 to 700 pounds per square inch, with a value of 400 pounds per square inch as an average assumption. Before final designs are initiated for major structures, the values for internal coefficient of friction and unit cohesion are determined by actual tests of the foundation rock. The minimum value for the shear-friction factor has been set at 4.0 under normal loading conditions. With regard to overturning, the structure will be stable if the concrete stresses are within allowable limits under extreme loading conditions and the quality and strength of the concrete and foundation are satisfactory.

During the 1920–30 decade it became evident that the Bureau would be called upon to construct dams in increasing numbers and of unprecedented proportions. The usual "gravity" method of analysis was suitable only for small gravity dams, and no method had yet been developed for the economical design of an arch dam. It was realized that improved design techniques would have to be developed to assure structures that would be economical as well as stable. Comprehensive studies by outstanding engineers and mathematicians over a period of several years led to the development of the trial-load method for the analysis of gravity and arch dams. Although the trial-load method involves some approximations and does not necessarily, in all cases, give the final stresses within a dam, it makes it possible to calculate the stresses for a given loading about as closely as they can be measured under the most closely controlled laboratory conditions. Through use of the trial-load method, dams of economical proportions can be designed, resulting in large savings of materials. The chief limitations of the method are its complexity and the amount of time and labor required to make an analysis. These are overcome by trained men, automatic calculating machines, and simplified short-cut methods which are constantly being developed.

The trial-load method is used for the design of important gravity dams since it takes into account the stresses caused by twisting and beam action, which are of considerable magnitude near the abutment walls if the walls are steep. The dam may be considered to be made up of a series of vertical cantilever elements that diminish in length from the center of the structure toward the abutments. These cantilevers will deflect by varying amounts due to the applied waterload, and the differential deflections will cause torsional moments or twists in the structure. In the trial-load method of analysis, the dam is assumed to be divided into vertical cantilevers, horizontal beams, and twisted elements, and the waterload is divided between the systems so that deflections and rotations are equal at conjugate points. As the name of the method implies, the division of the load is obtained by trial. Unit-load deformations are computed for the horizontal and vertical elements, and load patterns for each trial are made up of the unit loads multiplied by trial-load factors. After agreement in deflections and rotations is reached, the stresses are computed. Often several different designs are studied, and that one is selected which uses the least concrete and satisfies all design criteria, including stress limitations. Occasionally, a slight shift in position of a dam with respect to the canyon walls will improve the stress distribution. It

has been the practice of Bureau designers, especially on large structures, to specify that the reservoir be filled to some predetermined level dictated by design experience prior to welding the dam into a monolithic mass by grouting the contraction joints. This procedure in effect nullifies to a certain extent the bending in the horizontal elements and reduces the likelihood of tensile stresses occurring at the upstream face of the structure near the abutment sections.

The design of arch dams by the trial-load method is discussed in detail in "Trial Load Method of Analyzing Arch Dams," Bulletin 1 of Part V, Boulder Canyon Project Final Reports (1938). In this method, the dam is considered as being divided into systems of arches and cantilevers (see fig. 13), and the load is divided between these systems so that deflections of the arches and cantilevers are in agreement at various conjugate points throughout the dam. As with gravity dams, the division of the load is determined by trial. Each trial is carried out by a series of three steps or adjustments; namely, radial, tangential, and twist. Unit loads are employed as described for gravity dams. Each of the three adjustments is dependent upon the others and requires readjustment as trial loads are varied for the others. When the adjustments are completed— that is, when all deformations are in agreement— stresses in the arch and cantilever elements are computed, and also principal stresses at the abutments parallel to the upstream and downstream faces. (See figs. 14 and 15.)

A simplified or abridged trial-load method of analysis has been developed by the Bureau for use in the preliminary design of large arch dams and sometimes in the final design of small arch dams located in symmetrical canyons. The method makes use of tables and charts for determining the deflections of the arches and requires only a fraction of the time required for the complete trial-load method previously described. For the abridged method of analysis, symmetry of the dam is assumed and differences in elasticity of the foundation rock and the concrete are neglected, as is also the angle that the plane of the abutment makes with a vertical plane. For a massive arch dam where foundation movements are relatively large, the canyon walls of variable slope, and the dam unsymmetrical, the results obtained by the abridged method may vary greatly from the results of a complete analysis; hence the abridged method should be used with caution.

The two types of buttress dams that have been constructed by the Bureau are the flat-slab type and the multiple-arch type. The two types are similar in principle, each having the principal structural members of a sloping upstream deck supported by piers or buttresses. For the flat-slab type of structure the deck consists of a simply supported flat slab spanning between

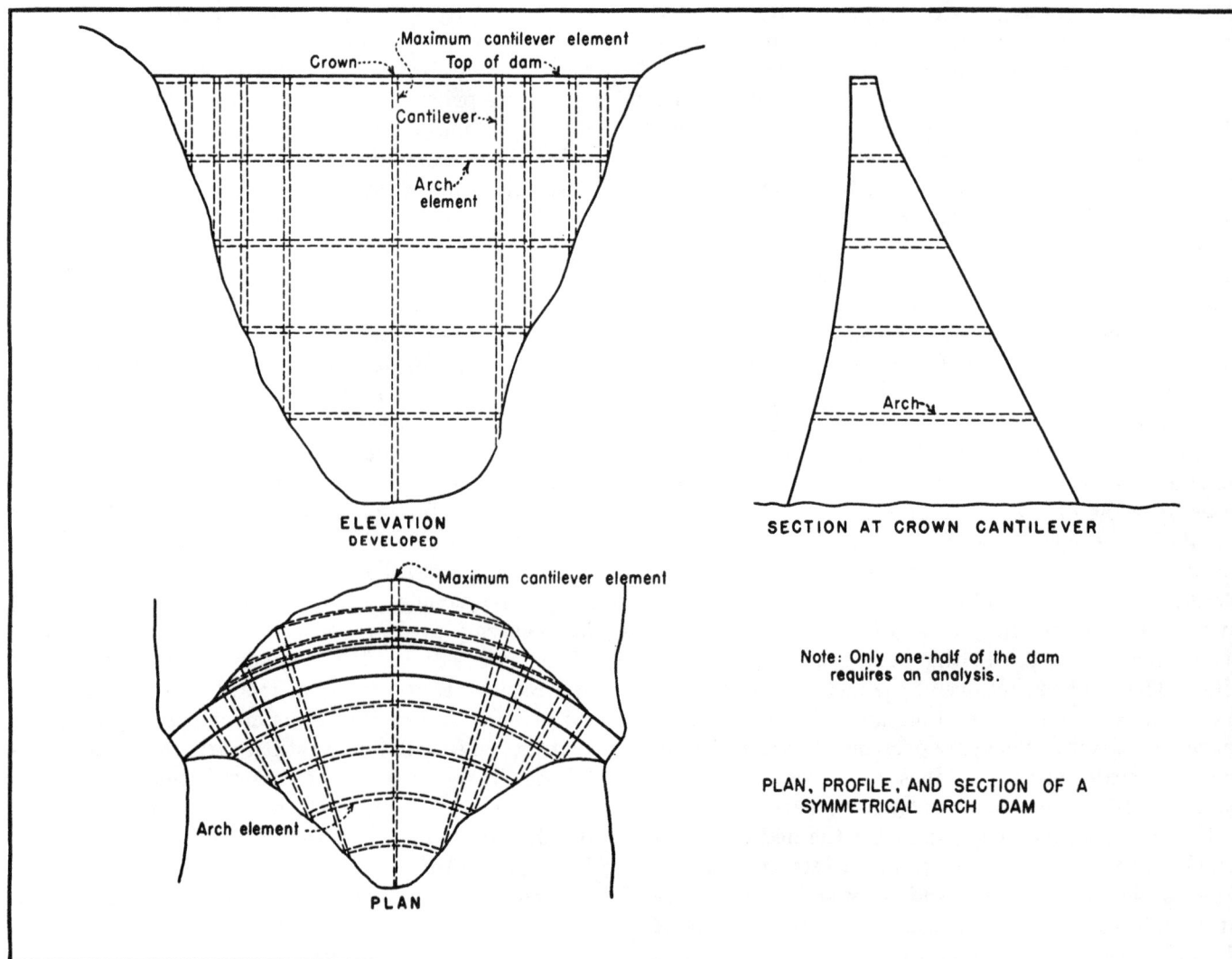

Figure 13.—Symmetrical arch dam showing selected arch and cantilever elements. Water load is distributed by trial between these elements so that deformations of arches and cantilevers at points of intersection are equal.

the buttresses, each unit being structurally independent. The multiple-arch type is a continuous structure with the deck made up of a series of arches, usually 180° and circular in shape. The multiple-arch type generally is preferred for higher buttress dams, while the flat-slab type is usually preferred for a long and low structure because of its simplicity and economy of construction. Although the Bureau has constructed very few buttress-type dams, the methods for economical designs have been worked out in detail. Curves and formulas have been devised whereby the most economical slope of the deck may be determined, together with the most economical spacing of the buttresses. Photoelastic studies have been made of the corbels that transfer the deck load to the buttresses in order to determine the concentrations of stresses. For buttress dams of unusual heights special studies may be necessary for the analysis of the deck. For Bartlett Dam,

a high multiple-arch structure, the trial-load method of analysis was used for the arch barrels.

SOME FACTORS AFFECTING LAYOUT OF CONCRETE DAMS

Beginning with the construction of Hoover Dam, the Bureau has considered temperature control of the mass concrete as one of the most important problems to be considered in the design and construction of major dams. Temperature control is a definite requirement since volume changes in the concrete due to uncontrolled temperature variations would produce tensile stresses that would crack the concrete, thereby affecting the appearance, durability, and even the stability of the dam. The accumulation and removal of excess heat due to heat of hydration or any other cause can be controlled to produce temperature variations and distributions that fall within previously determined allowable limits. Much of the accumulation of excess heat may be prevented by the use of special cements and reduced cement content, precooling aggregates and other materials used in the concrete, and regulating construction procedures as by limiting the thickness of placement lifts and exposure of lifts after placement.

Usually, however, excess heat must be removed from the concrete by artificial cooling methods if the volume is to be reduced to a minimum and contraction joints are to open sufficiently for grouting during the construction period.

The cooling system specified by the Bureau is made up of a number of embedded coils of 1-inch, thin-wall

Figure 14.—Arch and cantilever stresses obtained by trial-load analysis for Hungry Horse Dam.

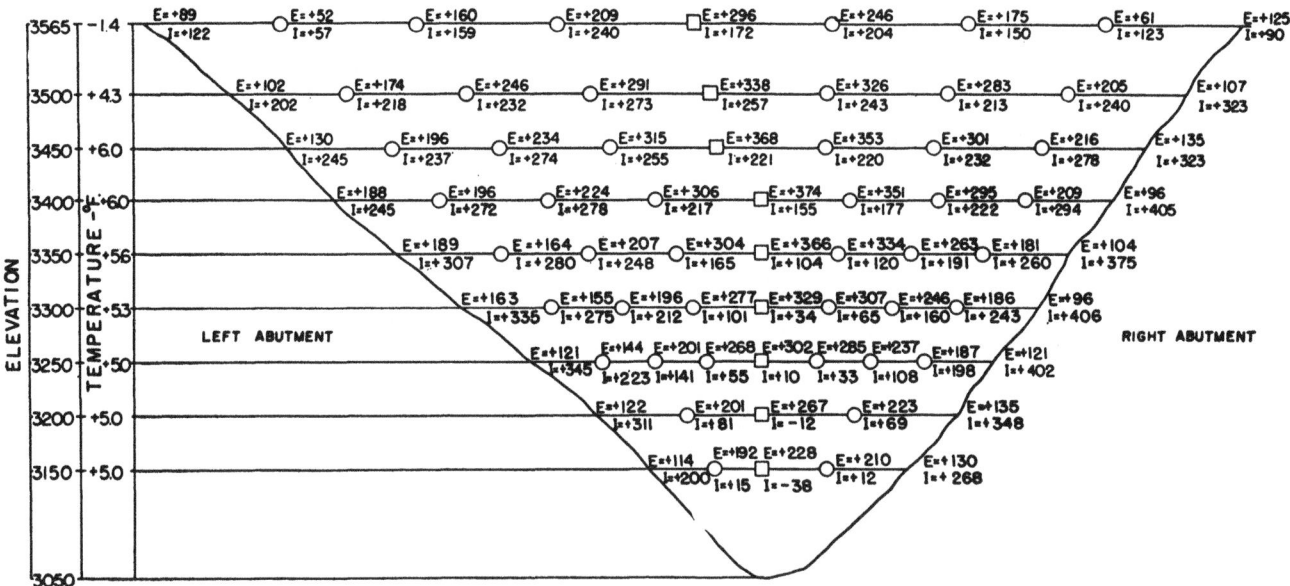

ARCH STRESSES

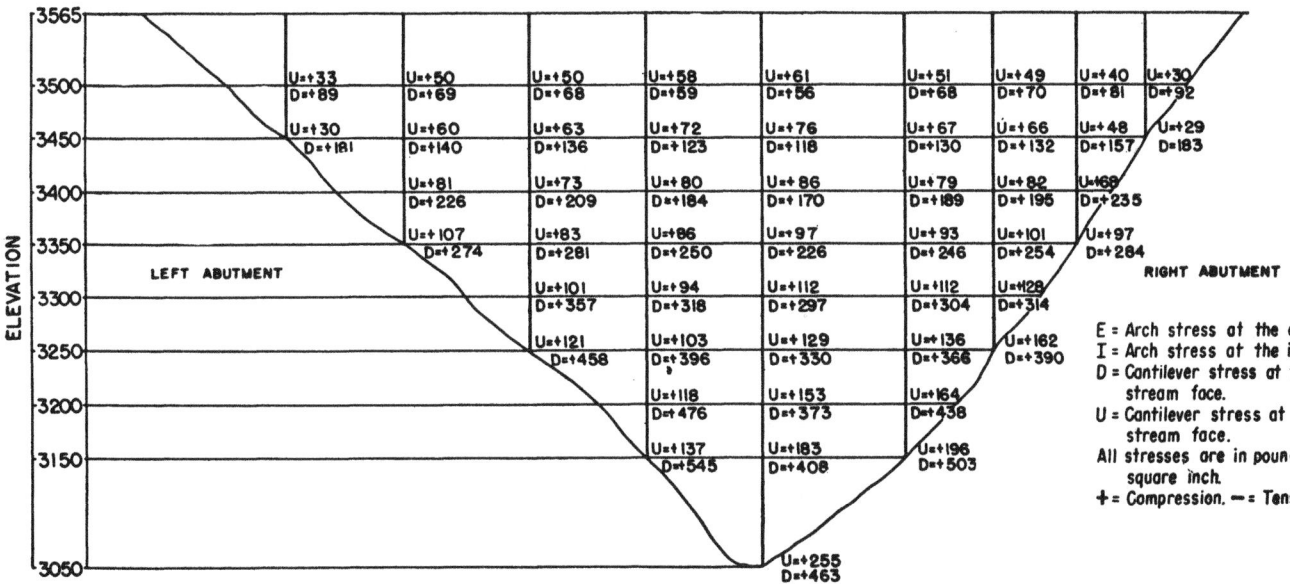

Elevation / Temperature:

Elev. 3565, Temp. −1.4: E=+89 I=+122; E=+52 I=+57; E=+160 I=+159; E=+209 I=+240; E=+296 I=+172; E=+246 I=+204; E=+175 I=+150; E=+61 I=+123; E=+125 I=+90

Elev. 3500, Temp. +4.3: E=+102 I=+202; E=+174 I=+218; E=+246 I=+232; E=+291 I=+273; E=+338 I=+257; E=+326 I=+243; E=+283 I=+213; E=+205 I=+240; E=+107 I=+323

Elev. 3450, Temp. +6.0: E=+130 I=+245; E=+196 I=+237; E=+234 I=+274; E=+315 I=+255; E=+368 I=+221; E=+353 I=+220; E=+301 I=+232; E=+216 I=+278; E=+135 I=+323

Elev. 3400, Temp. +6.0: E=+188 I=+245; E=+196 I=+272; E=+224 I=+278; E=+306 I=+217; E=+374 I=+155; E=+351 I=+177; E=+295 I=+222; E=+209 I=+294; E=+96 I=+405

Elev. 3350, Temp. +5.6: E=+189 I=+307; E=+164 I=+280; E=+207 I=+248; E=+304 I=+165; E=+366 I=+104; E=+334 I=+120; E=+263 I=+191; E=+181 I=+260; E=+104 I=+375

Elev. 3300, Temp. +5.3: E=+163 I=+335; E=+155 I=+275; E=+196 I=+212; E=+277 I=+101; E=+329 I=+34; E=+307 I=+65; E=+246 I=+160; E=+186 I=+243; E=+96 I=+406

Elev. 3250, Temp. +5.0: E=+121 I=+345; E=+144 I=+223; E=+201 I=+141; E=+268 I=+55; E=+302 I=+10; E=+285 I=+33; E=+237 I=+108; E=+187 I=+198; E=+121 I=+402

Elev. 3200, Temp. +5.0: E=+122 I=+311; E=+201 I=+81; E=+267 I=−12; E=+223 I=+69; E=+135 I=+348

Elev. 3150, Temp. +5.0: E=+114 I=+200; E=+192 I=+15; E=+228 I=−38; E=+210 I=+12; E=+130 I=+268

Elev. 3050

LEFT ABUTMENT   RIGHT ABUTMENT

CANTILEVER STRESSES

Elev. 3500: U=+33 D=+89; U=+50 D=+69; U=+50 D=+68; U=+58 D=+59; U=+61 D=+56; U=+51 D=+68; U=+49 D=+70; U=+40 D=+81; U=+30 D=+92

Elev. 3450: U=+30 D=+181; U=+60 D=+140; U=+63 D=+136; U=+72 D=+123; U=+76 D=+118; U=+67 D=+130; U=+66 D=+132; U=+48 D=+157; U=+29 D=+183

Elev. 3400: U=+81 D=+226; U=+73 D=+209; U=+80 D=+184; U=+86 D=+170; U=+79 D=+189; U=+82 D=+195; U=+68 D=+235

Elev. 3350: U=+107 D=+274; U=+83 D=+281; U=+86 D=+250; U=+97 D=+226; U=+93 D=+246; U=+101 D=+254; U=+97 D=+284

Elev. 3300: U=+101 D=+357; U=+94 D=+318; U=+112 D=+297; U=+112 D=+304; U=+128 D=+314

Elev. 3250: U=+121 D=+458; U=+103 D=+396; U=+129 D=+330; U=+136 D=+366; U=+162 D=+390

Elev. 3200: U=+118 D=+476; U=+153 D=+373; U=+164 D=+438

Elev. 3150: U=+137 D=+545; U=+183 D=+408; U=+196 D=+503

Elev. 3050: U=+255 D=+463

LEFT ABUTMENT   RIGHT ABUTMENT

E = Arch stress at the extrados.
I = Arch stress at the intrados.
D = Cantilever stress at the downstream face.
U = Cantilever stress at the upstream face.
All stresses are in pounds per square inch.
+ = Compression. − = Tension.

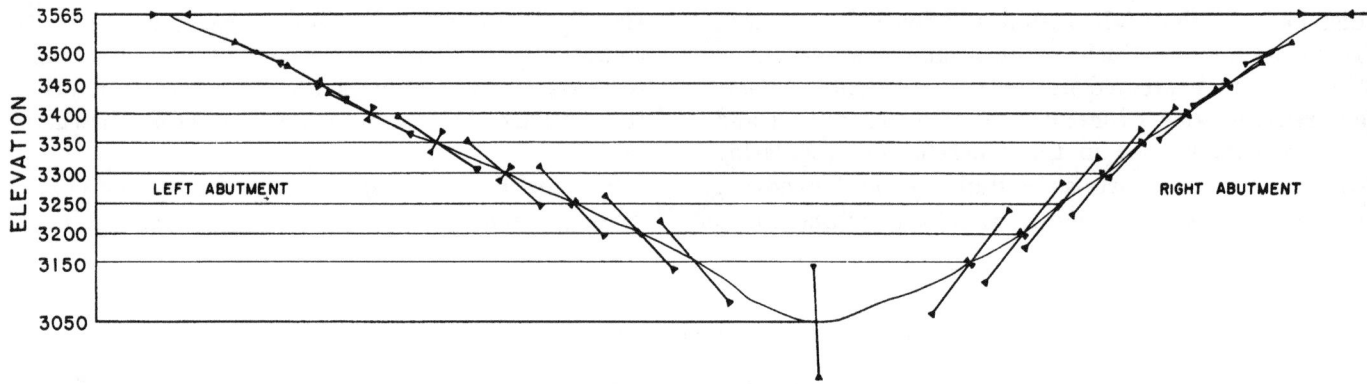

PRINCIPAL STRESSES-UPSTREAM FACE

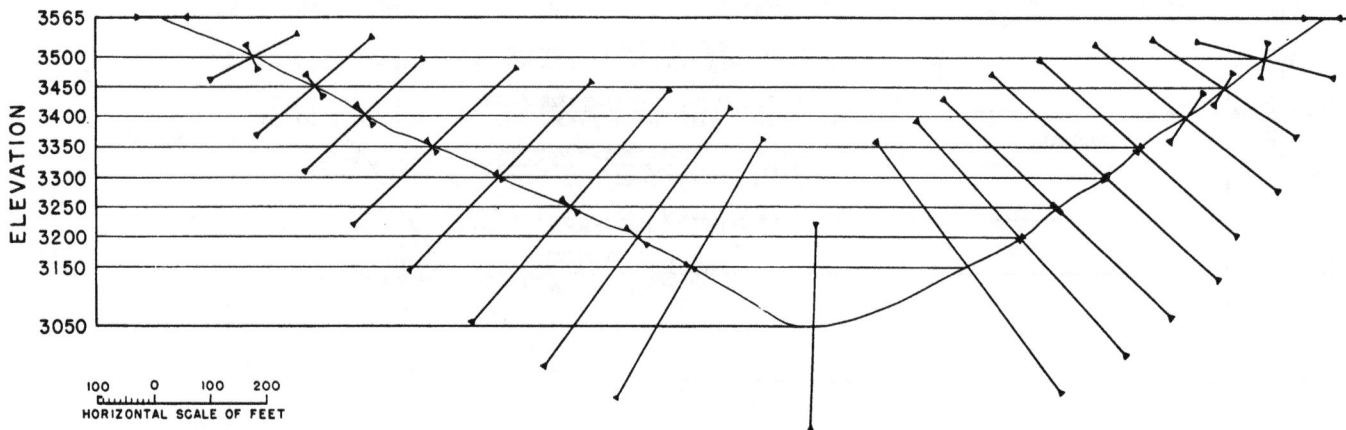

HORIZONTAL SCALE OF FEET
100   0   100   200

SCALE OF STRESSES IN LB. PER SQ. IN.
200 100 0   500   1000

→← Compression
←→ Tension

PRINCIPAL STRESSES-DOWNSTREAM FACE

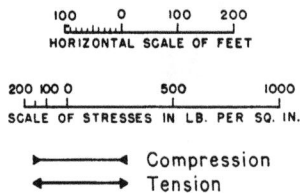

Figure 15.—Principal stresses obtained by trial-load analysis for Hungry Horse Dam.

tubing, served by supply and return headers. Coils are from 800 to 1,200 feet long and are placed on the rock foundation and on each subsequent 5-foot lift after the concrete has hardened. (See fig. 16). The horizontal spacing of the tubing ranges between 2 and 6 feet, depending on the temperature variations and the amount of control required. Cooling water is pumped through the embedded coils at the rate of about 4 gallons per minute for as many days as is required to lower the temperature of the concrete to the desired state. It is customary to specify that the flow through the coils be reversed periodically, usually once each day, so that all of the concrete is cooled equally.

The geographical location of the dam may have a pronounced effect upon the cooling requirements. For Hungry Horse Dam, which is located on a stream near the Canadian border that is fed by melting snow, the

Figure 16.—Cooling pipe was installed on top of each 5-foot lift at Friant Dam. Workmen are washing sandblasted surface preparatory to pouring another lift of concrete. Note vertical keyed construction joint between adjacent blocks.

mass concrete was initially cooled for a minimum period of 16 days by water supplied from a nearby tributary stream. Final cooling was then deferred until the winter months when the temperature of the river water was low enough so that the concrete could be subcooled to the desired temperature of 38° F. prior to the grouting of the contraction joints. Hoover Dam (see fig. 17) presented an entirely different problem since it is located in semidesert country where temperatures of both air and water are abnormally high. The initial cooling was accomplished by air-cooled river water, but the use of a refrigeration plant to cool the water was necessary for subcooling the concrete to the average desired temperature of 56°.

Contraction joints are formed in the mass to relieve the tensile stresses resulting from restrained contraction of the concrete, and thus prevent uncontrolled cracking. The spacing of the joints is usually dependent upon the following factors: physical features of the dam site, details of appurtenant structures to the dam, temperature control requirements, and size of blocks that will not require excessive plant capacity for handling and placing concrete.

All concrete dams have a system of transverse joints that are constructed normal to the axis of the dam. The spacing of these joints is dependent on the mixing-plant capacity since the concrete in the 5-foot lifts is placed in 15- to 20-inch increments in depth across the

Figure 17.—Schematic layout of cooling system at Hoover Dam. First-stage cooling used river water and cooling tower, second-stage used refrigerated water.

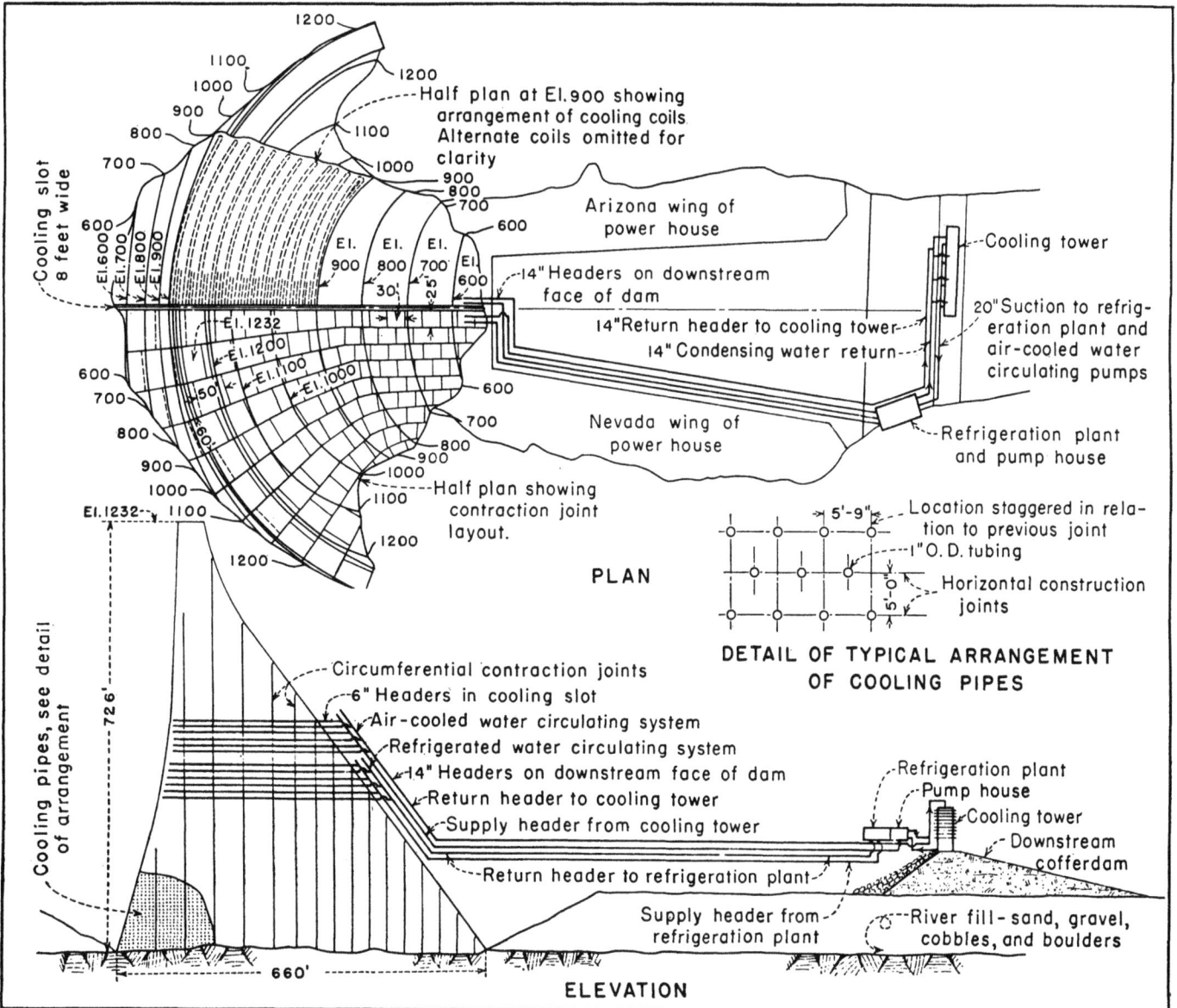

Figure 18.—Construction blocks at Shasta Dam. Note that longitudinal joints in alternate rows are staggered.

block, and the concrete in a previously placed increment must still be soft when the succeeding increment is placed. In the past, the practice has been to space the transverse joints about 50 feet apart in the dam, but recently a maximum spacing of 80 feet has proved feasible. The joints, as shown in figure 16, are constructed with keys that serve to minimize the leakage of water through the joints and also to provide shearing resistance. The dimensions of these keys have been standardized for all solid concrete dams constructed by the Bureau, being 6 inches deep, 2 feet 6 inches wide at the joint line, and 6 inches wide at the back of the key.

As the size of a dam, especially a gravity type, is increased, the base thickness approaches a limiting dimension beyond which there is a tendency for vertical cracking across the blocks due to temperature and shrinkage effects. To avoid this cracking, longitudinal contraction joints are constructed parallel to the axis of the dam and offset at the transverse joints, thus forming longitudinal rows of staggered columns. (See fig. 18.) In the construction of major Bureau dams the longitudinal joints were formerly spaced about 50

feet apart, but adoption of the practice of more positive temperature control of the concrete has permitted increasing the limiting dimension to about 200 feet. This has reduced by about 75 percent the number of longitudinal joints required. Shear keys are provided in the longitudinal joints with the key faces inclined so that they conform approximately with the lines of principal stress for the full water load on the dam.

After the dam has been cooled so that maximum shrinkage has taken place, the contraction joints are grouted to bind the entire structure into one continuous mass. A thin grout mixture of portland cement and water is forced into the opened joints under pressure so that adjacent blocks are bonded by a dense film of cement. The grout mixture is pumped through an embedded pipe system, usually consisting of $1\frac{1}{2}$-inch diameter supply and return headers from which $\frac{1}{2}$-inch branches lead to outlet units spaced uniformly over the face of the joints. In order to insure complete grouting before the grout mixture begins to set and to eliminate excessive pressures in the system, the joints are grouted in sections about 50 feet in height. The grouting may either be done from the downstream face of the dam or from galleries and shafts located in the interior of the mass, depending on the type of dam and spillway.

Galleries, shafts, and chambers (see fig. 19) are formed in the mass of a concrete dam for any one or a combination of the following purposes:

1. To provide drainage for water percolating through the upstream face or through the foundation.

2. To provide space from which to drill grout and drainage holes and grout the high-pressure curtain.

3. To provide space for headers and equipment used for the artificial cooling of the dam and the grouting of the contraction joints.

4. To provide access to the interior of the dam for inspection.

5. To provide access to and room for necessary mechanical and electrical equipment.

The gallery system is dependent upon the size and type of dam, foundation conditions, and location and type of appurtenant structures. The number of galleries in a dam may vary from only one foundation gallery to an elaborate system such as was formed in Shasta and Grand Coulee Dams, with inspection or gate-service galleries at each 50-foot level. For major structures it is customary to provide one or more elevator shafts for ready access between the bottom and top of the dam. All openings formed in the mass of a dam are reinforced for the tensions created by the change in direction of the compressive stresses at the openings.

DESIGN OF EARTH EMBANKMENTS

The objective in the design of an earth dam or embankment is to determine that cross section which,

Figure 19.—Grouting and drainage gallery in Friant Dam.

when constructed with the materials near at hand, will safely and most economically fulfill its required function. Because of the complexity and variability of the physical properties of soils under conditions of saturation and loading, an earth structure is not susceptible to exact mathematical analysis. Its design must be based on experience records of successful structures, together with results of examinations of foundation conditions, field and laboratory tests of the available construction materials, and a knowledge of the principles of soil mechanics. The usual procedure in designing an earth dam consists of making several trial designs and analyzing them for conformance with the following criteria:

1. The dam must be sufficiently watertight for the purpose intended.

2. The dam must be stable and hold its shape.

3. The dam must be safe from overtopping by floodwater.

4. The dam must be safe from damage by erosion of its surfaces.

The outer slopes selected for a dam depend on the height of the dam, the strength of its foundation, and the character of materials available for embankment construction. Slopes as steep as 2 to 1 may be satisfactory where optimum conditions prevail, but 4 to 1 or even flatter slopes are necessary for high dams where the foundation and materials are poor. Figure 20 shows sections of typical earth dams constructed by the Bureau of Reclamation.

In the past, many earth dams were constructed with a concrete core wall extending entirely through the embankment to stop percolation, but it has been adequately demonstrated that, with modern methods of construction, percolation can be effectively prevented by suitable foundation preparation, adequate control of moisture and compaction, and proper selection of embankment materials. Homogeneous dams are constructed throughout of one type of soil, which contains sufficient clay or silt to make the embankment impervious. Wherever suitable materials are available, the Bureau uses the zoned type of construction in which a compacted, impervious inner zone acts as a water barrier. It is recognized that even the tightest compacted earth fill is not impervious since there are pore spaces in the material which permit the passage of water. However, the quantity of flow through a properly compacted impervious zone of an earth dam is extremely small.

Most soils that are fine-grained and impervious enough for use in a water-barrier zone are structurally weak. (An exception to this is a well-graded gravel-sand mixture containing up to 25 percent of silt and clay which, when moistened and compacted, is both impervious and stable.) For this reason, the impervious

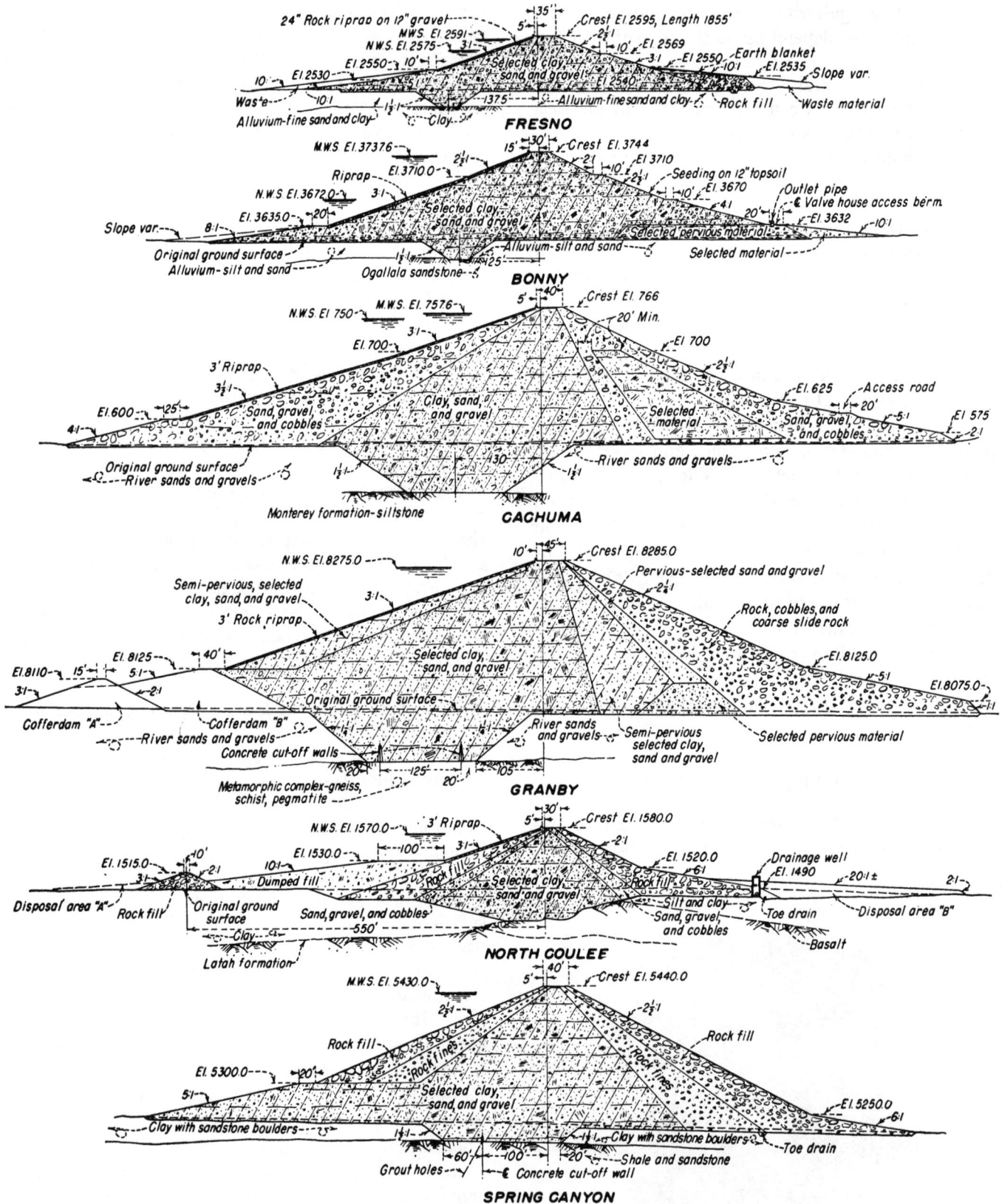

Figure 20.—Sections of typical earth dams constructed by the Bureau of Reclamation.

inner zone of a zoned type of dam is supported by zones of coarse-grained materials to secure a stable embankment with a minimum volume for the dam. These zones of coarse-grained soils are placed both upstream and downstream from the impervious core.

The stability of an embankment depends on the forces acting upon it and the shearing strength of the soils. Although several analytical methods of stability analysis have been developed and applied by Bureau engineers, variations of the Swedish slip-circle method are most commonly used. (See fig. 21.) This method is based on observations that embankment slides occur along nearly circular arcs, and that when a slide occurs the known tangential forces acting on the sliding mass of earth exceed the resisting forces along the arc. The cohesive property of soils and the strength developed from friction between the soil grains give rise to the resisting forces. The factor of safety is the ratio of resisting forces to the forces tending to make the material slide. The Swedish method requires several trial failure circles to determine the one with the smallest factor of safety. A minimum safety factor of 1.5 is commonly used in earth dam design. Stability analyses are made on a trial design for at least three conditions of loading:

1. For the condition during construction.
2. For the full reservoir condition.
3. For a condition of sudden draw-down.

A high earth embankment can fail during construction because of pore pressures in the foundation and in the impervious zone due to the superimposed weight of fill. The Bureau has made an intensive study of this phenomenon and has developed theoretical methods of estimating the magnitude of these pressures as well as devising and installing instruments to measure them. The information from field measurements on actual dams, supplemented by theory, has given a rational basis for controlling the construction of impervious zones to reduce these pore pressures and avoid failure of the fill during construction. Where there is danger of excessive pore pressures, the Bureau uses special construction control procedures, and provides an impervious core of minimum size, with outer zones of high-strength gravelly soil or rockfill to furnish lateral support for the core. A filter zone of intermediate-size material is

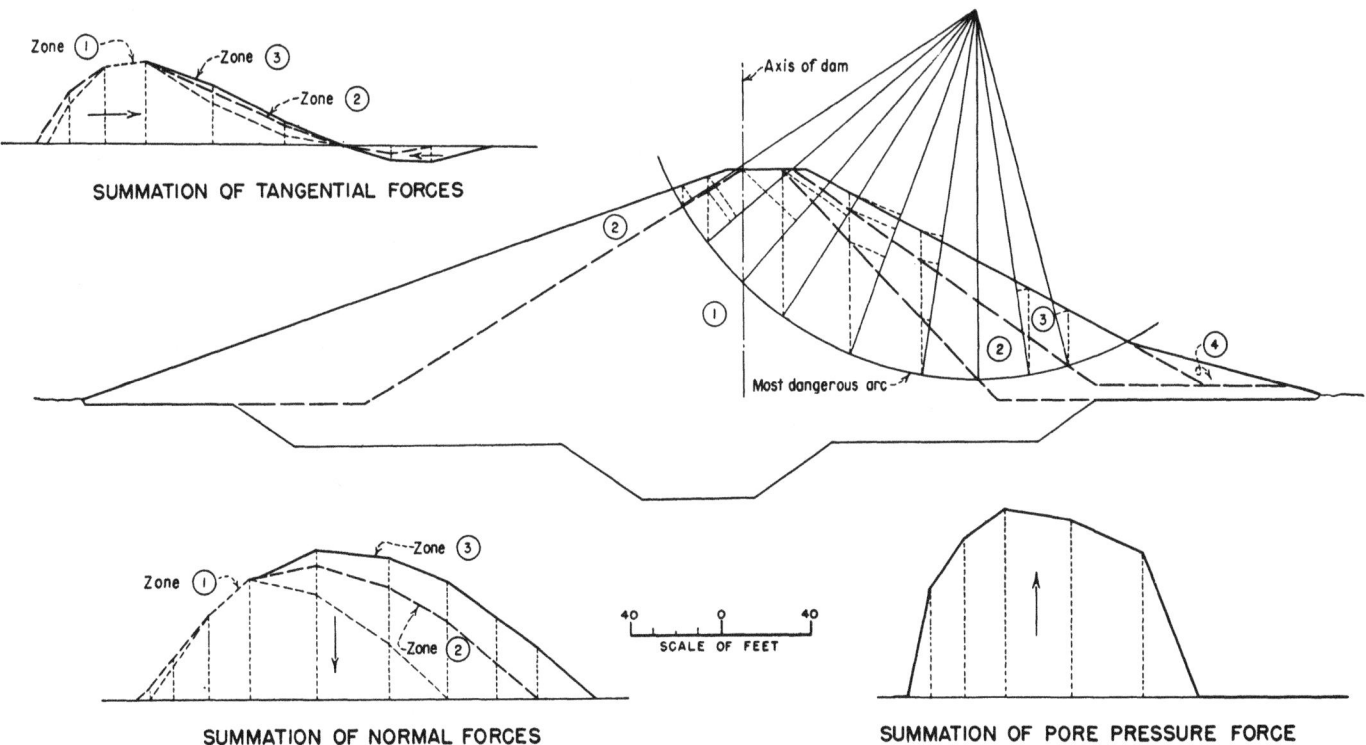

Figure 21.—Stability analysis of a zoned earth dam by the Swedish slip-circle method.

	ZONE ①	ZONE ②	ZONE ③
Wet density of materials, lbs./ft.³	127.9	137.5	105
Tan φ	0.65	0.70	0.70
Cohesion, lbs./ft.²	800	0	0

Normal force on arc in Zone 1 = 4,020 x 127.9 + 1,593 x 137.5 + 687 x 105 = 805,000 lbs.
Normal force on arc in Zone 2 and 3 = 615 x 137.5 + 1,086 x 105 = 198,500 lbs.
Tangent force on arc = 2,280 x 127.9 + 326 x137.5 - 72.4 x 105 = 328,700 lbs.
Pore pressure force = 7,820 x 62.5 = 488,000 lbs.
Length of arc in Zone 1 = 157.2 ft.
$$S.F. = \frac{157.2 \times 800 + (805,000 - 488,000) \, 0.65 + 198,500 \times 0.7}{328,700}$$
S.F. = 1.45

provided adjacent to the core to prevent fine particles from being carried into the coarse material by seepage water.

If special provisions were not made in design to avoid it, all homogeneous impervious embankments would begin to fail when water from the reservoir finally seeped through the structure and reached the downstream slope of the dam. The exit pressure of the water, regardless of the quantity seeping, would soon loosen the soil and cause the slope to slough. In addition, the flow of water through the dam would create an appreciable force in the direction of flow, which would act to decrease the factor of safety in the downstream portion of the dam. Seepage water may be prevented from reaching the downstream slope in a homogeneous dam by providing drains in the downstream portion of the embankment. Trial designs are analyzed for adequate safety against such failure; for this purpose a flow net is drawn and the slip-circle method of anlysis is used. Construction of a zoned-type dam on an adequate foundation virtually precludes the possibility of seepage failure since the line of seepage never reaches the downstream slope but is caused to drop at the boundary between the impervious core and the downstream pervious zone.

Irrigation reservoirs are usually emptied each year to serve their primary purpose of furnishing water. When the water surface is lowered, seepage forces act to decrease the stability of the upstream slope. The time involved in emptying the reservoir may be weeks or months, yet in view of the low permeability of the impervious fill the draw-down may be considered as "instantaneous." This instantaneous draw-down condition is analyzed by use of the flow net and the slip-circle method.

Special precautions are required to prevent overtopping of earth dams because of the destructive erosion and possible failure that would result. The freeboard, or the vertical distance from the maximum water surface to the top of the dam, is determined from the characteristics of the spillway and the height of waves that may occur in the reservoir. The heights of waves depend on the depth of the reservoir, length of open water for wave build-up (termed fetch), and wind velocity. For reservoirs up to 6 miles in length and wind velocities up to 100 miles per hour, it has been found that a freeboard of 6 feet for an ungated spillway on an earth dam is satisfactory. When gate-controlled spillways are used, the freeboard is usually increased somewhat to provide an additional margin of safety against possible malfunction of the gates.

The width of the crest of an earth dam is often determined by the requirements for a road across the structure. Where no roadway is required, the width will vary from a minimum of about 20 feet for low dams to 40 feet for high structures.

The action of waves against the upstream face of an earth dam would soon wash away large quantities of soil; hence, the upstream slope must be protected. The most satisfactory kind of protection is dumped rock riprap, which consists of hard, angular rock fragments graded in size from about one-half cubic foot to one-half cubic yard. With medium-sized reservoirs, the riprap is usually placed in a 3-foot layer on top of an 18-inch-thick filter layer of crushed rock or gravel. The rough rock surface tends to dissipate the energy of the waves and prevent them from riding up the slope of the dam. Concrete and asphalt paving have been used for protection of the upstream slope where riprap was expensive, but generally they have not been as satisfactory as riprap. The downstream slope of an earth dam must be protected against erosion by wind and rainfall. Usually a coarse gravel cover of 12-inch minimum thickness is used, but where gravel or rock is not economically available the downstream slope is seeded with hardy grasses. When no roadway is required over the dam, the crest is usually protected by a layer of gravelly material.

CONSTRUCTION OF EARTH EMBANKMENTS

Bureau earth dams are built by private contractors, who, to the greatest possible extent, are permitted to use their enterprise and ingenuity in producing the desired results. This policy has resulted in the development of efficient equipment and methods which have consistently reduced the cost of earthwork, while producing entirely satisfactory structures. Construction control is maintained by the Bureau's construction engineer and his inspection and field laboratory staffs.

The operations of excavation, transportation, placing, and compaction of the materials for the various zones of the dam embankment vary considerably depending on the nature and quantities of materials involved and the type of equipment the contractor elects to use. If appreciable amounts of rock larger than about 5 inches are present in the material to be placed in the rolled impervious fill, some sort of separation plant is used to remove this oversized rock in order to prevent interference with compaction. On small jobs, power shovels of 2 to 3 cubic yards capacity are frequently used to excavate materials from the borrow area, and trucks of 15 cubic yards capacity or greater transport the earth to the embankment where it is dumped, spread by bulldozers in layers of specified thickness, and compacted. Compaction of impervious fill is permitted only after the material has been brought to the proper moisture content, and is done with very heavy tamping, or sheepsfoot, rollers which compact

Figure 22.—Special tamping rollers compacted the earth embankment at Anderson Ranch Dam.

excavating equipment is slowly moved by tractors and the trucks are loaded while in motion, resulting in considerable saving of time. Two excavating units were used at Bonny Dam, each capable of excavating 1,500 cubic yards of earth per hour. Materials were hauled to the fill, spread by large bulldozers, and rolled to 6-inch compacted lifts by 6-foot diameter, dual-drum tamping rollers weighing over 40,000 pounds. More than 1,000,000 cubic yards of rolled fill per month were placed by this procedure. Other large jobs use 5-cubic-yard shovels with similar transporting, placing, and compacting equipment. Clay for the impervious section of Anderson Ranch Dam was excavated by electric shovels and loaded into the hopper of a special "pendulum" type loader that deposited the material on a conveyor belt for transportation to the dam. (See fig. 24.)

each layer of soil from the bottom up. The Bureau has developed a special roller with improved tamping characteristics. Figure 22 shows such a roller in operation. Compaction of material in pervious zones is done by the vibration and weight of crawler-type tractors after each layer of specified thickness has been placed and thoroughly wetted. Rock-fill zones are not especially compacted but are placed in layers and subjected to equipment travel.

On large projects where conditions of the borrow area are suitable, excavation of earth materials is often done by elevating graders equipped with vertical cutter blades which obtain a mixture of material from the face of a cut in the borrow area. The equipment forces the material onto a conveyor belt which raises it to a sufficient height whence it is dropped into large bottom-dump trucks. Figure 23 shows such equipment at Bonny Dam making a cut in the borrow area. The

Figure 24.—Clay for impervious section of Anderson Ranch Dam was loaded into hopper of "pendulum" by electric shovels for transportation to the dam over belt conveyor.

Figure 23.—Elevating graders excavated and loaded materials for Bonny Dam. Vertical cutter blades obtained a mixture of materials from the face of the cut, and the equipment loaded the trucks while in motion.

The shovels made long, vertical cuts on the embankment face, which served to mix any stratified materials.

On many projects where conditions are suitable, use is made of various types of self-loading scrapers which rapidly and efficiently perform all the operations of excavating, loading, hauling, and spreading the material. These scrapers are especially adaptable to embankment construction where borrow pits of uniform, unstratified material are located fairly near the site. When used in stratified deposits they have a disadvantage of excavating in horizontal layers, which permits little or no mixing of material. This can be overcome by using other equipment to perform the initial excavating and mixing. Self-loading scrapers are frequently used in stripping of the topsoil at a dam site.

Where the moisture content of material in the impervious borrow area is considerably less than that required for proper compaction, additional water must be added. In such instances, it has been found that much better results are obtained when the water is added to the material by irrigating the borrow pits rather than by sprinkling the fill. Ponding the area or use of a sprinkler system is the method usually employed, and it must be done well in advance of excavation to permit uniform penetration of the water. Where the borrow-pit material is too wet, drainage ditches are used to reduce the moisture content of the soil.

CONSTRUCTION CONTROL FOR EARTH EMBANKMENTS

There is no one scheme of construction control applicable to all earth dams or embankments. For each major structure, the criteria for processing, placing, and compacting the materials must be determined after careful consideration of the design requirements and the characteristics of the soils to be used. In controlling the construction of the impervious zone of a dam, the following criteria must be met:

1. The material must be formed into a homogeneous mass, free from any potential paths of percolation through the zone or along the contacts with the foundation or concrete structures.

2. The soil must be sufficiently impervious to preclude excessive water loss through the dam.

3. The material must not consolidate excessively under the weight of the fill above it.

4. The soil must be placed so as to develop and maintain its maximum practicable shearing strength.

5. The material must not soften excessively on saturation by the reservoir.

All of these requirements are interdependent and can be obtained by proper choice of materials, compactive effort, and moisture control. The Bureau has found that for impervious soils, heavy compactive effort must be accompanied by control of moisture to obtain the most desirable strength of material. Studies are made in the earth laboratory to determine the allowable range of placement conditions for the soils to be used, and the field inspection and laboratory personnel make tests of the soil in the borrow pit and on the fill during the entire construction period to insure that the criteria for successful construction are met. (See figs. 25 and 26.) The control tests include moisture determinations prior to compaction; tests of density and moisture in the compacted fill, which are correlated with the standard density and optimum water content as determined in the laboratory; and tests of permeability, gradation, and consolidation. The field laboratory (see fig. 27) is equipped to perform the control tests on the job.

Figure 25.—Determining density of rolled embankment material on Bonny Dam. Material was carefully removed and weighed and the hole filled with sand of known density. From the weight of soil removed and the weight of sand, the volume and hence the density of embankment material could be determined.

For pervious zones in an earth dam these attributes are needed:

1. The material must be formed into a homogeneous mass free from large voids.

2. The soil must be free draining.

3. The material must not consolidate excessively under the weight of the superimposed fill.

4. The soil must have high shearing strength.

These criteria can be met by selecting a material of suitable characteristics and compacting it to the maximum practicable density. The material must be such that, when compacted, it will be many times as pervious as the central impervious core of the dam. This permits any seepage water coming through the core to be drained quickly through the pervious zone, thus keeping the line of saturation well away from the downstream slope. The strength and permeability of pervious zones are checked periodically during construction by analyses of gradation for the soils used, density tests on the fill, and laboratory permeability tests.

In the construction of each zone of the embankment, so far as practicable the materials are placed so that the coarsest and most pervious soils are located toward the outer slopes of the dam. The fill is kept fairly level, except where a portion of the channel is being used for diversion, and all zones are built up together as far as practicable. Orderliness on a fill is a good indication of efficient operation and careful construction. Figure 28 shows a typical view of an embankment under construction, and figure 29 shows a completed embankment.

Figure 26.—Testing moisture content of rolled embankment on Long Lake Dam with a penetration resistance tester. In this method, the needle is inserted in a can of material that has been screened and compacted according to standard procedure, and the moisture condition is estimated from the penetration resistance. (Less than 1,000 pounds per square inch—too wet for proper compaction; more than 2,000 pounds per square inch—too dry.)

Figure 27.—Testing laboratory used for construction control at Enders Dam. View shows percolation test apparatus, electric oven, soil hydrometers, and gradation test apparatus.

Filter layers are often provided underneath the riprap on the upstream slope of the dam and on the foundation near the downstream toe. Careful selection of material for gradation and permeability characteristics is necessary for filter materials. The selection and placing of rock riprap is carefully controlled to insure that a well interlocked, rough face of durable rock is presented to the action of the waves.

DIVERSION DAMS

The term diversion dam, as used by the Bureau of Reclamation, applies to a dam the sole purpose of which is to divert water for irrigating land. Diversion dams are subjected to the same hydraulic forces as storage dams, and most of the same problems, as well as others, are encountered in their design. In fact, some dams serve both the purposes of storage and diversion. Hence the procedures of investigation, design, and construction as discussed previously for dams in general, apply whether a dam is for the purpose of storage or diversion. However, there are certain problems and requirements in the design and construction of diversion dams that warrant special attention, and these are discussed here.

The choice of site for a diversion dam is usually very limited in comparison with that for a storage dam. The elevation of the land to be served and the intervening topography determine in general the canal design

Figure 28.—Embankment construction operations on impervious section of Anderson Ranch Dam. At left material is deposited in bottom-dump buckets, in center rolling is in progress, and at right material is being smoothed after rolling. Shovel at extreme left loads trucks from stock pile at discharge end of conveyor.

Figure 29.—Panoramic view of nearly completed Davis Dam embankment. Note riprap on upstream face, and forms for construction of concrete parapet on roadway.

and location, and these in turn determine the general location of the dam. The canal grade is usually much flatter than the grade of the river, which limits the final choice of site to a relatively short stretch of river. The length of canal and height of dam may sometimes be varied within limits to obtain the most economical combination. Factors affecting the selection of sites for dams in general are given in another portion of this article.

Because of the limitations in choice of sites for diversion dams, they must frequently be constructed on very inferior foundations. Diversion dams are often founded on alluvial fill, requiring special provisions to assure stability and protect against percolation beneath the structure. These provisions include the use of lightweight construction to reduce the weight on the foundation, the use of sheet piling cut-off walls and long aprons to lengthen the path of percolation, and the use of weep holes with inverted filters to relieve uplift pressures in the foundation. Diversion dams are comparatively low in height, since they ordinarily provide no storage and raise the water surface only a sufficient height for diversion into a canal. This permits use of structures light in weight, which compensates to a great extent for the poor foundation conditions often encountered.

A typical diversion dam (see fig. 30) consists of an overflow section, a sluiceway, a headworks structure, upstream and downstream wing walls, a nonoverflow section, and stilling pools as required. The overflow section, or weir, serves to pass the river flow and is

usually designed to pass a flood of 50- or 100-year frequency. If the foundation for the weir is of rock, a concrete gravity section is preferred. Types of weir construction which have been used on earth foundations, in addition to gravity sections, include concrete slab-and-buttress sections and rock-fill sections with a sheet-piling cut-off wall. The operating water surface behind the weir is determined from the required water surface in the canal and the loss of head through the headworks. The crest of the weir is set about 0.5 foot above the operating water surface to minimize the loss of water over the weir due to wave action during periods

Figure 30.—Completed Roza Diversion Dam. Headworks is at left of photograph.

when all of the water is being diverted and to provide proper regulation of the pool level. The length of the weir is determined by economic studies of several plans, and it may be controlled by the allowable discharge per foot of crest length (as limited to prevent scour of the river bed), by the maximum allowable elevation of the upstream water surface, or by the width of the river. Hydraulically, the overflow section of a diversion dam is essentially the same as an overflow spillway, and the reader is referred to the portion of this article on spillways and outlet works for discussions of methods of control and energy dissipation. The crest of an overflow section is usually shaped parabolic to correspond roughly with the lower surface, or "nappe," of the overflowing sheet of water during maximum discharge. This improves the discharge characteristics and prevents destructive negative pressures from developing under the nappe.

Diversion dams are provided with sluiceways and sluice gates to regulate the pool level above the dam, to provide controlled release of water to satisfy downstream requirements, and to sluice deposited sand and silt from the area near the headworks. The discharge capacity may be determined by the quantity needed for sand sluicing or by the quantity required to fulfill rights of downstream users below the dam. The openings should be made large enough to pass floating debris. The sluiceways on many existing dams are simply round or square conduits passing through the dam. Recent trends in sluiceway design favor a separate sluiceway structure joined to the overflow weir on one side and to the headworks structure on the other. This type has a sill or floor at the elevation of the stream bed or slightly above, vertical sidewalls, and an operating deck. Radial gates or slide gates are used to control the size of the sluiceway opening. Radial gates are preferred because they can be raised above the high water level with a minimum of structure height, thus leaving a free water surface for floating debris. Energy dissipation is accomplished by a stilling pool, the same as for an overflow spillway. A training wall is provided to separate the flow through the sluiceway from that over the weir and to confine the sluiceway flow to insure the formation of a hydraulic jump when the sluiceway is operating without flow over the weir.

The headworks structure regulates the flow of water into the diversion canal. Water is usually diverted through a single headworks located on the bank at one end of the dam, but a headworks is sometimes required on each bank of the river. The centerline of the headworks structure should make an acute angle with the centerline of the sluiceway structure. There are no general rules for determining this angle and the best arrangement can be determined only by model studies.

The flow through the headworks is controlled by one or more gates the size and type of which determine the form of the headworks structure. The size of gates and width of the gate openings depend on the required discharge, the pool elevation, the desired velocity through the gate structure, and the canal design. The sill of the headworks structure is placed somewhat higher than the sluiceway sill to reduce the amount of sediment taken into the canal. The gates may be operated manually or by power. Automatic controls are sometimes required. The type of gate depends largely on the characteristics of flow in the river. The open type radial gate is suitable where there is little fluctuation in the river water surface. Large fluctuations in the upstream water surface require the use of a submerged gate, which may be either a top-seal radial gate or a slide gate. The structure usually consists of a floor slab or sill with vertical sidewalls and vertical piers dividing the gate chambers. An operating deck is required for the open type structure to support the gate hoists and to provide access to the sluiceway and overflow weir. An access road may be required over the downstream part of the structure.

The headworks is designed to operate with a free water surface for normal operating conditions. When high water occurs in the river during the irrigating season, it is necessary to operate the gates at partial gate opening. The downstream end of the headworks structure includes a transition from the rectangular stilling pool to the canal cross section, which is usually trapezoidal. The length of the headworks structure will ordinarily be controlled by the lengths of the gate chamber, the stilling pool, and the outlet transition. The required length of percolation path is provided by use of cut-off walls and, if necessary, by increasing the length of the structure. For unlined canals a cut-off wall is provided at the downstream end of the headworks structure to prevent undermining. A short length of the canal is protected from scour by dry rock paving or riprap.

The use of trashracks will depend on the character of the stream, the amount of floating debris anticipated, and the type of gate used. Trashracks are not usually provided where the submerged type of gate is used. Stop-log grooves are provided so that the gate opening can be closed as quickly as possible by stop logs in the event of gate failure to prevent damage to the canal system. The stop logs are also useful for making inspections and repairs in the gate chamber. Screens are required in some localities to prevent fish from entering the irrigation canal. Their use will depend on the species of fish and their importance from the standpoints of recreation, industry, and conservation, and also on the legislation or ordinances governing fish control. Fish screens may be classified in three groups as stationary, mechanical, or electrical, and may involve the use of either bars or screens. Migratory fish require a fish ladder or other means for allowing them to pass the dam.

Wing walls are provided to connect one or more of the components of the diversion dam with the river banks or dikes. They are used upstream and downstream and serve principally to retain the embankment at the ends of the dam, to provide sufficient percolation path around the end of the dam, and to prevent erosion of the banks due to eddy action or high, jet velocities. The walls are designed for load conditions that take into consideration an estimated amount of saturation behind the wall and the water level in front of the wall.

After the elevation of the water surface in the pool has been established for the condition of maximum design flow, freeboard is added to obtain the elevation of the top of component parts of the structure which must not be inundated by high water. The elevation of the top of the sluiceway, headworks, upstream wing walls, and nonoverflow section is determined in this manner. The amount of freeboard provided will vary with design conditions. A minimum of 3 feet is used when a dam is designed purely on the basis of flood frequency studies.

The nonoverflow section serves to close off the rest of the flood plain at the required elevation. It may consist of an earth dike or a concrete gravity or buttressed section. Concrete sections are designed as nonoverflow dams or walls. Earth sections are designed as low earth dams with protection on the upstream side to prevent erosion from wave action or lateral currents. The downstream slope may also require protection from high water. The top of the nonoverflow section is sometimes used for access to the dam for purposes of control and maintenance.

Some Western streams contain excessive quantities of silt, which if not removed at the point of diversion would be carrried into the canals and laterals, necessitating removal at heavy expense to maintain the discharge capacities required. Desilting works of various types have been constructed as appurtenant features to diversion dams, varying from a simple settling basin from which the settled deposits are flushed periodically to a complex system of settling basins, silt scrapers, and sluicing devices such as used at Imperial Dam to remove silt from water entering the All-American Canal. (A special article on Imperial Dam is included in this publication.) Sometimes it is possible to eliminate the need for a desilting works by so arranging the headworks and sluiceway that most of the bed load of the stream is directed past the headworks and into the sluiceway. This procedure was used for the Superior-

Courtland Diversion Dam. The best arrangement of headworks, sluiceway, and guide walls is determined by model studies.

SPILLWAYS AND OUTLET WORKS

The spillway structure serves as a safety valve for a dam, providing means for the passage of flood flows without damage to the dam or appurtenant structures. In Bureau procedure, the maximum flood to be provided for is determined on the basis of field investigations and studies of rainfall and run-off records. The spillway capacity is determined by a series of studies that consider the comparative economies of passing and storing various portions of the design flood. If the top few feet of a reservoir contain a large storage capacity, it is sometimes economically possible to construct the dam with sufficient freeboard that much of the flood can be stored in the freeboard space, thereby permitting considerable reduction in the design capacity of the spillway. For example, the design capacity of the spillway for Hungry Horse Dam was reduced from 80,000 to 50,000 cubic feet per second by constructing the dam an additional 5 feet in height. On the other hand, it was obviously impossible economically to take advantage of any freeboard storage for such a dam as Grand Coulee, where the tremendous flood designed for overshadowed the reservoir storage.

Spillways are of two general types: the overflow type, which is constructed as an integral part of the dam, and the channel type, which is an independent structure from the dam. Overflow spillways are generally most economical for large capacities, and are readily adaptable to gravity and buttress dams. They are seldom used for arch dams except in small quantities and low overflow heads since arch action cannot be considered to exist above the spillway crest. Some type of open channel or tunnel-type spillway located in the abutment is usually found most suitable for an arch dam. In some instances diversion channels or tunnels are planned for use as part of the spillway. Spillways for earth dams are usually located in an abutment. Typical spillways are shown in figures 31 and 32.

Both overflow and channel-type spillways may be constructed with either controlled or uncontrolled crest, dependent on requirements for reservoir operation. That is, gates may or may not be installed on the crest to provide a means of controlling the discharge over the spillway. Most dams constructed by the Bureau have been multiple-purpose structures, and the various requirements for the reservoir operations have usually dictated the use of control gates on the spillway crests. Spillway gates are discussed in the following article.

The construction of a hydraulic scale model is essential for the design of a spillway of any importance,

Figure 31.—These side-channel spillways and intake towers at Hoover Dam discharge through tunnels in the abutments that formerly served as diversion tunnels.

since by its use the designer is able to study the spillway as it will operate and obtain answers to problems that would be extremely difficult to solve by analytical methods. In the operation of the hydraulic model the following are some of the items that may be determined:

1. The proper design for an inlet channel to a channel-type spillway for improved delivery of flow to the control structure.

2. The proper design for a pier and the crest shape to give maximum discharge with least disturbance of the flow.

3. The necessary height of training walls for adequate freeboard, or the size of tunnel for proper flow conditions.

4. Various types of energy dissipators and their effectiveness for varying discharges and tailwater elevations.

5. The possibility of scour in a stream bed and the consequent formation of bars that might affect tailwater.

Overflow spillways are so designed that the water passes over the curved crest and directly down the downstream face of the dam. The flow is confined to the spillway section by training walls constructed on the downstream face of the dam at either end of the spillway. The length of the spillway crest is usually determined by the width of the stream bed and the location of appurtenant structures that may be required at the toe of the dam. Generally the most economical design is a long spillway crest with shallow depth of flow, since with this design there is less energy to dissipate per foot length of spillway at the toe of the dam, the concrete in the crest is not so costly, and the gates are smaller and less expensive.

Perhaps the most important consideration involved in the design of an overflow spillway is the method of dissipating the energy generated by the high velocities at the toe of the dam. The method most favored by Bureau designers has been the construction of a sloping apron downstream from the toe of the dam. The length of apron is determined sufficient to enable the flow to form a hydraulic jump while still on the apron, thus reducing the velocity to about that of the original stream. The length of the apron will usually be about five times the depth of the tailwater at the lowest point of the apron. For dams of moderate height where the velocity of the water is not too great, baffle piers may be constructed in the path of the high-velocity jet, thereby reducing the required length of apron to some extent. For large flows at high velocities with deep tailwater, the apron may reach such a length as to be economically infeasible. Under these conditions a deeply submerged "bucket" may be designed and located at the toe of the dam so that the energy is dissipated by rolling the water upward through the deep tailwater. The spillway bucket at Grand Coulee Dam

is an example. Some spillway buckets have been designed so that they are located above the tailwater, discharging the water up and outward from the toe of the dam, but this design is not usually suitable except in extremely favorable locations since the fall of the water subjects the stream bed to severe erosion.

The channel-type spillway discharges from the control structure through an open channel or tunnel called the discharge carrier. The control structure may be any one of a number of different designs, dependent on the surrounding topography. The simplest and most widely used type of control structure is the normal-flow control structure, where the water approaches the control structure normal to the crest and enters the discharge carrier with no general change in direction. A different type of control structure is the side channel

Figure 32.—Spillway at Anderson Ranch Dam during construction. Stilling pool is in foreground.

type, such as used at Hoover Dam. (See fig. 31.) In this design the water flows over the crest of the control structure into a side channel which directs the flow downstream and at a right angle to the flow over the crest. While this type is not efficient hydraulically, it may be economically justified where the side channel can be located to lie along the canyon wall, thus avoiding excessive excavation. Another type of control structure sometimes adaptable to dam sites with steep canyon walls is the drop-inlet or morning-glory type, such as used at Hungry Horse Dam. In this design, the water flows over a circular crest and drops directly into a tunnel-type discharge carrier.

Whether channel-type spillways are designed with open-channel or tunnel-type discharge carriers is usually governed by economic studies. The final design of an open-channel discharge carrier is determined by a comparison of costs for several tentative layouts with various cross sections and grades. The grades of a tunnel-type discharge carrier are usually established at the beginning of the design, since it is important that the tunnel be dropped in elevation as rapidly as possible, thereby increasing the velocity of the water and decreasing the required size of the tunnel. Present-day tunnels are lined with concrete to reduce friction losses, and the tunnel section is designed to flow about three-quarters full. Many channel-type spillways for concrete dams require no energy dissipation at the outlet end, since the flow can be discharged into a rocky river bed a considerable distance from the dam. Special precautions must sometimes be taken, however, to dissipate the energy by various methods or to throw the water clear of the outlet end by a deflector.

Reservoir outlet works are required at most dams for the controlled withdrawal of water for irrigation or other purposes. Reservoir outlets serve to regulate the river flow and consequently the reservoir elevation. The number and size of outlets must be such that the discharge requirements for the reservoir at various elevations will be satisfied, and may vary from one conduit of small diameter to a considerable number of large-size conduits. The largest number of outlets that have been installed in a structure is at Grand Coulee Dam, where 60 conduits were installed, each 102 inches in diameter. Where an overflow-type spillway is constructed on a dam, as is usually the case for concrete gravity dams, the outlets may be located in the spillway section and discharge onto the sloping face of the spillway or into the stilling basin at the toe of the dam. For this type of installation, the flow is usually controlled from galleries and chambers by gates situated near the entrance to the conduit. Grand Coulee outlets are of this type. Locating the outlets in the spillway, where possible, is generally advantageous from an economic viewpoint since the length of the conduit is

usually a minimum, and the necessity of a valve house and perhaps a stilling basin is avoided.

Outlets for concrete dams, whether gravity, arch, or buttress, may consist of conduits extending through the dam, with control valves located in galleries or in valve houses at the toe of the dam; or the outlets may be located in concrete-lined tunnels driven through the abutments (frequently use is made of diversion tunnels) and controlled by valves in valve houses near the downstream toe of the dam (see fig. 32). The latter location is preferred for large-capacity outlets in thin arch dams since these dams are not well adapted to large openings through them.

Outlets for earth dams are preferably located in concrete-lined tunnels driven through the abutments, but may be located in trenches excavated in the foundation under the dam. (See figs. 33 and 34, respectively.)

Figure 33.—Left abutment at Anderson Ranch Dam showing portion of upstream face and intake to outlet structure. Borrow area is in background.

If no other location is suitable, the outlets may be installed in conduits through the earth fill, provided adequate precautions are taken. When outlet works are located in rock, the rock is always pressure-grouted upstream from the axis of the dam and along the cut-off trench grout curtain to provide a continuous, impermeable barrier to the flow of water. When outlets are located through or under an earth fill, special precautions must be taken to prevent leakage of water along the outside of the conduit. Cut-off collars are placed at intervals along the conduits, and the embankment is carefully compacted around them. For earth dams constructed by the Bureau, the preferred practice has been to install the operating gates in a gate chamber located about midway of the length of the tunnel

Figure 34.—Outlet conduit for Bonny Dam, looking upstream. Conduit upstream from gate chamber is now under hydrostatic pressure, and entire conduit is covered by earth embankment.

or conduit, rather than at the upstream end, as illustrated in figure 34. This provides simplicity in construction and obviates the necessity of locating an outlet valve house on the sloping upstream face of the dam extending to the reservoir water surface, but it exposes the upstream one-half of the conduit to hydrostatic pressure, requiring special precautions to prevent seepage. It is essential that outlet works be located on foundations that will provide adequate support since settlement might cause rupture of the conduits, with disastrous results. The structure must be designed to withstand the superimposed embankment load without cracking.

MEASUREMENTS OF BEHAVIOR

The Bureau of Reclamation installs a variety of testing and measuring devices in all dams of major proportions. Some of the devices give information which is used in controlling the construction operation, and others provide for the continued observation of the behavior of the structure after construction is completed. By this continued observation, not only can the safety of the structure be ascertained at all times, but also much actual behavior data can be compiled which will be useful in the design of future dams.

Most devices used for determining the behavior of concrete dams consist of electrical measuring instruments embedded in the concrete and include the following:

Resistance thermometers for measurement of the temperature of the concrete at various locations.

Strainmeters for measurement of deformation in mass concrete or in reinforcement steel around open-

ings in the dam, thereby enabling the stress in the concrete and reinforcement to be computed.

Joint meters for measurement of contraction joint openings.

Hydrostatic pressure gages for measurement of uplift pressures underneath the dam.

Nonelectric instruments not embedded in the concrete include plumb lines and tiltmeters that are used to determine deflections and tilting of the structure, also precise traverse instruments and levels to determine deflections and settlements at various locations. (See fig. 35.)

The Bureau installs instruments in its high earth dams to measure pore pressures, consolidation, and surface movements. The pressure measuring devices, called piezometers, consist of tubing filled with water connecting a porous disk, which is placed in the fill or the foundation, with a pressure gage located in a concrete manhole in the downstream slope of the dam. During construction any fluid pressures developed at the piezometer tips in the fill or foundation are recorded by the pressure gages and are compared with the design requirements of pore pressure and stability. After completion of the structure and filling of the reservoir, these piezometers serve to trace the actual flow through the dam and serve as warning signals in case of distress. A companion system to the piezometers is the internal settlement device, which consists of a series of vertical telescoping pipes and cross arms free to move independently of each other in every 5 feet of depth. By means of a specially designed plumb bob, readings are taken which record the actual movement of the fill as it consolidates under the weight of superimposed material during construction. These measurements of the behavior of the interior of the dam have shown that pore pressure and consolidation in soils are related. This has given Bureau engineers a better understand-

Figure 35.—Deflection measuring equipment in Friant Dam.

ing of soil action which has contributed to safer and more economical designs. In addition to the above two systems of measurement, surface points along the crest of the dam and on the downstream slope and toe

are checked periodically by surveying methods to record any movements during the life of the structure which may indicate trouble. Figures 36 and 37 are examples of typical instrument installations.

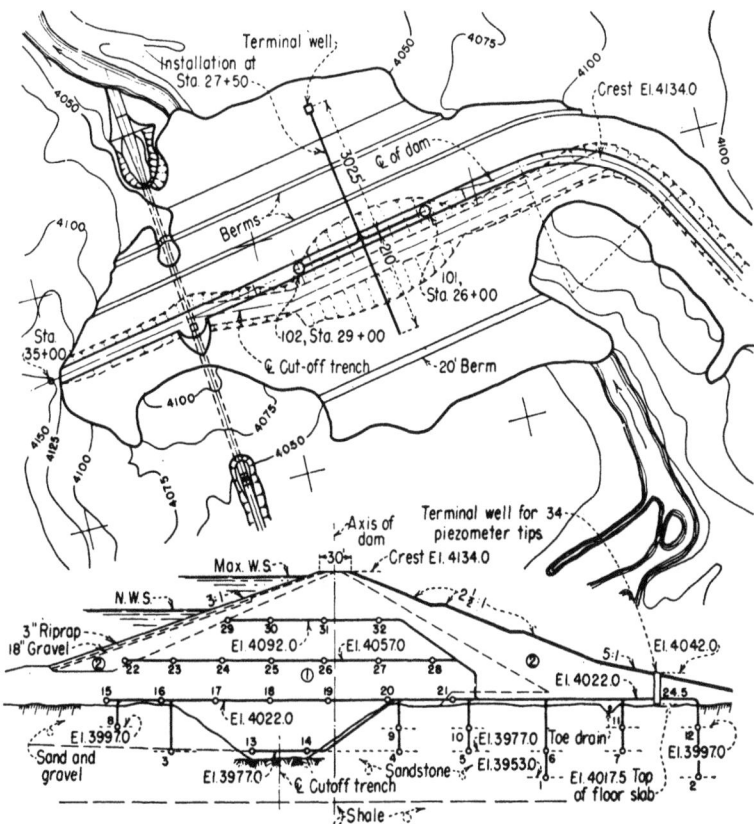

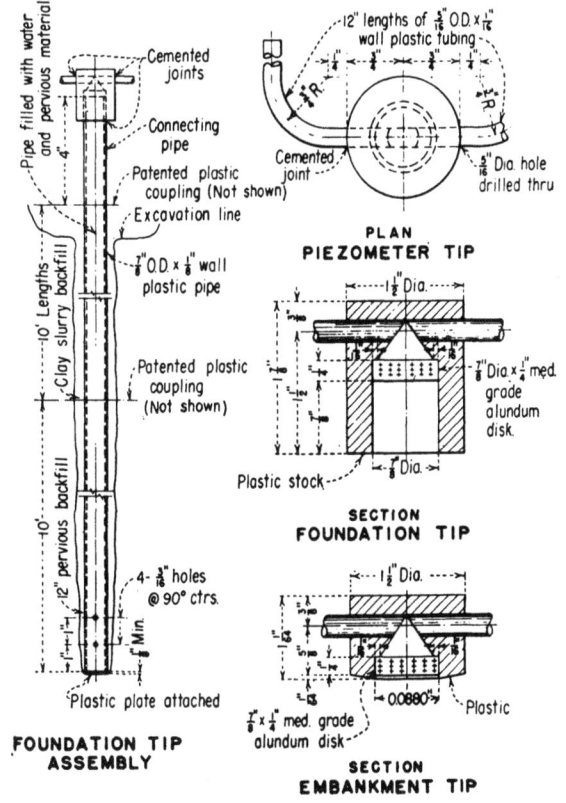

Figure 36.—Installation of piezometers for measuring fluid pressures in Keyhole Dam and foundation, and detail of piezometer.

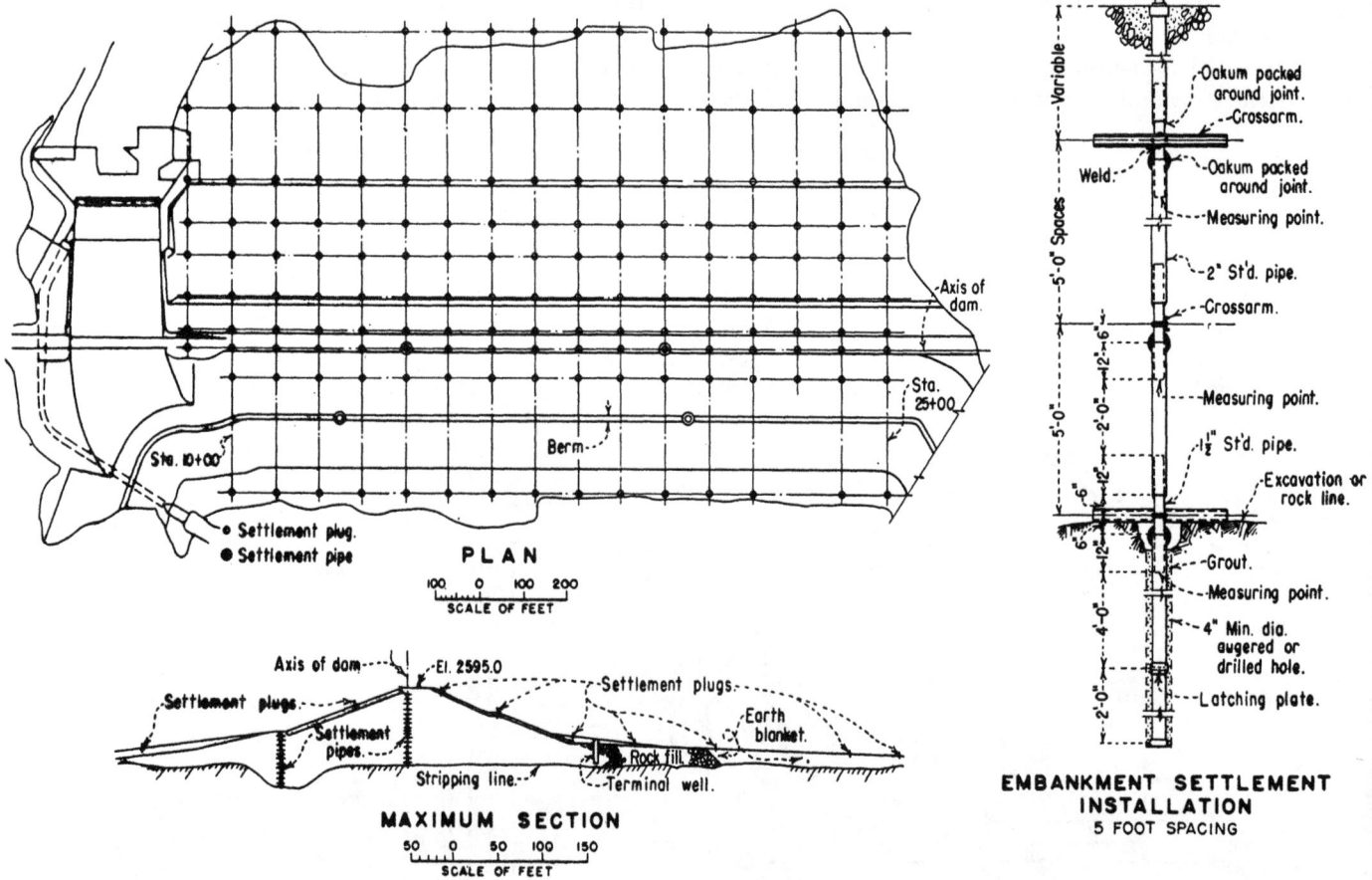

PLAN

SCALE OF FEET
100 0 100 200

MAXIMUM SECTION

SCALE OF FEET
50 0 50 100 150

EMBANKMENT SETTLEMENT
INSTALLATION
5 FOOT SPACING

Figure 37.—Installation of cross-arm settlement devices for measuring embankment settlement at Fresno Dam, and detail of settlement device.

RESERVOIR OUTLETS AND SPILLWAY GATES

By Arthur W. Tschannen

❯❯

Modern dams are frequently of immense size, requiring the control of great volumes of water under high pressure head. The energies involved are ofttimes tremendous, whether the discharge be through outlet pipes, through penstocks, or over the dam spillway. For example, when the spillway of Shasta Dam in California is operating at capacity, the falling water generates kinetic energy at the rate of nearly 7¼ million horsepower which must be controlled and dissipated without damage to the structure. The problems involved in the development of suitable control works for modern dams are many and complex, and the efforts of many Bureau engineers—as well as others both here and abroad—have contributed toward their solution.

Spillways and outlet works in general are discussed in the previous article. This article describes the types of outlet gates and valves and spillway gates generally used by the Bureau of Reclamation, including their development, uses and limitations. The variety of gates and valves available to the designer of modern spillways and outlets is diverse enough for him to select the type which will best fit the particular conditions. In the interest of hydraulic efficiency, simplicity, and economy, modifications and new designs of existing types of gates and valves are made as a matter of course. Every dam is unique, and the control works for that dam are to a great extent tailor-made—designed to give, in the judgment of the designer, the best possible performance.

RESERVOIR OUTLETS

The form of a reservoir outlet works will vary considerably with the type of dam and the purpose of the reservoir, as heretofore described. Basically, however, each installation consists of a conduit through or under the dam or through the abutment, with suitable control valves or gates to regulate the flow. Near the entrance or at some point along the conduit upstream from the control valve or gate, an emergency or guard gate is installed to permit unwatering the conduit for maintenance and to allow inspection and repair of the control gate or valve.

The usual method of flow regulation, particularly at the smaller structures, makes use of one or more outlet pipes with gates or valves which can be set to discharge any volume within their capacity. Gates and valves satisfactory for this kind of service have been developed from early Bureau designs. In some unusual cases of flow regulation at large dams where large discharges are expected, a number of outlets have been provided so that the gates and valves may be operated fully open, thus avoiding the damage frequently caused by operating gates or valves at partial openings. The outlets may be arranged at different levels, as at Grand Coulee Dam, so that as the water level fluctuates, the valves at the different levels may be operated at full gate opening. The following brief discussion outlines the use of control and emergency gates and valves at typical Bureau installations.

At Enders Dam, the typical earth structure shown in figure 1, a circular reinforced concrete conduit leads from a cylindrical trashrack inlet to the gate chamber which houses a high-pressure emergency gate. From the gate chamber, a plate steel pipe extends in a reinforced concrete conduit to a valve house at the downstream toe of the dam where it terminates in a wye leading to two hollow-jet control valves discharging into the open air. A walkway in the conduit between the gate chamber and the valve house provides access to the emergency gate.

Figure 2 illustrates the outlet works at Hungry Horse Dam, a typical concrete structure. Three outlet pipes lead from trashracks on the upstream face of the dam to ring-follower emergency gates, thence to hollow-jet control valves in the valve house. The air inlet pipes, which extend from the top of the dam to the bellmouth

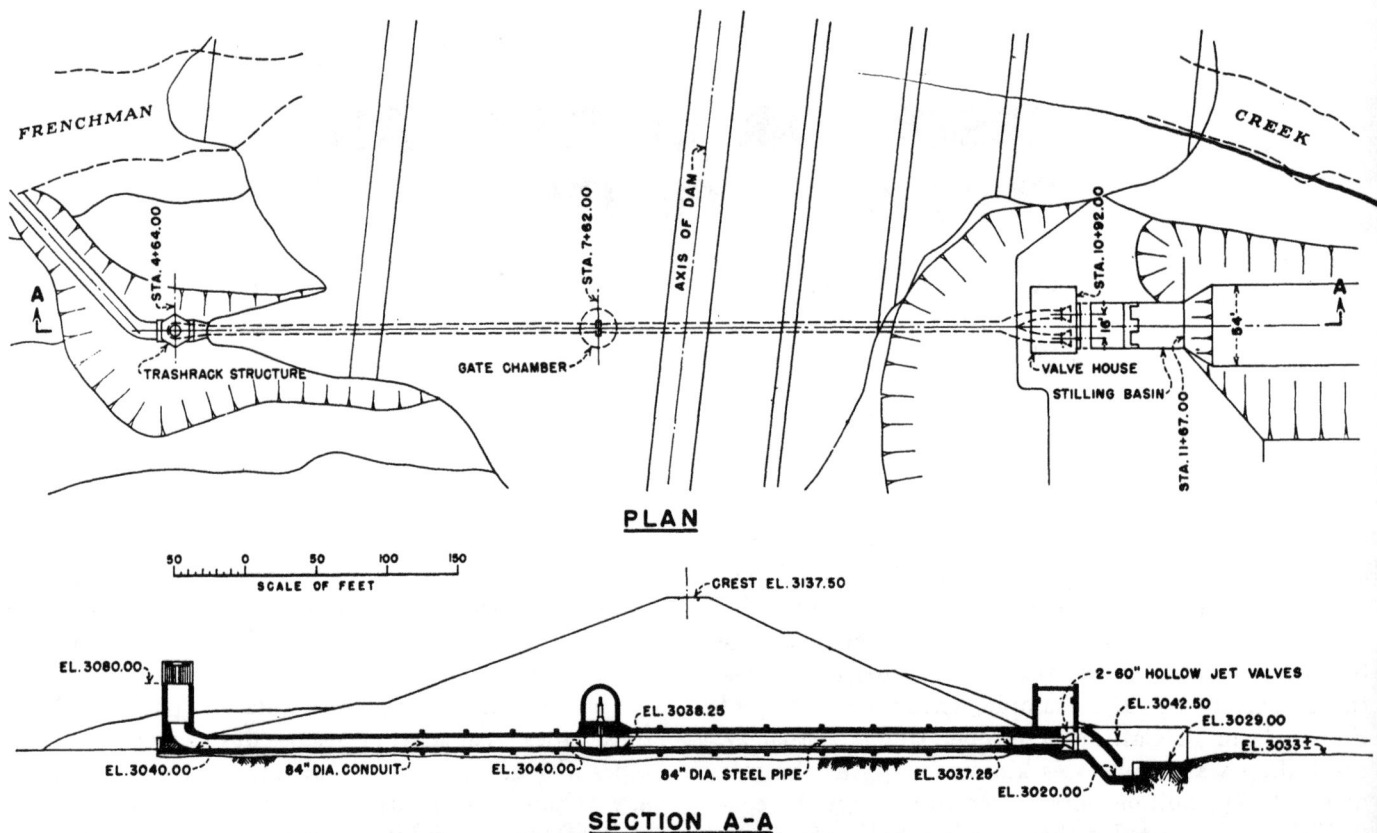

Figure 1.—The outlet works at Enders Dam is typical of those used in earth dams built by the Bureau. The gate chamber under the structure houses a high-pressure emergency gate, and the 84-inch diameter steel pipe terminates at the valve house in two hollow-jet control valves.

at the intake, minimize the erosional effects of high-velocity water flow in the closed conduit. Further description of this outlet system is given in the special article on Hungry Horse Dam.

Several types of gates and valves have dual applications: they may be installed for regulation of flow through outlet works, as described above, or they may serve for emergency closures in either penstocks or outlets. Figure 3 shows the intake and penstock arrangement for each of the four turbines at Hungry Horse Dam. Water for the turbine passes through the trashrack into the bellmouth entrance to the penstock, which can be closed for unwatering the penstock by lowering the gate on the face of the dam. The penstock entrances are vented to the open air, similar to the outlet conduits. The plate steel penstocks are embedded in the dam concrete and extend downward through the dam to the turbines in the power plant.

Control Gates

According to the definition used by Reclamation engineers, gates in general differ from valves in that they are so constructed that the water passage through the gate is not obstructed when the gate is in the fully opened position. Control gates for reservoir outlets used by the Bureau fall into five main categories: slide gates, high-pressure gates, radial gates, jet-flow gates, and cylinder gates.

(a) The *slide gate* is one of the simplest types of controls used for reservoir outlets. As generally defined, the slide gate is a type of gate having a continuous bearing surface under compression all around the opening to provide the supporting and sealing surface for the leaf. The operation of the gate consists of moving the leaf over a circular or rectangular opening to provide the desired discharge. Movement of the leaf is usually by means of a screw or hydraulic hoist. Although this type of gate is commonly used for irrigation distribution systems and for emergency closures, it has been installed in very limited numbers for the controlling of reservoir discharges. The more generally used control gates are described in paragraphs b, c, and d which follow.

(b) The *high-pressure gate*, by Bureau definition, is a special slide-type gate in which the rectangular leaf is encased in a body and bonnet and the gate is equipped with a hydraulic hoist for moving the leaf. Figure 4 is a cross section through a typical high-pressure gate. The leaf is raised vertically into the bonnet when oil is forced into the hydraulic cylinder above it, at pressures between 750 and 1,500 pounds per square inch, raising the piston to which the leaf is connected. Both the housing and the leaf have wearing surfaces of special bronze which also act as seals for the gate. Normally, these gates are used for controlling relatively low-head discharges (100-foot head or less) through conduits or pipes. High-pressure gates are often placed in tandem so that the downstream gate regulates the flow and the other is reserved for emergency use. They are frequently used as emergency or guard gates in combination with other types of regulating gates or valves. To conserve material, the cast conduit liner and the bonnet are not designed to withstand full water pressure, but are embedded in reinforced concrete which takes part of the load.

The gate frame and housing are strongly ribbed to carry the loads exerted by the hoist during opening and closing. Since the size of the leaf and the height of water above the gate influence the sliding friction between the leaf and the seal seats, loads on the frame and housing of large installations may be very high. Use of a special nonseizing bronze in the seats and seals permits movement of the leaf under loads which in some cases exceed 1,000,000 pounds.

In a high-pressure emergency gate installation, the gate may be held in the open position by a semi-automatic hanger which can be released by hand; but a high-pressure control gate usually requires a fully automatic hanger which will hold the gate in any predetermined position. When a high-pressure gate is used for regulating or throttling, a large supply of air is required in the conduit downstream from the gate leaf to minimize pressure fluctuations and to inhibit cavitation erosion below the gate.

(c) *Radial gates*, as they are called by the Bureau, are more frequently known as taintor gates. Their most common use, both on Bureau projects and elsewhere, is for regulating flow over a spillway crest, as discussed in another portion of this article. However, radial gates have been used successfully as control gates in closed conduits which withstand heads of over 100 feet, and some have been designed for heads approaching 200 feet. The present discussion will be confined to their use in closed conduits.

Early radial gates were often made of wood, but modern ones are made wholly of steel. Basically, the gate consists of a section of a cylinder connected to, and supported by, two side beams which are connected by hinge pins to the supporting structure. The hinge pins

Figure 2.—Outlet conduits for Hungry Horse Dam extend through the dam and discharge into the river channel. Flow is controlled by 96-inch hollow-jet valves, and upstream from each valve is a 96-inch ring-follower emergency gate.

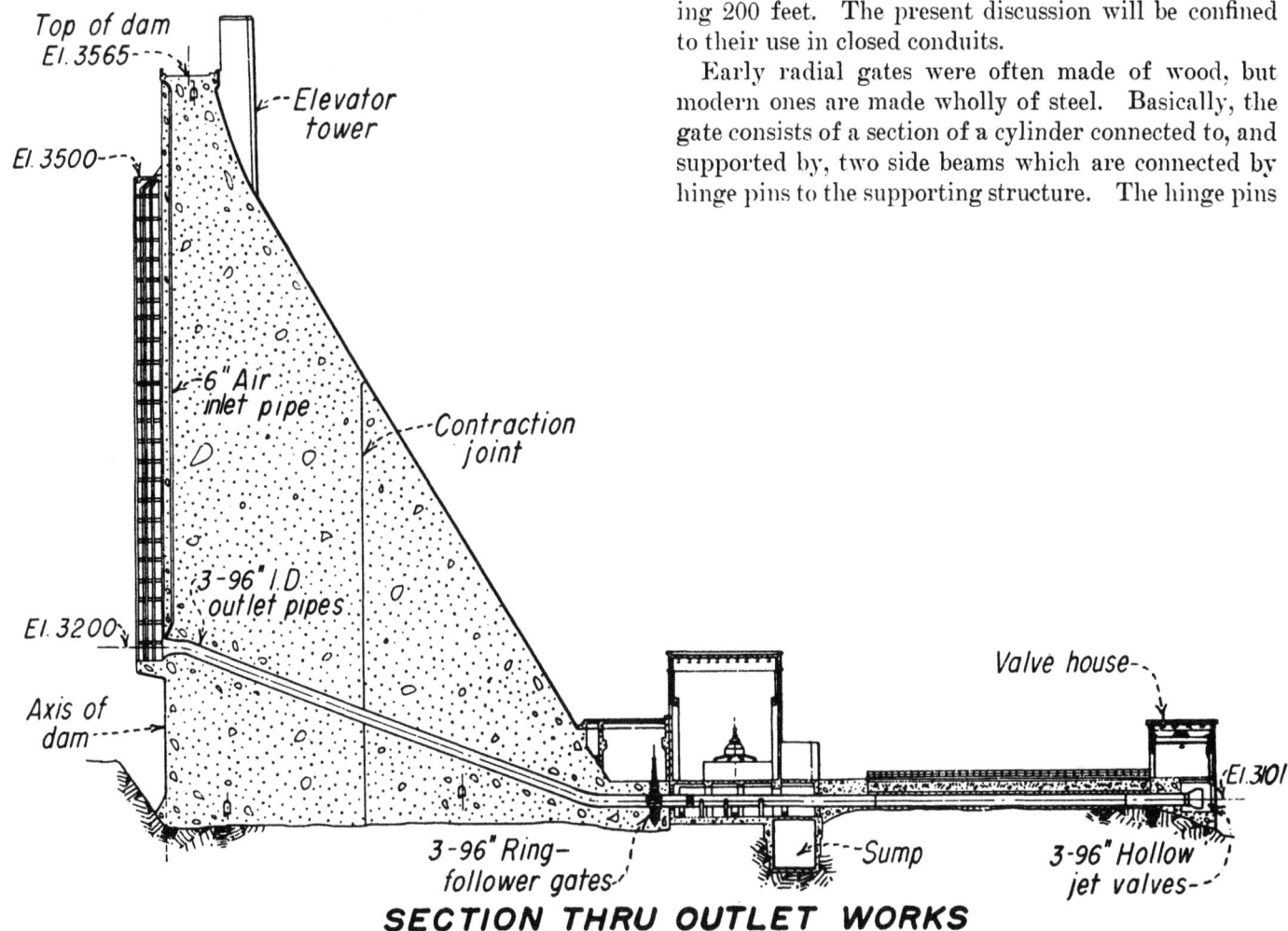

SECTION THRU OUTLET WORKS

P.I. El. 3585
Nor. H.W.S. El. 3560
Top of dam El. 3565
400'R
24" Air inlet pipe
Contraction joints
El. 3319
13'-6" I.D. penstock
Axis of dam
Foundation gallery
Drainage gallery
4-75,000 K.va. Generators
Max. T.W.S. El. 3101
Min. T.W.S. El. 3074.5
5:1
El. 3080
El. 3048
355'-6" On line of centers
44'-6"
Ɛ Units
4-105,000 H.P. Turbines

MAXIMUM SECTION THRU PENSTOCK AND POWERPLANT

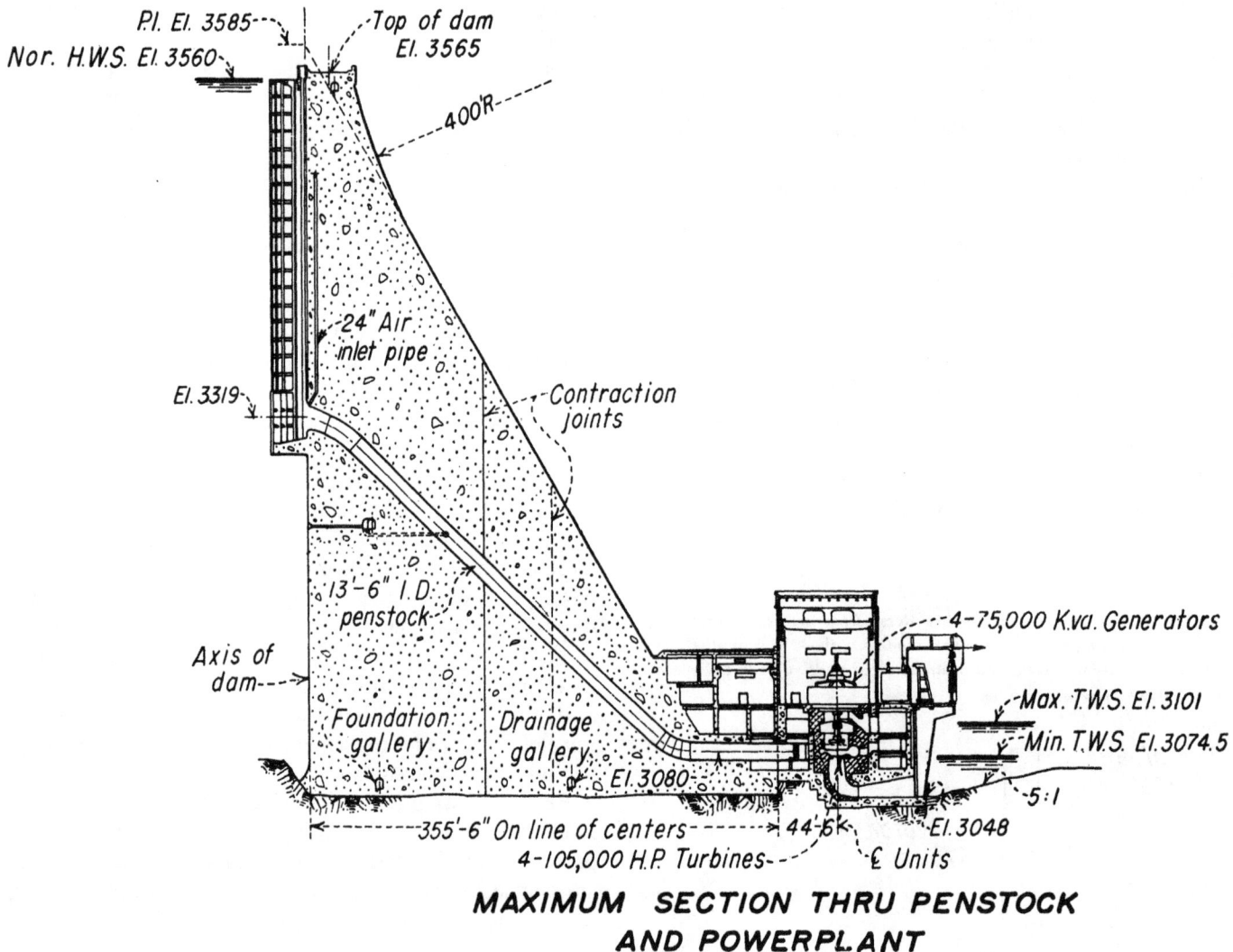

Figure 3.—The penstocks at Hungry Horse Dam and power plant may be unwatered by closing bulkhead gates on the upstream face of the dam.

are located on the axis of the cylinder. Figure 5 illustrates the installation of a radial gate in a closed conduit. In this particular arrangement, a slide gate can be lowered in front of the radial gate to permit repair and maintenance. When a radial gate is installed in a closed conduit, close-fitting seals are required across the top of the gate as well as at the sides and bottom.

High-head radial gates are usually operated by means of a screw lift, which permits opening the gate and holding it at any desired opening, allowing variable discharges. No guides or seals protrude into the waterway with this type of gate, thus imparting good hydraulic characteristics to the water passages. Discharges do not fill the conduits or tunnels below the gates, and an abundant supply of air admitted through large vents assures minimum fluctuation of the water at all openings.

(d) *Jet-flow gates* are designed for service as control gates. They may be installed at the discharge end of,

or at any point in, a conduit. The operating element of the gate consists of a wheel-mounted leaf moved vertically by a motor-driven screw hoist. The leaf and the housing surrounding the water passage are shaped to form an orifice which will cause the water to issue in a jet at any gate opening. The mechanical drive permits close and constant regulation of flow through the gate.

Figure 6 illustrates the design of a jet-flow gate. From a point about one diameter upstream from the gate leaf, the conduit flares slightly until its diameter is about 20 percent larger than the jet orifice. From this point the conduit contracts sharply to the orifice diameter, on an angle of 45° with the centerline of the conduit, forming the nozzle. The restriction produces a jet which springs free of the wheel slots at the side of the gate leaf and imparts a lift to the bottom of the jet to decrease impingement of the high-velocity water on the bottom portion of the conduit.

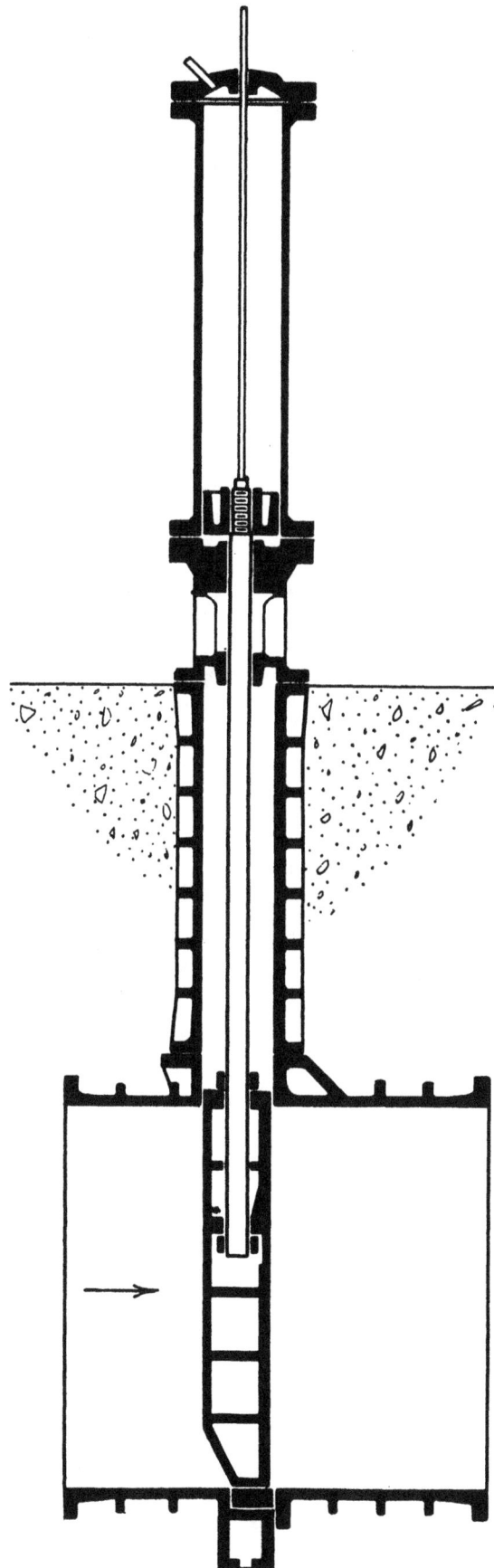

Figure 4.—The high-pressure leaf gate is used both as an emergency gate and for control or throttling purposes. It is raised and lowered by means of the hydraulic cylinder above the floor of the gate chamber.

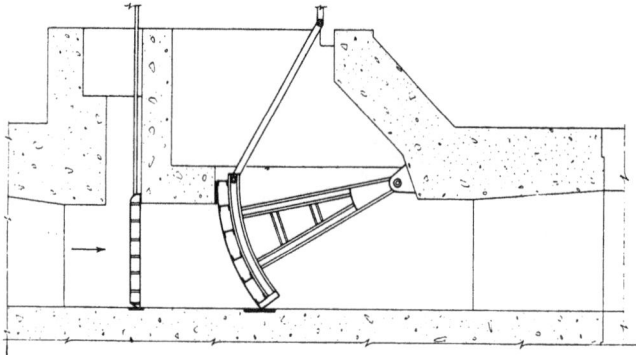

Figure 5.—When a radial gate is installed in a closed conduit, provision is frequently made for a bulkhead gate which may be lowered to permit maintenance or repair of the radial gate. Tight-fitting seals are necessary at the top as well as at the sides and bottom of the radial gate.

The seal, which provides complete closure of the opening when the leaf is lowered, consists of a bronze wearing-ring vulcanized to a rubber diaphragm which is clamped to the downstream surface of the nozzle. Contact of the seal with the gate leaf is maintained by hydrostatic pressure of the water behind the seal. Large air inlets supply air to the jet discharge, preventing vibration of the gate and cavitation erosion that might be present were air excluded.

(e) The *cylinder gate* is one of the least common control-type gates installed on Bureau projects. It is mentioned chiefly because of its novel features. This type of control is normally installed in an intake tower upstream from the dam. Water at high pressure is discharged into the intake tower through the gate, which consists of a movable cylinder sealing cylindrical seats. The released water then flows through a gravity tunnel. Hoisting is usually accomplished by means of screw lifts or hydraulic cylinders located in the top of the intake tower above water level. Lifting effort is transmitted from the hoists through long steel stems to the gates below. Cylindrical gates are also sometimes used for emergency closure of penstocks and outlets.

Control Valves

According to the definition used by Reclamation engineers, valves—as distinguished from gates—are so constructed that the closing member remains in the water passageway for all operating positions. Four types of control valves have been used on Bureau projects. They are, in order of development, the Ensign valve, the needle valve, the tube valve, and the hollow-jet valve. All of these are variations or modifications of the balanced-needle type in which the hydraulic

thrust on the regulating element, or needle, is maintained at a minimum or is automatically balanced at all openings of the valve. In a fully balanced needle-type valve, only sliding friction need be overcome in setting the valve for different openings; the hydraulic pressure, rather than opposing the motion of the regulating element, actually may provide the operating force.

(a) The *Ensign valve*, although it is no longer specified for Bureau projects, has rendered and is still rendering dependable service at many of the early dams. It was invented in 1908 by O. H. Ensign, then chief engineer of the Bureau. It was the first of the balanced-needle type valves to be installed, and the other valves mentioned were developed in a direct line of descent. Model studies in the hydraulic laboratories were of invaluable aid in determining desirable modifications for the successive developments.

Figure 7 is a section through an Ensign valve and a portion of the discharge conduit. Parts of the valve are a stationary cylinder, called the body; a hollow cylindrical plunger, with a needle tip and seal ring at its downstream end; a ribbed support ring holding the valve in position in front of the discharge conduit and forming the seat; and a discharge throat liner connected to the conduit. At the outer diameter of the

Figure 6.—The jet-flow gate, by means of the steeply sloping constriction just upstream from the gate leaf, produces a jet of water springing free of the gate guides and walls of the conduit when the gate is fully opened. Large quantities of air are admitted to the conduit just downstream from the gate.

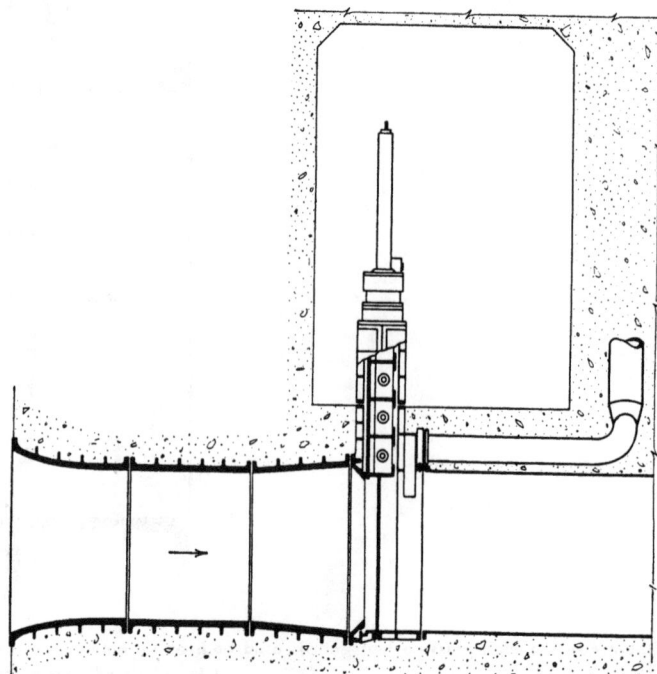

plunger and about midlength is a piston-ring arrangement, called a bull ring, which fits the stationary cylinder.

An Ensign valve is usually mounted on the upstream face of the dam, below the waterline, so that it operates completely submerged. Water flows between the support ribs and is diverted into the discharge conduit by the shape of the needle tip. Flow is regulated by varying the position of the plunger. The bull ring fits the cylinder with sufficient clearance to equalize pressure inside and outside the cylinder when a valve in the control pipe is closed. With the pressure thus equalized, the plunger will move to the closed position because the area at the back of the plunger exposed to the pressure is greater than the effective area of the needle tip. Opening the valve in the control pipe reduces the pressure in the cylinder. The unbalanced pressure then causes the needle to withdraw, opening the valve and allowing the discharge of water.

Ensign-type valves have not been entirely satisfactory, partly because of cavitation erosion which damaged the needle tip and throat liner and partly because it is necessary to lower the reservoir water level to gain access to the valve for maintenance or repair.

(b) The *needle valve* for reservoir outlet service was

developed in the early 1920's in an effort to overcome the difficulties which had been experienced with Ensign valves. Needle valves are designed to operate at the downstream end of a conduit, where discharge into the open air is possible, thus eliminating the opportunity for cavitation within the conduit and other destruction to the dam.

A considerable evolution has taken place in the design of needle valves, each redesign being an attempt to improve the hydraulic characteristics, reduce the size, or reduce the cost of the valve. Though frequently modified in design, a needle valve remains essentially a globular body, supporting and enclosing a stationary cylinder mating with a movable needle which effects the closure. An annular water passage surrounds the needle and cylinder. Movement of the needle, by which waterflow is regulated, is effected by changing the water pressure within the hollow cylinder as discussed for the Ensign valve.

As shown in figure 8, the construction of a needle valve is such as to guide the waterflow smoothly through the valve. Water flowing in the direction of the arrow passes the pointed end of the stationary cylinder, flows smoothly around the cylinder and movable needle, and issues into the open air at the nozzle.

Figure 7.—The Ensign valve is mounted on the upstream face of the dam, below the water surface. Water pressure on the face of the needle is balanced by water pressure in the interior of the valve, between the cylinder and the plunger.

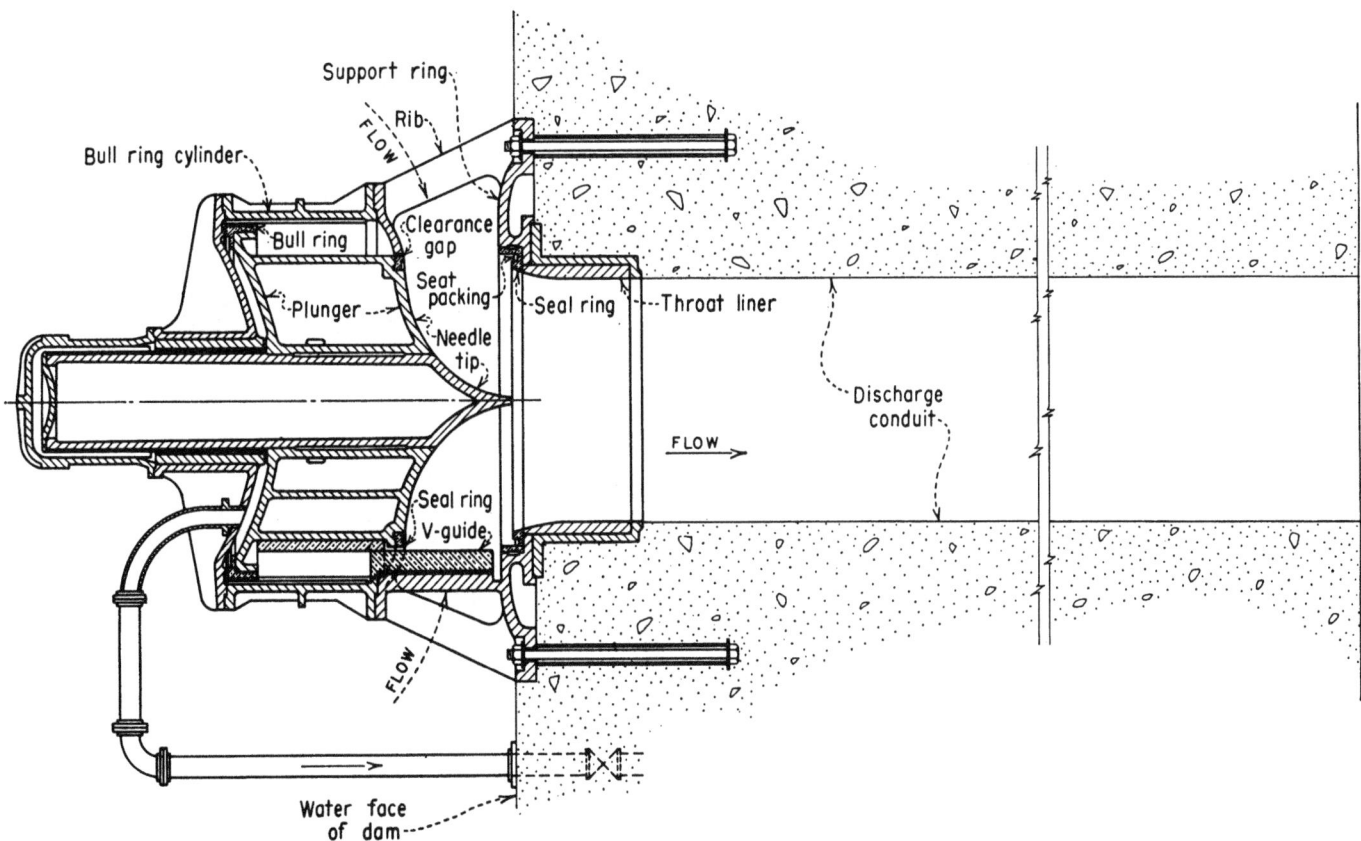

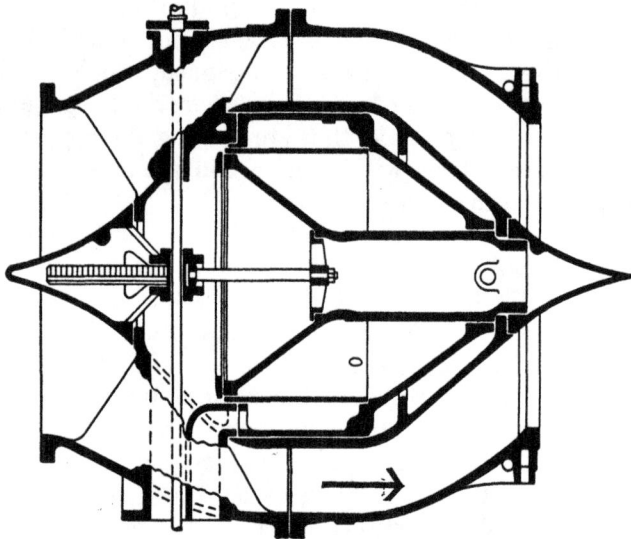

Figure 8.—The needle valve is designed so that water flows through the annular space surrounding the needle, issuing in a jet. Motion of the needle is brought about by water pressure exerted upon the inner surface of the needle, causing it to move downstream toward the seat.

Needle valves 96 inches in diameter have been installed by the Bureau, in some cases operating at heads in excess of 550 feet. The advantages of a needle valve, in addition to those already cited, are that the water issuing from the valve is in the form of a jet, whatever the valve opening, and control is positive and easy to achieve. Needle valves are usually protected by emergency gates in the conduit upstream from the valve.

(c) The *tube valve* is essentially a needle valve with the tip of the downstream or movable needle eliminated. It was designed to minimize the cavitation erosion which developed at the downstream end of the needle valve. Modifying the needle valve in this manner also resulted in a saving in weight and permitted its use in a conduit with less danger of damage to the conduit from cavitation erosion.

The controlling tube is usually moved by an actuating screw driven by an electric motor through bevel gears. In the closing operation, the tube moves against the seat at the downstream end of the valve. Hydraulic characteristics of the tube valve are essentially the same as those of the needle valve, except that at valve openings of less than 30 percent the jet is rough and unstable—a feature which limits application of the valve. At greater openings, a smooth, efficient jet issues. When a tube valve is installed within a conduit, rather than at the downstream end, a longer body is used to increase the capacity. When the valve is in this location special provision is required for aerating

the jet as it leaves the valve and enters the conduit downstream. Figure 9 is a section through a tube valve. The Bureau has installed valves of this type as large as 102 inches in diameter both within and at the down stream ends of conduits.

(d) The *hollow-jet valve* is the Bureau's most recent and outstanding development in control valves. It is essentially a needle valve with the movable or closing needle pointed upstream and the downstream portion of the body eliminated. The hollow-jet valve is characterized by its relatively low cost and high discharge coefficient. For a given size, the valve will discharge 25 percent more water than a needle valve of the latest improved design. The body of the hollow-jet valve is not subject to the pressure of the water in the reservoir and can therefore be of lighter construction than other designs which close at the downstream end of the valve. The hollow-jet valve derives its name from the shape of its discharge—a hollow or annular jet dispersed over a wide area. Figure 10 shows the general proportions and working arrangement of the valve.

The water passages through the body of a hollow-jet valve are designed to minimize cavitation erosion within the valve. Although the valve is not of the fully balanced design, water from the conduit is permitted to pass through small openings in the upstream face of the needle, thereby materially reducing the mechanical effort necessary for operation. Force for mov-

Figure 9.—The tube valve resembles the needle valve with the downstream tip of the needle cut off. The tube is moved upstream or downstream mechanically by a large screw to open or close the valve.

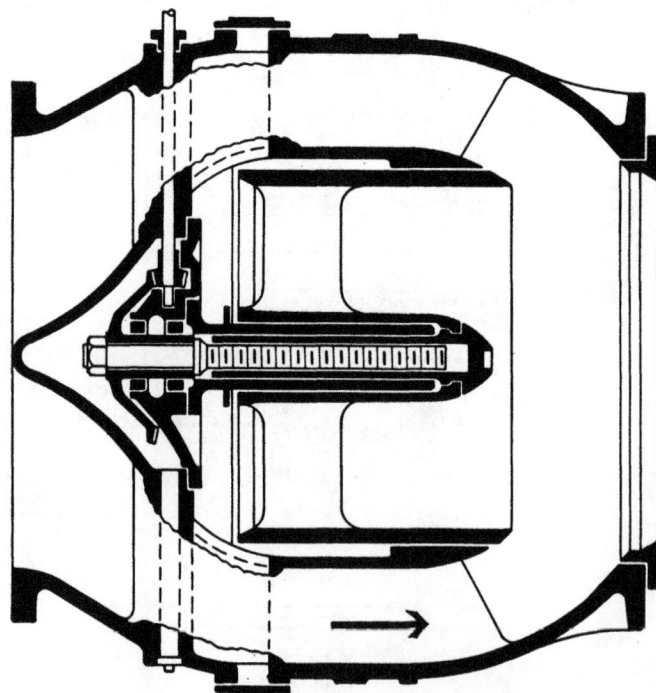

ing the needle is transmitted through a vertical shaft to a bevel gear and pinion arrangement turning a threaded stem which moves the needle to the desired position. Figure 11 is a photograph of the downstream end of one of the 96-inch hollow-jet valves installed at Hungry Horse as it appeared when assembled for shop testing.

Emergency Guard Gates and Valves

As indicated previously, conduits for penstock or outlet works operating under high pressures are usually provided with a gate or valve permitting closure for servicing the conduit, turbine, or regulating valve. It is usually designed for rapid emergency closure in event of failure of some part of the system. This emergency gate or valve may be at the entrance or within the conduit itself, and may be any one of the following types according to the particular requirements: slide gates, wheel-mounted (fixed-wheel) gates, roller-mounted (coaster) gates, ring-follower and ring-seal gates, and butterfly valves.

(a) Simple *slide gates* have been used for emergency closure of outlets, but have generally been restricted to

Figure 11.—Large gates and valves are assembled in the shop for testing. The 96-inch hollow-jet valve for Hungry Horse Dam dwarfs the workmen. Photograph courtesy Hardie-Tynes, Birmingham, Ala.

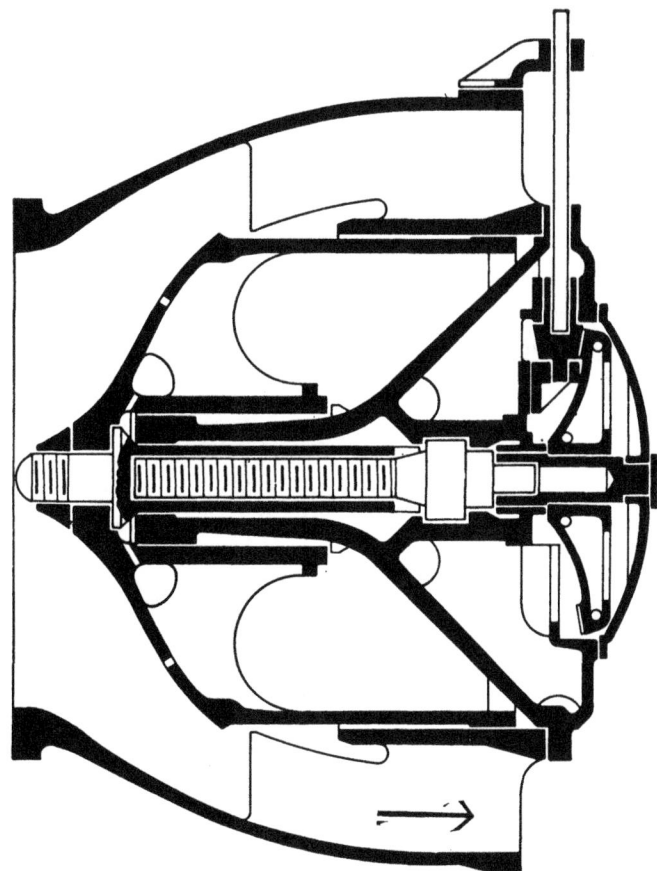

Figure 10.—The closing member of the hollow-jet valve moves upstream against the flow of water to shut off the flow. This valve is designed primarily to discharge into the open air.

small installations operating under low heads. Slide gates are described in the portion of this article discussing control gates.

(b) *Wheel-mounted gates*, referred to as *fixed-wheel gates* by the Bureau, are constructed of structural steel and mounted on wheels. They usually operate as closure gates for relatively high head penstocks or outlet conduits. The usual construction consists of a skin plate supported on horizontal beams which transmit their loads through vertical girders to the wheels and tracks. Structural steel gate frames, which are anchored to the concrete of the dam structure, support the tracks and the gate seal seats.

Figure 12 shows three different applications of the fixed-wheel gate: the surface type, or water-level installation, usually used for controlling spillway releases; the tunnel type, or midconduit installation; and the face type, used at the face of a dam to close the intake to a penstock or outlet conduit. Either mechanical or hydraulic gate-lifting devices may be used with the fixed-wheel gate.

Leakage around a fixed-wheel gate when it is in the closed position is a continuing problem, the solution of which is engaging the attention of Bureau engineers. The seal much used at the present time can be seen in the slot sections of the tunnel and face types of gate in figure 12. It is a rubber strip resembling a music note in cross section, affixed to the gate and pressing against the seal seat when the gate is closed.

Depending on wheel loads, fixed-wheel gates are fitted with either flanged or unflanged wheels. The wheels are supported on corrosion-resistant axles by self-lubricating bronze bushings. Axle supports on the

TUNNEL TYPE

FLOW

SLOT SECTION

SURFACE TYPE

FLOW

SLOT SECTION

FACE TYPE

FLOW

SLOT SECTION

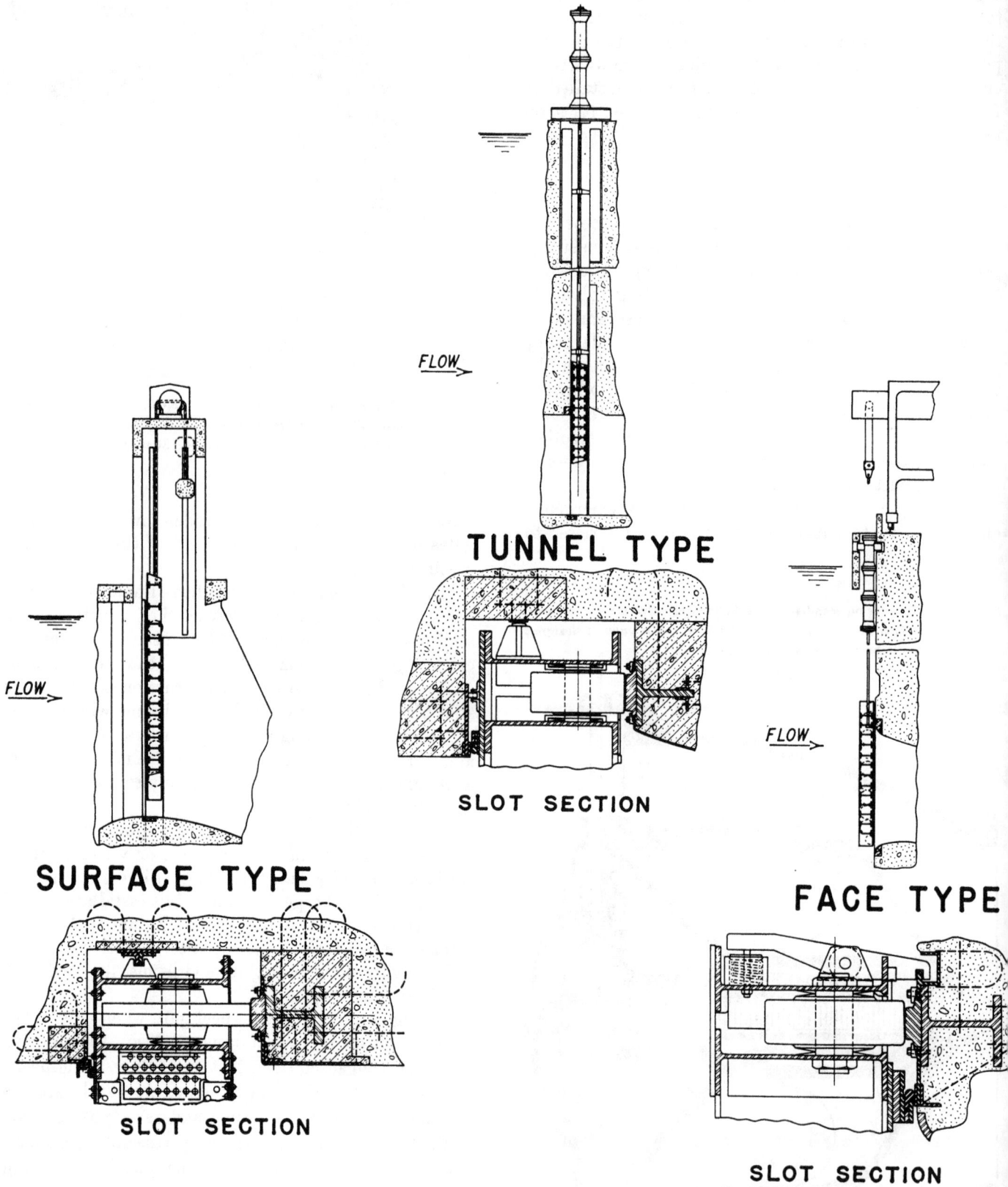

Figure 12.—Fixed-wheel gates are adaptable to a variety of installations. The slot sections illustrate different methods of mounting wheels, rails, guides, and seals for various applications.

gate are so mounted that, by means of an eccentric arrangement, individual wheels can be brought into bearing on the track, insuring that all wheels carry equal loads. Fixed-wheel gates as large as 17.5 by 34.66 feet have been designed for installation on the upstream face of a dam. Gates of this size may be moved into position by a gantry crane operating from the top of the dam, or hydraulic hoists for gate operation may be installed. The fixed-wheel gate may be used as a bulkhead gate under balanced conditions for the purpose of unwatering the conduit for inspection, maintenance, or other needs. The term "balanced conditions" implies that the gate is raised or lowered only when no water is flowing and the hydrostatic loads on either side of the gate leaf are equal.

(c) Leaf gates of the *roller-mounted* type, designated by the Bureau as *coaster gates*, are likewise used for emergency closure of a conduit or penstock. The coaster gate is similar in construction to the fixed-wheel gate, except that endless chains of rollers are substituted for the wheels. The coaster gate is chosen where unusually high heads or large sizes make the fixed-wheel gate impractical due to its greater frictional resistance. Although the coaster gate is designed for sufficient strength to be closed under emergency or unbalanced conditions, it is usually operated under balanced pressures to effect closure and unwatering of the conduit for inspection or servicing of turbines, valves, or other equipment.

Like the fixed-wheel gate, the coaster gate is built of structural steel in the form of a skin plate supported by beams and girders. A hydraulic hoist is usually employed for raising the gate. Placed above the water level, the hydraulic hoist is connected to the gate by a long stem. Since both fixed-wheel and coaster gates are designed so that the weight of the gate is sufficient to lower it, the gate stems carry only a tensile load. Coaster gates 15 by 29.65 feet in size have been installed at Grand Coulee Dam for the penstock intakes.

(d) *Ring-follower* and *ring-seal gates* are modified slide gates designed for emergency closure at high heads. (See figs. 13 and 14.) In both of these types the leaf is extended to include an opening equal in diameter to that of the conduit, forming an unobstructed passageway when the leaf is in the raised or open position. This avoids the turbulence caused by the vertical guide slot as in an ordinary slide gate. The leaf of the ring-follower gate slides on bronze seats and is ordinarily moved by means of a hydraulic cylinder. The leaf of the ring-seal gate is usually carried on rollers or wheels and is moved vertically by an electric motor operating a pair of threaded stems. When the ring-seal gate closes, a movable ring, actuated hydraulically by water taken from the conduit, moves laterally to form a watertight seal between the leaf and the

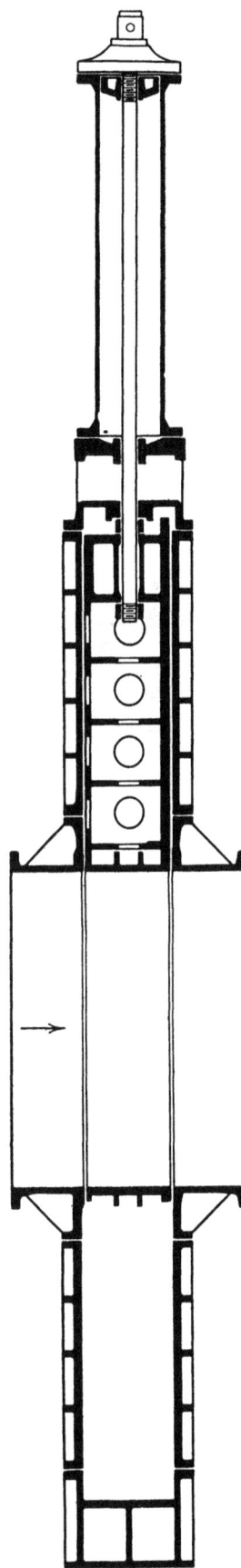

Figure 13.—Ring-follower and ring-seal gates are similar in that the lower end of the leaf forms an open section in the water passage, continuous with the conduit, which avoids the turbulence induced by the gate-guide slots in the usual design of leaf gate.

Figure 14.—Gate frames for the 96-inch ring-follower gates at Hungry Horse Dam were erected in the field before encasement in concrete.

valve body. Butterfly valves are occasionally installed as regulating controls for low-head discharges, but their primary use is as service or guard gates in power penstocks immediately upstream from the turbine or in outlet works, to provide an emergency shutoff. Another application is in a pipeline on either or both sides of a pumping unit so that reverse flow may be prevented in the event of a power failure or a pump shutdown.

When used for emergency closure, the butterfly valve operates either in the fully open or fully closed position and never remains at intermediate positions. Also, for this application the leaf is designed to have minimum hydraulic resistance to flow through the valve. The leaf is moved either by a hydraulic rotor or a hydraulic cylinder. Sealing of the leaf with the valve body when in the closed position has been a problem to the designing engineer because hydraulic forces tend to aid sealing on one side of the leaf while deflecting the other side away from the sealing surface. To assist in the solution of the problem, adjustable seals are provided, sometimes on the leaf and sometimes on the valve body.

The largest butterfly valves installed on a Bureau project are the 168-inch diameter valves in the main unit penstocks at Hoover Dam. Including the hydraulic operating rotor, each of these units is approximately 28 feet high and weighs 190 tons.

SPILLWAY GATES

Although spillways are of many types, their primary purpose is to protect the dam from being overtopped and to discharge the excess reservoir inflow in such a manner that neither the dam and its appurtenant works,

body. The ring is usually located in an annular recess in the gate housing and is placed concentric with and around the opening in the gate body. Some gates, however, have been designed with the sealing ring contained in the leaf, rather than in the housing. Figure 14 shows the gate frame and bonnet of one of the 96-inch ring-follower gates at Hungry Horse Dam erected in the field before being encased in concrete.

(e) The *paradox gate* has been installed at several Bureau projects, but has been superseded by the ring-follower and ring-seal gates. In use, the paradox gate functions in the same locations and under the same conditions as the latter types. Its design differs from that of other types of leaf gates in that the leaf is mounted on a pair of roller trains, so arranged on wedge-shaped tracks that water pressure on the leaf assists in sealing the gate; in opening, the wedge moves first, releasing the leaf and reducing the friction force and the hoisting load. High manufacturing and maintenance costs were among the factors which led to abandonment of the paradox gate in later installations.

(f) The *butterfly valve*, shown in figure 15, is essentially a circular leaf, slightly convex in form, mounted on a transverse shaft carried by two bearings in the

Figure 15.—Butterfly valves are designed for quick closure in the event of failure of some part of the system. Note that the diameter of the valve body is larger than the conduit to compensate for the resistance of the leaf to flow.

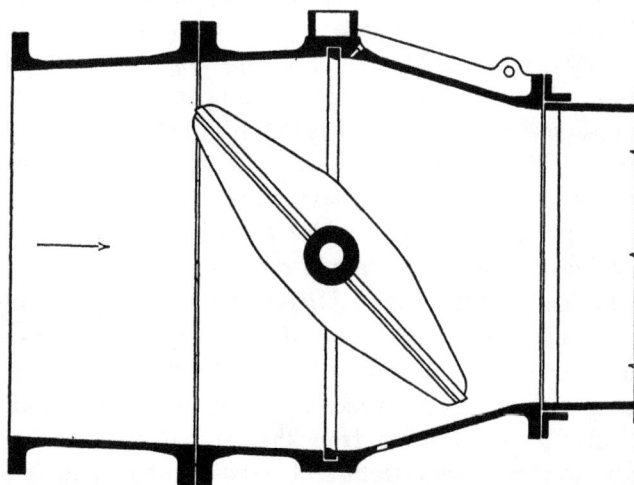

nor the developed areas downstream from the reservoir, are endangered. Except for the uncontrolled, or "free-flow" spillway, which is designed merely to discharge excess water, releases are controlled by various types of gates, chosen for capacity, structural limitations, hydraulic characteristics, or economic reasons. Radial gates, vertical wheel-mounted gates, or drum gates are usually chosen for Bureau of Reclamation installations, but special conditions have dictated the selection of ring-type or roller-type gates.

(a) Use of *radial gates* for regulation of flow under submerged conditions and high heads has already been discussed. As a spillway gate, the radial gate is designed for open-channel flow. Sealing of a radial spillway gate is relatively simple. The usual type of end seal is a rubber "music note" strip bolted to the sides of the gate and bearing against steel wall plates. A rubber strip bolted to the bottom of the gate forms a seal when the gate is lowered. When the gate is in the open, or raised, position the water passageway is unobstructed, resulting in excellent hydraulic characteristics for the installation. Figure 16 shows a typical radial gate installation on a spillway crest. Radial gates 51 feet wide by 34.5 feet high have been designed for installation at Canyon Ferry Dam.

(b) *Vertical leaf gates*, when used for spillway control, are similar to the emergency gates of the surface type, illustrated in figure 12. They are usually of either the wheel- or roller-mounted type, traveling on vertical tracks anchored in guide slots in the spillway structures. Motor-driven hoists housed above the gates lift the gates to any desired position by means of link chains. To reduce the load on the hoist, reinforced concrete counterweights are usually attached to the hoist chains and arranged to move vertically in guide slots.

(c) *Drum gates* have been installed on the spillway crests of large Bureau dams. Drum gates are noteworthy because discharges are released over the gate, affording an unobstructed water passage, and because the curved surface of the lowered gate forms the spillway crest, improving the hydraulic characteristics of the spillway at maximum discharges. In the raised, or closed, position the gates effectively increase the storage capacity of the reservoir, making possible the use of the upper levels of the lake for flood storage with simultaneous increase in operating head for the power turbines. The gates may be lowered to draw down the reservoir in anticipation of an approaching flood.

In cross section, the drum gate resembles the Greek capital letter delta, with two curved sides one of which forms the spillway crest when the gate is lowered into the gate recess chamber in the top of the spillway. The gate is hinged along its upstream edge to the upper edge of the gate chamber. Since the gate is a hollow, water-

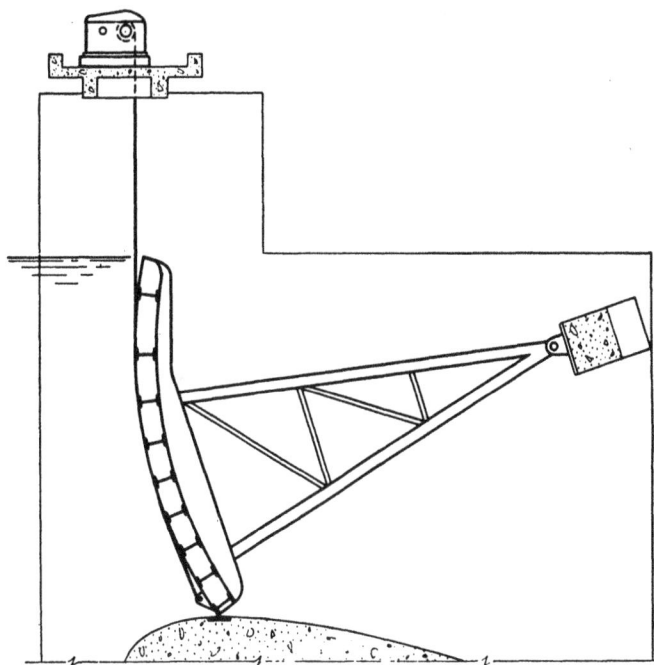

Figure 16.—Radial gates are especially suitable for control of overflow spillways. Gate hoists mounted on the operating bridge lift the gate allowing water to flow beneath in the desired amount.

tight, buoyant vessel, it is supported by the body of water within the gate recess chamber and may be made to assume a desired position without the necessity for a mechanical hoist. Releasing water from the gate recess chamber, thereby lowering the supporting water surface, will lower the gate. Conversely, allowing water to enter the gate recess chamber will raise the water surface and the gate. The cross section of the gate, figure 17, shows the general arrangement. Seals are provided at both the upstream and downstream edges of the gate, as well as at the ends, to prevent leakage into the gate recess chamber. A drain hose permits removal of water from the interior of the gate. Admission and release of water into or out of the gate chamber is controlled manually or by automatic float valves.

As with other types of spillway gates, a drum gate installation may be designed to regulate the reservoir water elevation by controlling the discharges over the spillway. Other installations, not designed for regulation, are for the primary purpose of increasing the maximum reservoir storage capacity.

(d) The *ring gate* is another gate of the buoyant type used in the control of spillway releases. It is essentially a ring- or annular-shaped steel drum, operating in a recess or gate chamber in the spillway crest and controlled in a manner similar to that described for the drum gate. Figure 18 is a simplified cross section through the crest of a ring gate showing the general

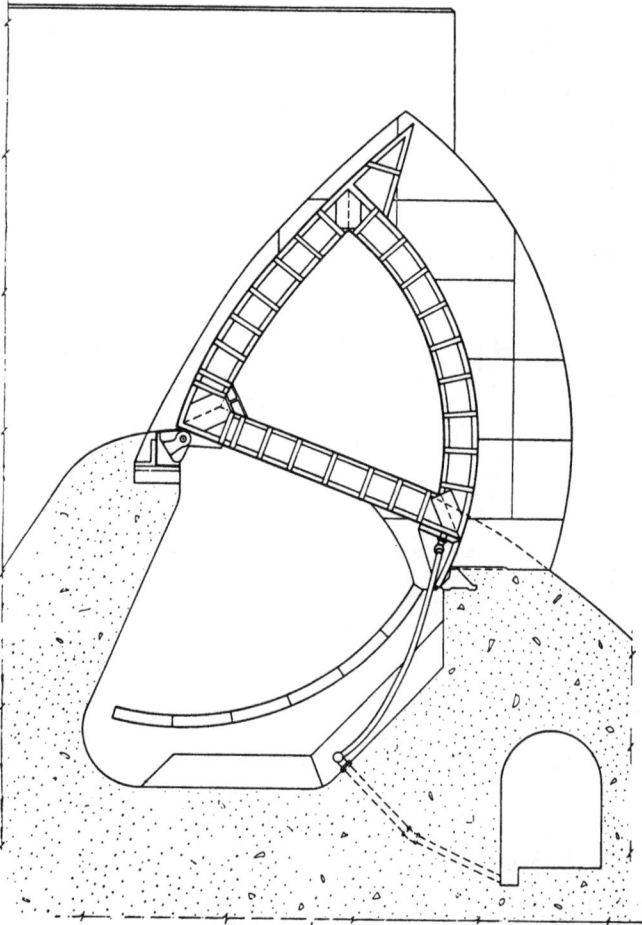

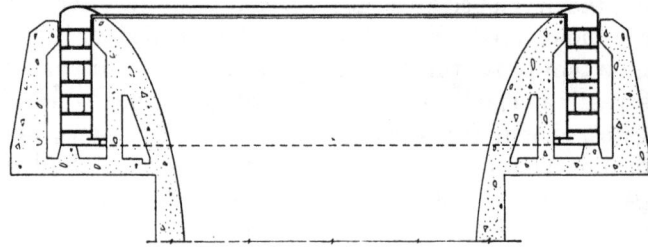

Figure 18.—The ring gate is especially adapted for "morning glory" spillways. Like the drum gate, the ring gate floats in a chamber. Changes in the gate position are made by raising or lowering the water surface in the chamber.

Figure 17.—Drum gates are frequently installed in large concrete dams. When lowered, the gate forms a continuous smooth crest for the spillway, capable of passing large floods containing trash and debris.

debris. Some roller gates have been designed to operate submerged, allowing floating material to pass over the top; but the usual design provides for lifting the gate free of the crest, allowing flow beneath the gate. The section through a typical roller gate, figure 19, illustrates the design and operation. The hollow, plate-steel gate cylinder has an extended lip along the lower edge, which seats against a seal in the spillway crest. Teeth on a portion of the ends of the drum engage in geared racks, and the gate is rolled up or down inclined supporting rails by a hoist through sprocket chains around the ends of the cylinder. The weight of the gate is sufficient to lower and seal it when the chain is released. The roller gates at Roza Diversion Dam, on the Yakima project in Washington, are 110 feet long and weigh 233 tons each.

Hoists for roller gates are installed on the piers. Where several gates are required, two hoists are usually installed in one hoist house to serve adjacent gates, minimizing the number of hoisting structures since the gates need be driven from one end only.

arrangement. The ring gate is particularly adaptable to the morning-glory type of spillway, in which the water plunges down a shaft and discharges through a tunnel. As the figure shows, the crest of the gate is rounded in a smooth contour, continuous with the bellmouth of the spillway crest. The gate is kept level during raising and lowering by means of a leveling device, consisting of a number of geared shafts extending around the periphery of the gate, with the gears meshing vertical gear racks anchored to the wall of the gate chamber.

The first ring gate installation on a Bureau project was at Owyhee Dam in eastern Oregon, and the most recent is at Hungry Horse Dam in western Montana. Both are about 60 feet in diameter and 12 feet high.

(e) *Roller gates* were developed in Europe and were first used by the Bureau on the Colorado River Dam near Grand Junction, Colo. These gates are particularly suitable for controlling discharges at low-head dams where wide, unobstructed passageways between piers are desired to permit passage of ice or floating

Figure 19.—Roller gates are adapted to spillways where floating ice or other debris is a problem. With the gate raised, a free passageway for water is left beneath the gate.

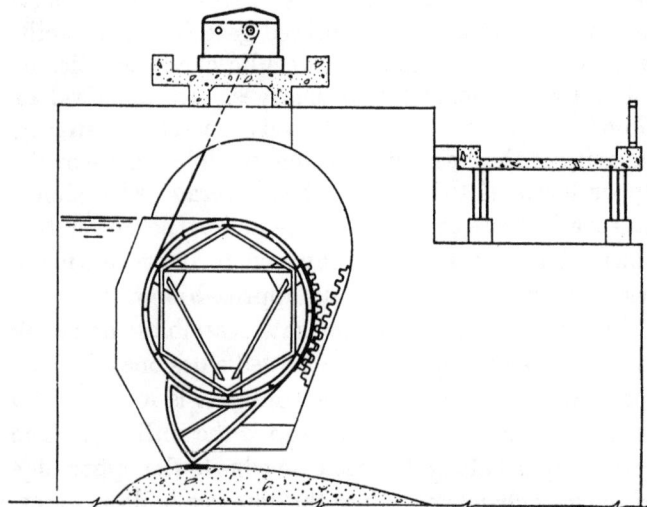

www.ingramcontent.com/pod-product-compliance
Lightning Source LLC
Chambersburg PA
CBHW061325190326
41458CB00011B/3894